Nonconvex Optimization and Its Applications

Volume 77

GENERALIZED CONVEXITY, GENERALIZED MONOTONICITY AND APPLICATIONS

Proceedings of the 7th International Symposium on Generalized Convexity and Generalized Monotonicity

Edited by

ANDREW EBERHARD
RMIT University, Australia

NICOLAS HADJISAVVAS
University of the Aegean, Greece

DINH THE LUC
University of Avignon, France

Springer

eBook ISBN: 0-387-23639-2
Print ISBN: 0-387-23638-4

Created in the United States of America

Visit Springer's eBookstore at: http://ebooks.kluweronline.com
and the Springer Global Website Online at: http://www.springeronline.com

Contents

Preface

In recent years there is a growing interest in generalized convex functions and generalized monotone mappings among the researchers of applied mathematics and other sciences. This is due to the fact that mathematical models with these functions are more suitable to describe problems of the real world than models using conventional convex and monotone functions. Generalized convexity and monotonicity are now considered as an independent branch of applied mathematics with a wide range of applications in mechanics, economics, engineering, finance and many others.

The present volume contains 20 full length papers which reflect current theoretical studies of generalized convexity and monotonicity, and numerous applications in optimization, variational inequalities, equilibrium problems etc. All these papers were refereed and carefully selected from invited talks and contributed talks that were presented at the 7th International Symposium on Generalized Convexity/Monotonicity held in Hanoi, Vietnam, August 27-31, 2002. This series of Symposia is organized by the Working Group on Generalized Convexity (WGGC) every 3 years and aims to promote and disseminate research on the field. The WGGC (http://www.genconv.org) consists of more than 300 researchers coming from 36 countries.

Taking this opportunity, we want to thank all speakers whose contributions make up this volume, all referees whose cooperation helped in ensuring the scientific quality of the papers, and all people from the Hanoi Institute of Mathematics whose assistance was indispensable in running the symposium. Our special thanks go to the Vietnam Academy of Sciences and Technology, the Vietnam National Basic Research Project "Selected problems of optimization and scientific computing" and the Abdus Salam International Center for Theoretical Physics at Trieste, Italy, for their generous support which made the meeting possible. Finally, we express our appreciation to Kluwer Academic Publishers for including this volume into their series. We hope that the volume will be useful for students, researchers and those who are interested in this emerging field of applied mathematics.

ANDREW EBERHARD

NICOLAS HADJISAVVAS

DINH THE LUC

I

INVITED PAPERS

Chapter 1

ALGEBRAIC DYNAMICS OF CERTAIN GAMMA FUNCTION VALUES

J.M. Borwein*
Research Chair, Computer Science Faculty,
Dalhousie University, Canada

K. Karamanos
Centre for Nonlinear Phenomena and Complex Systems,
Université Libre de Bruxelles, Belgium

Abstract We present significant numerical evidence, based on the entropy analysis by lumping of the binary expansion of certain values of the Gamma function, that some of these values correspond to incompressible algorithmic information. In particular, the value $\Gamma(1/5)$ corresponds to a peak of non-compressibility as anticipated on a priori grounds from number-theoretic considerations. Other fundamental constants are similarly considered.

This work may be viewed as ah invitation for other researchers to apply information theoretic and decision theory techniques in number theory and analysis.

Keywords: Algebraic dynamics, symbolic dynamics.

MSC2000: 94A15, 94A17, 37Bxx, 11Yxx, 11Kxx

1. Introduction

Nature provides us with a wide variety of symbolic strings ranging from the sequences generated by the symbolic dynamics of nonlinear systems to RNA and DNA sequences or DLA patterns (*diffusion limited*

*email:jborwein@cs.dal.ca

aggregation patterns are a classical subject in Nonlinear Chemistry); see Hao (1994); Nicolis et al (1994); Schröder (1991).

Entropy-like quantities are a very useful tool for the analysis of such sequences. Of special interest are the *block entropies,* extending Shannon's classical definition of the entropy of a single state to the entropy of a succession of states (Nicolis et al (1994)). In particular, it has been shown in the literature that *scaling* the block entropies by length sometimes yields interesting information on the structure of the sequence (Ebeling et al (1991); Ebeling et al (1992)).

In particular, one of the present authors has derived an entropy criterion for the specialized, yet important algorithmic property of *automaticity* of a sequence. We recall that, a sequence is called *automatic* if it is generated by a finite automaton (the lowest level Turing machine). For more details about automatic sequences the reader is referred to Cobham (1972), and for their role in Physics to Allouche (2000).

This criterion is based on entropy analysis by *lumping. Lumping* is the reading of the symbolic sequence by 'taking portions' (see expression (1)), as opposed to *gliding* where one has essentially a 'moving frame'. Notice that gliding is the standard approach in the literature. Reading a symbolic sequence in a specific way is also called *decimation* of the sequence.

The paper is articulated as follows. In Section two we recall some useful facts. In Section three we present the mathematical formulation of the entropy analysis by lumping. In Section four we present our intuitive motivation based on algorithmic arguments while in Section five we present a central example of an automatic sequence, taken from the world of nonlinear Science, namely the Feigenbaum sequence. In Section six we present our main results. In Section seven we speak about automaticity and algorithmic compressibility measures. In section eight we analyse $\exp(\pi/\sqrt(2))$. Finally, in Section nine we draw our main conclusions and discuss future work.

2. Some definitions

We first recall some useful facts from elementary number theory. As is well known, *rational* numbers can be written in the form of a fraction p/q, where p and q are integers and *irrational* ones cannot take this form. The k**-ary** expansion of a rational number (for instance the decimal or binary expansion) is periodic or eventually periodic and conversely. Irrational numbers form two categories: *algebraic irrational* and *transcendental,* according to whether they can be obtained as roots of a polynomial with rational coefficients or not. The k**-ary** expansion of an irrational number is necessarily *aperiodic.* Note that transcendental numbers are *well approximated* by fractions. In 1874 G. Cantor showed that 'almost all' real numbers are transcendental.

A *normal* number in base $k \geq 2$ is a real number x such that, for each integer $d \geq 1$, each block of length d occurs in the k-**ary** expansion of x with (equal) asymptotic frequency $1/k^d$. A rational number is never normal, while there exist numbers which are normal and transcendental, like Champernowne's number. This number is obtained by concatenating the decimal expansions of consecutive integers (Champernowne (1933))

$$0.12345678910111121314...$$

and it is simultaneously transcendental and normal in base 10.

There is an important and widely believed conjecture, according to which all algebraic irrational numbers are believed to be normal. But present techniques fall woefully short on this matter, see Bailey et al (2004). It seems that E. Borel was the first who explicitly formulated such a conjecture in the early fifties (Borel (1950)). Actually, normality is not the best criterion to distinguish between algebraic irrational and transcendental numbers. In fact, there exist transcendental numbers which are normal, like Champernowne's number (Champernowne (1933), Chaitin (1994), Allouche (2000)) and probably π (Schröder (1991), Wagon (1985) Allouche (2000)). One of the first systematic studies towards this direction dates back to ENIAC also some fifty years ago (Metropolis et al (1950); Borwein (2003)). No truly 'natural' transcendental number has been shown to be normal in any base, hence the interest in computation.

3. Entropy analysis by lumping

For reasons both of completeness and for later use, we compile here the basic ideas of the method of entropy analysis by lumping. We consider a subsequence of length N selected out of a very long (theoretically infinite) symbolic sequence. We stipulate that this subsequence is to be read in terms of distinct 'blocks' of length n,

$$\ldots \underbrace{A_1 \ldots A_n}_{B_1} \underbrace{A_{n+1} \ldots A_{2n}}_{B_2} \ldots \underbrace{A_{jn+1} \ldots A_{(j+1)n}}_{B_{j+1}} \ldots$$

We call this reading procedure *lumping.* We shall employ lumping throughout the sequel. The following quantities characterize the information content of the sequence (Khinchin (1957); Ebeling et al (1991)).

i) The *dynamical (Shannon-like) block-entropy* for blocks of length n is given by

$$H(n) := - \sum_{(A_1,...A_n)} p^{(n)}(A_1,\ldots,A_n)\cdot \ln p^{(n)}(A_1,\ldots,A_n) \quad (1.1)$$

where the probability of occurrence of a block $A_1 \ldots A_n$, denoted $p^{(n)}(A_1,...,A_n)$, is defined (when it exists) in the statistical limit as

$$p^{(n)}(A_1,\ldots,A_n) = \frac{\#\text{of blocks } A_1\ldots A_n \text{ found when lumping}}{\text{Total } \# \text{ of blocks found}} \quad (1.2)$$

starting from the beginning of the sequence, and the associate entropy per letter

$$h^{(n)} = \frac{H(n)}{n}. \quad (1.3)$$

ii) The *conditional entropy* or entropy excess associated with the addition of a symbol to the right of an n-block

$$h_{(n)} = H(n+1) - H(n). \quad (1.4)$$

iii) The *entropy of the source* (a topological invariant), defined as the limit (if it exists)

$$h = \lim_{n\to\infty} h_{(n)} = \lim_{n\to\infty} h^{(n)} \quad (1.5)$$

which is the discrete analogue of metric or Kolmogorov entropy.

We now turn to the selection problem, that is to the possibility of emergence of some preferred configurations (blocks) out of the complete set of different possibilities. The number of all possible symbolic sequences of length n (*complexions* in the sense of Boltzmann) in a K-letter alphabet is

$$N_K = K^n. \quad (1.6)$$

Yet not all of these configurations are necessarily realized by the dynamics, nor are they equiprobable. A remarkable theorem due to McMillan (see Khinchin (1957)), gives a partial answer to the selection problem asserting that for stationary and ergodic sources the probability of occurrence of a block $(A_1,\ldots,A_n)$ is

$$p^{(n)}(A_1,\ldots,A_n) \sim e^{-H(n)} \quad (1.7)$$

for almost all blocks $(A_1, \ldots, A_n)$. In order to determine the abundance of long blocks one is thus led to examine the scaling properties of $H(n)$ as a function of n.

It is well known that numerically, block entropy is underestimated. This underestimation of $H(n)$ for large values of n is due to the simple fact that not all words will be represented adequately if one looks at long enough samples. The situation becomes more and more prominent for calculating $H(n)$ by 'lumping' instead of 'gliding'. Indeed in the case of 'lumping' an exponentially fast decaying tail towards value zero follows after an initial plateau.

Since the probabilities of the words of length m are calculated by their frequencies, i.e. $p_n = N_1/N_{[sample]}$ where $N_{[sample]}$ is the size of the available data-sample i.e. the length of the 'text' under consideration, then as $N_1 \to 0$ for long words, the block entropy calculated will reach a maximum value, its *plateau,* at

$$H_{MAX} = \log_{[K]}(N_{[sample]})$$

where K the length of the alphabet. Indeed, this corresponds to the maximum value of the entropy for this sample, given when

$$p^{(n)} = 1/N_{[sample]}.$$

This value corresponds also to an effective maximum word length

$$n_{max} = \ln N_{[sample]}$$

in view of eqs. (1), (6) and (7).

For instance, if we have a binary sequence with 10,000 terms, of course $b = 2$ and $N_{[sample]} = 10^4$. This way, the value of H_{MAX} can determine a safe border for finite size effects. In our case

$$H_{MAX} = n_{max} = \ln\left(10^4\right) = 9.2\ldots, \tag{1.8}$$

so that $n_{max} = 9$ and we can safely consider the entropies until $n = 8$.

After this small digression, we recall here the main result of the entropy analysis by lumping, see also Karamanos (2001b); Karamanos (2001c). Let m^k be the length of a block encountered when lumping, $H(m^k)$ the associated block entropy. We recall that, in view of a result by Cobham (Theorem 3 of Cobham (1972)), a sequence is called ***m-automatic*** if it is the image by a letter to letter projection of the fixed point of a set of substitutions of constant length m. A substitution is called *uniform* or *of constant length* if all the images of the letters have the same length. For instance, the Feigenbaum symbolic sequence can in an equivalent manner be generated by the Metropolis, Stein and Stein algorithm (Metropolis et al (1973); Karamanos et al (1999)), or as the fixed point $(\sigma^F)^\infty(R)$ of the set of substitutions of length 2: $\sigma^F(R) = RL, \sigma^F(L) = RR$ starting with *R*, or by the finite automaton of Figure 1 (see also Section five).

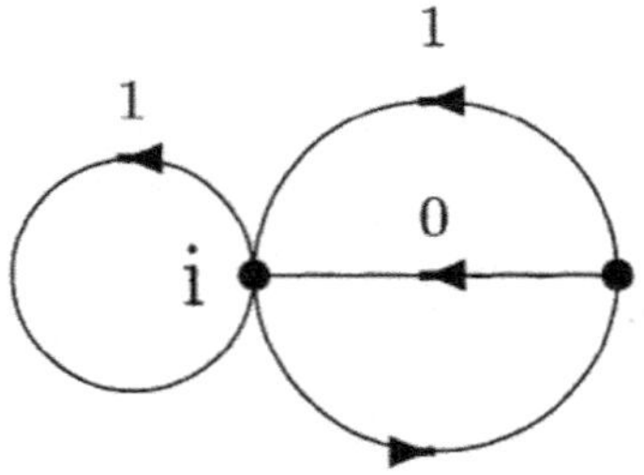

Figure 1.1. Deterministic finite automaton described by Cobham's algorithmic procedure. This automaton contains two states: i and a and to each state corresponds by the function of exit F a symbol; either $F(i) = R = 1$ or $F(a) = L = 0$. To calculate the nth term of the 2^∞ sequence we first express the number n in its binary form and then we start running the automaton from its initial state, according to the binary digits of n. In this trip we read the symbols contained in the binary expansion of n from the left to the right following the targets indicated by the letters. For instance $n = 3 = (11 \text{ base } 2)$ gives the run $i \to i \to i$ so that $u(3) = R = 1$, while $n = 9 = (1001 \text{ base } 2)$ gives the run $i \to i \to a \to i \to i$ so that $u(9) = R = 1$.

The term 'automatic' comes from the fact that an automatic sequence is generated by a finite automaton.

The following properties then holds:

- If the symbolic sequence $(u_n)_{n \in \mathcal{N}}$ is m-automatic, then

$$\exists \quad k_o \in \{0,1\}, \quad m \in \mathcal{N}^*, \quad \forall \quad k \geq k_o: \quad H(m^{k_o}) = H(m^k) \tag{1.9}$$

  when lumping, starting from the beginning of the sequence.

The meaning of the previous proposition is that for m-automatic sequences there is always an envelope in the diagram $H(n)/n$ versus n, falling off exponentially as $\sim m^{-k}$ for blocks of a length m^k, $k = 1, 2, ...$ For infinite ergodic strings, the conclusion does not depend on the starting point. Similar conclusions hold if instead of a one-to-one letter projection we have a one-to-many letters projection of constant length. In particular, we have the following result.

- If the symbolic sequence $(u_n)_{n\in\mathcal{N}}$ is the image of the fixed point of a set of substitutions of length m by a projection of constant length μ, then

$$\exists \quad k_o \in \{0,1\}, \quad m \in \mathcal{N}^*, \quad \forall \quad k \geq k_o : \quad H(\mu \cdot m^{k_o}) = H(\mu \cdot m^k) \tag{1.10}$$

  when lumping, starting from the beginning of the sequence.

Our propositions give an interesting *diagnostic* for automaticity. When one is given an unknown symbolic sequence and numerically applies entropy analysis by lumping, then if the sequence does not obey such an invariance property predicted by the propositions, it is certainly non-automatic. In the opposite case, if one observes evidence of an invariance property, then the sequence is a good candidate to be automatic.

For stochastic automata, the following proposition also holds (see Karamanos (2004)).

- If the symbolic sequence $(u_n)_{n\in\mathcal{N}} \in \{0,1\}$ is generated by a *Cantorian stochastic automaton,* then (see Karamanos (2004))

$$\forall \quad k \geq 1 : \quad H(m^k) = k \cdot H(m) \tag{1.11}$$

  when lumping, starting from the beginning of the sequence.

4. The example of the Feigenbaum sequence

Before proceeding to the analysis of binary expansions of the values of the gamma function $\Gamma(1/n)$ (which as we shall see presently seems not to be automatic) we first give an example of entropy analysis by lumping of a 2-automatic sequence: the period-doubling or *Feigenbaum sequence,* much studied in the literature (Grassberger (1986); Ebeling et al (1992); Karamanos et al (1999)).

The Feigenbaum symbolic sequence can in an equivalent manner be generated by the Metropolis, Stein and Stein algorithm (Metropolis et al (1973); Karamanos et al (1999)), or as the fixed point $(\sigma^F)^\infty(R)$ of the set of substitutions of length 2: $\sigma^F(R) = RL, \sigma^F(L) = RR$ starting with *R,* or by the finite automaton of Fig.1. According to our first proposition, this sequence satisfies

$$H(1) = H(2) = \cdots = H(2^k) \tag{1.12}$$

when lumping, while for any integer r,

$$H(2 \cdot r) = H(r) \tag{1.13}$$

as is shown in Karamanos et al (1999).

Thus, the Feigenbaum sequence appears to be extremely compressible from the viewpoint of algorithmic information theory—memorizing the finite automaton (instead of memorizing the full sequence) lets one reproduce every term and so, the complete sequence. We say that the information carried by the Feigenbaum sequence is 'algorithmically compressible'.

The period-doubling sequence, is the only one for which an exact functional relation between the block-entropies when *lumping* and when *gliding* exists in the literature, so that it is an especially instructive example.

5. Motivation for the Gamma function

The basis of reduced complexity computation of Gamma function values is illustrated by the cases of $\pi = \Gamma^2(1/2)$, and of $\Gamma(1/4)$, and $\Gamma(3/4)$. These algorithms are discussed at length in Borwein et al (1987) and related material is to be found in Borwein (2003). Their origin is very classical relying on the early elliptic function discoveries of Gauss and Legendre but they do not appear to have been found earlier.

Algorithm. Let $x_0 := \sqrt{2}, \pi_0 = 2 + \sqrt{2}$ and $y_1 := 2^{1/4}$, compute

$$x_{n+1} := \frac{\left(\sqrt{x_n} + 1/\sqrt{x_n}\right)}{2}$$

for $n \geq 0$ and

$$y_{n+1} := \frac{\left(y_n\sqrt{x_n} + 1/\sqrt{x_n}\right)}{y_n + 1},$$

$$\pi_n := \frac{x_n + 1}{y_n + 1}\pi_{n-1}$$

for $n > 0$. Then

$$0 < \pi_n - \pi < 10^{-2^{n+1}}.$$

Hence

$$(2+\sqrt{2})\prod_{n\geq 1}\frac{1+x_n}{1+y_n} = \Gamma^2\left(\frac{1}{2}\right)$$

while

$$(1+1/\sqrt{2})\prod_{n\geq 1} x_n\frac{1+x_n}{1+y_n} = \Gamma^4\left(\frac{3}{4}\right),$$

and

$$(1+1/\sqrt{2})^3\prod_{n\geq 1} x_n^{-1}\left(\frac{1+x_n}{1+y_n}\right)^3 = \frac{1}{16}\Gamma^4\left(\frac{1}{4}\right),$$

provide corresponding quadratic algorithms for $\Gamma(1/2), \Gamma(3/4)$ and $\Gamma(1/4)$, see Borwein et al (1987), pp. 46–51.

There are similar algorithms for $\Gamma(1/6), \Gamma(1/3), \Gamma(2/3)$ and $\Gamma(5/6)$, and related elliptic integral methods for $\Gamma(k/24)$ for all positive integer k are given by Borwein et al (1992). For example,

$$\mathrm{K}\left(\frac{\sqrt{3}-1}{2\sqrt{2}}\right) = \frac{3^{-1/4}}{4}\beta\left(\frac{1}{3}, \frac{1}{6}\right) = \frac{3^{1/4}}{4^{1/3}}\frac{\Gamma^3\left(\frac{1}{3}\right)}{\pi}.$$

In consequence, since elliptic function values are fast computable, we obtain algorithms for $\Gamma(k/6)$.

No such method is known for other rational Gamma values, largely because the needed elliptic integral and Gamma function identities are too few and do not allow one to separate $\Gamma(1/5)$ and $\Gamma(2/5)$, for example, while they do allow for their product to be computed.

This does not rule out the existence of other approaches but it suggests that the algorithmic complexity of $\Gamma(1/5)$ should be greater than that of $\Gamma(1/5)\,\Gamma(2/5)$, and that the algorithmic complexity of $\Gamma(1/5)$ or $\Gamma(1/9)$ should be greater than that of $\Gamma(1/12)$. This in part motivates our analysis.

Similarly, we note that

$$\Gamma\left(\frac{1}{20}\right)\Gamma\left(\frac{3}{20}\right)\Gamma\left(\frac{7}{20}\right)\Gamma\left(\frac{9}{20}\right) = 160\sqrt{+\sqrt{5}-2}\,\pi\,\mathrm{K}\left(k_5\right)$$

where

$$k_5 = \sqrt{\frac{\sqrt{5}-1}{2} - \frac{\sqrt{10}-\sqrt{2}}{4}}.$$

Thus this Gamma product is fast computable, as are many others.

6. Results

In this work, we have considered the first 10,000 digits of the binary expansions of numbers of the form $\Gamma(1/n)$, where $n = 2, 3, 4, 5, 8, 9, 12, 24$. We have good statistics until a block length $n = 9$.

We can report the following results:

1 The binary expansion of $\Gamma(1/5)$ presents the maximum value of the entropy throughout almost the whole range.

2 The binary expansions of $\Gamma(1/5)\Gamma(2/5)$ and $\Gamma(1/2)$ present the minimum value of the entropy through almost the whole range. This corresponds to significant algorithmic compressibility.

3 The binary expansion of $\boldsymbol{\Gamma(1/3)}$ presents (within the limits of the numerical precision) non-monotonic behaviour of the block entropy per letter (not recorded below), indicating a deep and unanticipated algorithmic structure for this number.

4 The binary expansions of the other numbers present intermediate behavior.

There is now the question of the error bars. In any case, due to finite-sample effects the values of the entropy are underestimated, as we have already explain in Section three. To estimate the error of these computations, suppose that, for $k = 1$ there is an error in one digit over 10,000 digits. Then the corresponding error in the entropy H_1 by lumping will be

$$\delta H_1 = -\frac{1}{10^4} \ln \frac{1}{10^4} \sim 0.00092 \simeq 0.001, \tag{1.14}$$

while due to lumping there is an error for the entropy H_8 (at the limit of our numerical precision) of 1 block per $\frac{10^4}{8} = 1250$ blocks of length 8, leading to a corresponding error in the entropy H_8 by lumping

$$\delta H_8 = -\frac{1}{1250} \ln \frac{1}{1250} \sim 0.0057 \simeq 0.006, \tag{1.15}$$

so that we can keep three significant digits of the entropy in the whole range.

In particular, we have the following results for $H(k)/k$ for k from 1 to 9, 12 and 24.

$n = 2$	
k	$H(k)/k$
1	0.693
2	0.693
3	0.693
4	0.692
5	0.689
6	0.688
7	0.680
8	0.670

$n = 3$	
k	$H(k)/k$
1	0.693
2	0.693
3	0.693
4	0.692
5	0.692
6	0.690
7	0.688
8	0.679

$n=4$	
k	$H(k)/k$
1	0.693
2	0.693
3	0.693
4	0.692
5	0.692
6	0.690
7	0.685
8	0.681

$n=5$	
k	$H(k)/k$
1	0.693
2	0.693
3	0.693
4	0.693
5	0.692
6	0.692
7	0.690
8	0.687

$n=8$	
k	$H(k)/k$
1	0.693
2	0.693
3	0.693
4	0.692
5	0.692
6	0.689
7	0.688
8	0.679

$n=9$	
k	$H(k)/k$
1	0.693
2	0.693
3	0.693
4	0.693
5	0.691
6	0.690
7	0.687
8	0.680

$n=12$	
k	$H(k)/k$
1	0.693
2	0.693
3	0.693
4	0.693
5	0.692
6	0.690
7	0.687
8	0.681

$n=24$	
k	$H(k)/k$
1	0.693
2	0.693
3	0.693
4	0.693
5	0.692
6	0.689
7	0.687
8	0.681

The basic conclusion from these tables is that these Gamma function values correspond to little compressible information, as the entropy per letter $H(n)/n$ approaches in all cases its maximum value

$$h_{max}^{(n)} = \ln 2 = 0.693147.... \tag{1.16}$$

Furthermore, on inspecting the blocks that appear, one can check that (within the limits of our numerical precision), all possible blocks of letter occur in the binary expansions of these Gamma function values (as we would say in the language of the ergodic theory and dynamical systems, the system is "mixing"), a fact that validates both the statistics and the conclusions about the algorithmic incompressibility of the next Section.

We have also considered the first 5,000 digits of the binary expansion of $\Gamma(1/5)\Gamma(2/5)$. We have good statistics up to a block length $n = 9$. In particular, we obtain the following results for $H(k)/k$ for k from 1 to 8. This as conjectured shows significantly more compressibility.

k	$H(k)/k$
1	0.693
2	0.693
3	0.693
4	0.692
5	0.690
6	0.689
7	0.680
8	0.667

7. Automaticity measures

As we have already mentioned, when a symbolic sequence is generated by a deterministic finite automaton with m-states, then the block entropies measured by lumping respect an invariance property:

$$H(m) = H(m^2) = \cdots = H(m^j) = const.$$

for k integer, $k > 1$.

When this invariance property breaks, the sequence is not generated by a deterministic finite automaton with m-states. Still, one can still obtain a measure of algorithmic complexity (in particular of 'algorithmic compressibility') taking values from 0 % to 100 % the index: (in our notation)

$$A(j) = \left| \frac{h(m^{j+1}) - h(m^j)}{h_{max}(m^{j+1})} \right|, \tag{1.17}$$

properly normalized, on dividing by $h_{max}(m^{j+1})$.

To fix the ideas, let us consider the 2-states automaticity measure (so $m = 2$) of order $j = 1$, which can be expressed as

$$A(2) = \left| \frac{h(8) - h(4)}{h_{max}(8)} \right| = 0.032 = 3.2\% \tag{1.18}$$

In terms of 2-states automata, the variation of these indices is as follows:

n	$A(2)$
2	3.2
3	1.9
4	1.6
5	0.9
8	1.9
9	1.9
12	1.7
24	1.7

from which our conclusion about the algorithmic non-compressibility of $\Gamma(1/5)$ follows. Indeed, the more incompressible the sequence, the smaller the index $A(k)$. In confirmation of our earlier analysis, the corresponding value of $A(2)$ for $\Gamma(1/5)\Gamma(2/5)$ is 3.6%, indicating the highest algorithmic compressibility.

We arrive at exactly the same conclusions if we treat the values of $h(m)$ individually (instead of taking the absolute differences), searching directly for an alternative index of algorithmic compressibility

$$A' = \frac{h(i)}{h_{max}(i)}\%. \tag{1.19}$$

8. Entropy analysis of the constant $e^{\pi/\sqrt{2}}$

It has been shown (Contopoulos et al (1994); Contopoulos et al (1980); Heggie (1985)) that, for a wide class of Hamiltonian dynamical systems, the constant

$$\varrho := e^{\pi/\sqrt{2}}$$

plays the role that is played by the Feigenbaum constant δ for the logistic map and for dissipative systems in general (Nicolis (1995); Feigenbaum (1978); Feigenbaum (1979); Briggs (1991); Briggs et al (1998); Fraser et al (1985)). Thus, this constant (bifurcation ratio of period doubling bifurcations) is not universal, rather it depends on the particular dynamical system considered.

Recently, after the calculation of the Feigenbaum fundamental constants α and δ for the logistic map (quadratic non-linearity), to more than 1,000 digits by D. Broadhurst (Briggs (1991)), a careful statistical analysis of these constants has been presented (Karamanos et al (2003)), indicating the real possibility that these constants are non-normal (so probably transcendental) numbers.

Now, it is easy to show that the constant ϱ is transcendental (Waldschmidt (2004); Waldschmidt (1998a); Waldschmidt (1998b)). Indeed, according to the theorem of Gel'fond and Schneider—which resolved Hilbert's seventh problem—for a nonzero complex number λ and an irrational algebraic number b, one at least of the three numbers e^λ, $e^{b\lambda}$, b is transcendental. In our case, taking $\lambda = i\pi$ and $b = \sqrt{2}$, we easily obtain the transcendence of $e^{\frac{\pi}{\sqrt{2}}}$. As this constant is a combination of three fundamental constants π, e and $\sqrt{2}$, presumably all normal, it is reasonable to ask if ϱ also appears normal.

We first present an entropy analysis of the first 100,000 terms of the binary expansion of the constant $\varrho = e^{\pi/\sqrt{2}}$. We have reliable statistics for block lengths not exceeding $n = 10$.

Regarding the error bars now, we estimate the error of these computations as follows. Suppose that, for $k = 1$ there is an error in one digit over 100,000 digits. Then the corresponding error in the entropy H_1 by lumping will be

$$\delta H_1 = -\frac{1}{10^5} \ln \frac{1}{10^5} \sim 0.000115... \simeq 0.0001, \tag{1.20}$$

while due to lumping there is an error for the entropy H_{10} (at the limit of our numerical precision) of 1 block per $\frac{10^5}{10} = 10^4$ blocks of length 10, leading to a corresponding error in the entropy H_{10} by lumping

$$\delta H_8 = -\frac{1}{10^4} \ln \frac{1}{10^4} \sim 0.00092 \simeq 0.001. \tag{1.21}$$

For reasons of uniformity of our treatment, however, we keep three significant digits for the entropy per letter.

In particular, we record the following results for $H(k)/k$ as a function of k.

Length	Lumping of ϱ
1	0.693
2	0.693
3	0.693
4	0.693
5	0.693
6	0.693
7	0.693
8	0.692
9	0.691
10	0.688

This indicates serious evidence that ϱ is a normal number in base 2, since the entropy per letter approaches in all cases its maximum value $h_{max}^{(n)} = \ln 2 = 0.693....$

One should also notice that, all possible blocks of letters *(within the range computed)* appear in the binary expansions of ϱ (as we would say in the language of the ergodic theory and dynamical systems, the system is "mixing"), a fact that validates both the statistics and the conclusion about algorithmic incompressibility.

In order to observe the results of the change of the basis expansion, we also present here an entropy analysis of the first 100,000 terms of the decimal expansion of the constant $\varrho = e^{\pi/\sqrt{2}}$. We have reliable statistics for block lengths not exceeding $n = 5$.

For the error bars now, we estimate the error of these computations, suppose that, for $k = 1$ there is an error in one digit over 100,000 digits.

Then the corresponding error in the entropy H_1 by lumping will be

$$\delta H_1 = -\frac{1}{10^5} \ln \frac{1}{10^5} \ln 10 \sim 0.000265... \simeq 0.0003, \qquad (1.22)$$

while due to lumping there is an error for the entropy H_4 (at the limit of our numerical precision) of 1 block per $\frac{10^5}{4} = 25,000$ blocks of length 4, leading to a corresponding error in the entropy H_4 by lumping

$$\delta H_4 = -\frac{1}{25,000} \ln \frac{1}{25,000} \ln 10 \sim 0.000933 \simeq 0.001. \qquad (1.23)$$

For reasons of uniformity, we also decided to keep three significant digits for the entropy per letter. In particular, we record the following results for $H(k)/k$.

Length	Lumping of ϱ
1	2.303
2	2.302
3	2.298
4	1.952

This again indicates serious evidence that ϱ would be a normal number in base 10, since the entropy per letter approaches in all cases its maximum value $h^{(n)}_{max} = \ln 10 = 2.303....$ Again, we notice that, one can check that all possible blocks of letters appear, a fact that validates both the statistics and the conclusion about the algorithmic incompressibility.

Finally, we note that in terms of algorithmic complexity ϱ is one of the most accessible constants. The following algorithm, a precursor to those given above for $\Gamma(1/n)$ (Borwein et al (1987); Borwein (2003)) provides $O(D)$ good digits with $\log D$ operations.

```
epi:=proc(r,N) local k,n;
k:=r; for n to N
do k:=sqrt(1-k^2); k:=(1-k)/(1+k) od;
(k/4)^(-2^(-N));end;
```

Then $\mathrm{epi}(\sqrt{2}-1, N)$ returns roughly 2^N good digits of ϱ, while $\mathrm{epi}(1/\sqrt{2}, N)$ does the same for $\exp(\pi)$.

9. Conclusions and outlook

We have performed an analysis of some binary expansions of the values of the Gamma function $\Gamma(1/n)$ by lumping. The basic novelty of this method is that, unlike use of the Fourier transform or conventional entropy analysis by gliding, it gives results that can be related to algorithmic characteristics of the sequences and, in particular, to the property of automaticity.

In light of the paucity of analytic techniques for establishing normality or other distributional facts about specific numbers, such experimental-computational tools are well worth exploring further and refining more.

Acknowledgments

All the entropy calculations in this work have been performed using the program ENTROPA by V. Basios (see Basios (1998)) mainly at the Centre of Experimental and Constructive Mathematics (CECM) in Burnaby, BC, Canada and also at the Centre for Nonlinear Phenomena and Complex Systems (CENOLI) in Brussels, Belgium.

We first thank Professors G. Nicolis and J.S. Nicolis for useful discussions and encouragement. We should also like to thank M. Waldschmidt, G. Fee, N. Voglis, and C. Efthymiopoulos for fruitful discussions.

JB thanks the Canada Research Chair Program and NSERC for funding assistance. Financial support from the Van Buuren Foundation and the Petsalys-Lepage Foundation are gratefully acknowledged. KK has benefited from a travel grant Camille Liégois by the Royal Academy of Arts and Sciences, Belgium, from a grant by the Simon Fraser University and from a grant by the Université Libre de Bruxelles. His work has been supported in part by the Pôles d'Attraction Interuniversitaires program of the Belgian Federal Office of Scientific, Technical and Cultural Affairs.

References

Allouche, J.-P. (2000), Algebraic and analytic randomness, in *Noise, Oscillators and Algebraic Randomness,* M. Planat (Ed.), *Lecture Note in Physics, Springer* Vol. 550, pp. 345–356.

Bailey, D., Borwein, J.M., Crandall, R., and Pomerance, C. (2004), On the binary expansions of algebraic numbers, *J. Number Theory Bordeaux,* in press. [CECM Preprint 2003:204]

Bailey, D. H. and Crandall, R. E. (2001), On the random character of fundamental constant expansions, *Exp. Math.* Vol. 10(2), pp. 175.

Bai-Lin, H. (1994), *Chaos,* World Scientific, Singapore.

Basios, V. (1998), *ENTROPA* program in C++, (c) Université Libre de Bruxelles.

Borel, E. (1950), Sur les chiffres décimaux de $\sqrt{2}$ et divers problèmes de probabilités en chaîne, *C. R. Acad. Sci. Paris* Vol. 230, pp. 591–593.

Reprinted in: **Œuvres d'E. Borel,** Vol. 2, pp. 1203–1204. Editions du CNRS: Paris (1972).

Borwein, J. and Bailey, D. (2003), *Mathematics by Experiment: Plausible Reasoning in the 21st Century,* AK Peters, Natick Mass.

Borwein, J. M. and Borwein, P. B. (1987), *Pi and the AGM: A Study in Analytic Number Theory and Computational Complexity,* John Wiley, New York.

Borwein, J. M. and Zucker, I. J. (1992), Elliptic integral evaluation of the Gamma function at rational values of small denominator, *IMA Journal on Numerical Analysis* Vol. 12, pp. 519-526.

Briggs, K. (1991), A precise calculation of the Feigenbaum constants, *Math. Comp.* Vol. 57(195), pp. 435–439. See also

http://sprott.physics.wisc.edu/phys505/feigen.htm
http://sprott.physics.wisc.edu/phys505/feigen.htm

http://pauillac.inria.fr/algo/bsolve/constant/fgnbaum/brdhrst.html
http://pauillac.inria.fr/algo/bsolve/constant/fgnbaum/brdhrst.html

Briggs, K. M., Dixon, T. W. and Szekeres, G. (1998), Analytic solutions of the Cvitanovic-Feigenbaum and Feigenbaum-Kadanoff-Shenker equations, *Int. J. Bifur. Chaos* Vol. 8, pp. 347-357.

Chaitin, G. J.(1994), Randomness and Complexity in Pure Mathematics, *Int. J. Bif. Chaos* Vol. 4(1), pp. 3–15.

Champernowne, D. G. (1933), The construction of decimals normal in the scale of ten, *J. London Math. Soc.* Vol. 8, pp. 254–260.

Cobham, A. (1972), Uniform tag sequences, *Math. Systems Theory* Vol. 6, pp. 164–192.

Contopoulos, G., Spyrou N. K. and Vlahos L. (Eds.) (1994), , *Galactic dynamics and N-body Simulations,* Springer-Verlag; and references therein.

Contopoulos, G. and Zikides (1980).

Derrida, B, Gervois A. and Pomeau, Y. (1978), *Ann. Inst. Henri Poincaré, Section A: Physique Théorique* Vol. XXIX(3), pp. 305–356.

Ebeling W. and Nicolis, G. (1991), *Europhys. Lett.* Vol. 14(3), pp. 191–196.

Ebeling W. and Nicolis, G. (1992), *Chaos, Solitons & Fractals* Vol. 2, pp. 635.

Feigenbaum, M. (1978), Quantitative Universality for a Class of Nonlinear Transformations, *J. Stat. Phys.* Vol. 19, pp. 25.

Feigenbaum, M. (1979), The Universal Metric Properties of Nonlinear Transformations, *J. Stat. Phys.* Vol. 21, pp. 669.

Fraser S. and Kapral, R. (1985), Mass and dimension of Feigenbaum attractors, *Phys. Rev.* Vol. A31(3), pp. 1687.

Grassberger, P. (1986), *Int. J. Theor. Phys.* Vol. 25(9), pp. 907.

Heggie, D. C. (1985), *Celest. Mech.* Vol. 35, pp. 357.

Karamanos, K. and Nicolis, G. (1999), Symbolic dynamics and entropy analysis of Feigenbaum limit sets, *Chaos, Solitons & Fractals* Vol. 10(7), pp. 1135–1150.

Karamanos, K. (2000), From Symbolic Dynamics to a Digital Approach: Chaos and Transcendence, Proceedings of the Ecole Thématique de CNRS 'Bruit des Fréquences des Oscillateurs et Dynamique des Nombres Algébriques', Chapelle des Bois (Jura) 5–10 Avril 1999. 'Noise, Oscillators and Algebraic Randomness', M. Planat (Ed.), *Lecture Notes in Physics* Vol. 550, pp. 357–371, Springer-Verlag.

Karamanos, K. (2001), From symbolic dynamics to a digital approach, *Int. J. Bif. Chaos* Vol. 11(6), pp. 1683–1694.

Karamanos, K. (2001), Entropy analysis of automatic sequences revisited: an entropy diagnostic for automaticity, Proceedings of Computing Anticipatory Systems 2000, CASYS2000, *AIP Conference Proceedings* Vol. 573,*D. Dubois (Ed.),* pp. 278–284.

Karamanos, K. (2001), Entropy analysis of substitutive sequences revisited, *J. Phys. A: Math. Gen.* Vol. 34, pp. 9231–9241.

Karamanos K. and Kotsireas, I. (2002), Thorough numerical entropy analysis of some substitutive sequences by lumping, *Kybernetes* Vol. 31(9/10), pp. 1409–1417.

Karamanos, K. (2004), Characterizing Cantorian sets by entropy-like quantities, to appear in *Kybernetes.*

Karamanos, K. and Kotsireas, I. (2003), *Statistical analysis of the first digits of the binary expansion of Feigenbaum constants* α *and* δ, submitted.

Khinchin, A. I. (1957), *Mathematical Foundations of Information Theory,* Dover, New York.

Metropolis, N., Reitwisner, G. and von Neumann, J. (1950), Statistical Treatment of Values of first 2000 Decimal Digits of π and e Calculated on the ENIAC, *Mathematical Tables and Other Aides to Computation* Vol. 4, pp. 109–111.

Metropolis, N., Stein, M. L. and Stein, P. R.(1973), On finite limit sets for transformations on the unit interval, *J. Comb. Th.* Vol. A 15(1), pp. 25–44.

Nicolis, G. (1995), *Introduction to Nonlinear Science,* Cambridge University Press, Cambridge.

Nicolis, J. S. (1991), *Chaos and Information Processing,* Word Scientific, Singapore.

Nicolis G. and Gaspard, P. (1994), *Chaos, Solitons & Fractals* Vol. 4(1), pp. 41.

Schröder, M. (1991), *Fractals, Chaos, Power Laws* Freeman, New York.

Wagon, S. (1985), Is π normal? *Math. Intelligencer* Vol. 7, pp. 65–67.

Waldschmidt M. (2004), personal communication.

Waldschmidt, M. (1998) Introduction to recent results in Transcendental Number Theory, *Lectures given at the Workshop and Conference in number theory held in Hong-Kong,* June 29 – July 3 1993, preprint 074-93, M.S.R.I., Berkeley.

Waldschmidt, M. (1998) Un Demi-Siècle de Transcendence, in Développement des Mathématiques au cours de la seconde moitié du XXème Sièle, *Development of Mathematics 2000,* Birkhäuser-Verlag.

Chapter 2

(GENERALIZED) CONVEXITY AND DISCRETE OPTIMIZATION

Rainer E. Burkard*
Institut für Mathematik B, Graz University of Technology
Austria.

Abstract This short survey exhibits some of the important roles (generalized) convexity plays in integer programming. In particular integral polyhedra are discussed, the idea of polyhedral combinatorics is outlined and the use of convexity concepts in algorithmic design is shown. Moreover, combinatorial optimization problems arising from convex configurations in the plane are discussed.

Keywords: Integral polyhedra, polyhedral combinatorics, integer programming, convexity, combinatorial optimization.

MSC2000: 52Axx, 52B12, 90C10, 90C27

1. Introduction

Convexity plays a crucial role in many areas of mathematics. Problems which show convex features are often easier to solve than similar problems in general. This short survey based on personal preferences intends to exhibit some of the roles convexity plays in discrete optimization. In the next section we discuss convex polyhedra all of whose vertices have integral coordinates. In Section 3 we outline the concept of polyhedral combinatorics which became basic for solving $\mathcal{NP}$-hard

---

*This research has been supported by the Spezialforschungsbereich "Optimierung und Kontrolle", Projektbereich "Diskrete Optimierung".
email: burkard@tugraz.at

problems like the travelling salesman problem. In Section 4 we show some of the roles (generalized) convexity plays in the algorithmic design for combinatorial optimization problems. In the last section combinatorial optimization problems arising from convex geometric configurations will be discussed.

2. Convexity and integer programming

At the end of the 19th century Minkowski began to study convex bodies which contain lattice points. In 1893 he proved the following fundamental theorem (see also his monograph *Geometry of Numbers* of 1896):

Theorem 2.1 *Let C be a convex body in $\mathbb{R}^d$, symmetric with respect to the origin, and let the volume V(C) of C be $V(C) \geq 2^d$. Then C contains a pair of points with integral coordinates.*

In connection with the development of linear and integer programming this area of the *geometry of numbers* got a new relevance. The main theorem of linear programming states that the finite optimum of a linear program is always attained in an extreme point (vertex) of the set of feasible solutions. If we can derive a bound on the coordinates of vertices of the feasible set, even if the underlying polyhedral set is unbounded, then the feasibility and optimality of an integer program can be checked in finitely many steps. To be more precise, let us assume that A is an integral $m \times n$ matrix and let $b \in \mathbb{Z}^m$. We consider the points with integral coordinates in the convex polyhedral set

$$P := \{x \in \mathbb{R}^n | Ax \leq b, \ x \geq 0\}$$

and call $S := P \cap \mathbb{Z}^n$. The following theorem, see Nemhauser and Wolsey (1988) Theorem I.5.4.1., is basic that an integer programming problem can be solved by enumeration.

Theorem 2.2 *Let $K := \max_{1 \leq i \leq m, \ 1 \leq j \leq n}(|a_{ij}|, |b_i|)$. If x is an extreme point of conv(S), then*

$$0 \leq x_j \leq ((m+n)nK)^n \quad \textit{for } j = 1, 2, ..., n.$$

As a consequence of this result the feasibility and optimality problems in integer linear programming belong to the complexity class $\mathcal{NP}$. Bank and Mandel (1988) generalized this result to constraint sets described by quasi-convex polynomials with integer coefficients.

Since integer programming can be reduced to linear programming provided that all extreme points of the feasible region have integral coordinates, there is a special interest in convex polyhedral sets with integral vertices. A convex polyhedron

$$P := \{x \in \mathbb{R}^n | Ax \leq b, \ x \geq 0\}$$

is called *integral,* if all its vertices have integral coordinates. A nice characterization of integral polyhedral sets defined by arbitrary right hand sides b has been given by Hoffman and Kruskal (1956). A matrix A is called *totally unimodular,* if any regular submatrix of A has determinant ± 1. Now the following fundamental theorem holds:

Theorem 2.3 *(Hoffman and Kruskal, 1956)*
Let A be an integral matrix. Then the following two statements are equivalent:

1 $P(A,b) := \{x \in \mathbb{R}^n | Ax \leq b, \ x \geq 0\}$ *is integral for all* b *with* $P(A,b) \neq \emptyset$.

2 *A is totally unimodular.*

Important examples for problems with totally unimodular coefficient matrices are assignment problems, transportation problems and network flow problems. Seymour (1980) showed that totally unimodular matrices can be recognized in polynomial time.

If we specialize the right hand side in the constraint set to b with $b_i = 1$ for all i, we get the constraint sets of

- set packing problems: $Ax \leq 1, \ x \geq 0$,
- set partitioning problems: $Ax = 1, \ x \geq 0$,
- set covering problems: $Ax \geq 1, \ x \geq 0$.

For this kind of problems not only totally unimodular matrices, but even a larger class of matrices leads to integral polyhedra. We call a matrix A with entries 0 and 1 *balanced,* if it does not contain a square submatrix of odd order with row and column sums equal to 2. For example, the following 3×3 submatrix constitutes a forbidden submatrix:

$$\begin{pmatrix} 1 & 1 & 0 \\ 0 & 1 & 1 \\ 1 & 0 & 1 \end{pmatrix}$$

Fulkerson, Hoffman and Oppenheim (see Fulkerson et al (1974)) showed the following result.

Theorem 2.4 *If A is balanced, then the set partitioning problem*

$$\min\{c'x | \ Ax = 1, \ x \geq 0\}$$

has integral optimal solutions.

For many years the recognition of balanced matrices has been an open problem. In 1999, Conforti, Cornuéjols and Rao (see Conforti et al. (1999)) showed that balanced matrices can be recognized in polynomial time.

The following result of Berge (1972) with respect to set packing and set covering problems is more along the lines of the Hoffman-Kruskal theorem.

Theorem 2.5 *Let matrix A be without 0-row and 0-column. Then the following statements are equivalent:*

1 A is balanced.

2 $\{x|Ax \leq b,\ x \geq 0\}$ *is integral for all* b *with* $b_i = 1$ *or* ∞.

3 $\{x|Ax \geq b,\ x \geq 0\}$ *is integral for all* b *with* $b_i = 1$ *or* 0.

For a recent survey on packing and covering problems the interested reader is referred to Cornuéjols (2001).

3. Polyhedral combinatorics

In the following we consider combinatorial optimization problems which can be described by

- a finite ground set E,
- a class $\mathcal{F}$ of feasible solutions which are subsets $F \subseteq E$, and
- cost coefficients $c(e)$ for all elements $e \in E$.

The cost of a feasible solution F is defined by $c(F) := \sum_{e \in F} c(e)$. The goal is to find a feasible solution with minimum cost.

For example, the *travelling salesman problem* may be described by the ground set E consisting of all edges (roads) between n vertices (cities) of a graph. A feasible solution F corresponds to a tour through all cities. A *tour* is a subset of the edges which corresponds to a cyclic permutation ϕ of the underlying vertex set, i.e., F consists of all edges $[i, \phi(i)],\ \ 1 \leq i \leq |V|$. Less formally spoken, a tour visits all vertices of the graph starting from vertex 1 and does not visit any vertex twice. The length of a tour F is given by $c(F) := \sum_{e \in F} c(e)$. The objective is to find a tour with minimum length.

In order to model this problem with binary variables we introduce a 0-1 vector x with $|E|$ components. A feasible solution F corresponds to

$$x(F) := \begin{cases} 1, & \text{if } e \in F \\ 0, & \text{if } e \notin F. \end{cases}$$

The combinatorial optimization problem

$$\min_{F \in \mathcal{F}} \sum_{e \in F} c(e)$$

can be written as

$$\min c'x \text{ for } x \in \operatorname{conv}\{x(F),\ F \in \mathcal{F}\}.$$

This means that the linear function $c'x$ is to be minimized over the convex hull of finitely many points. *Polyhedral combinatorics* consists in describing the polytopes given as convex hull of all feasible points by linear inequalities. Let us discuss as examples matching problems and symmetric travelling salesman problems.

Matching problems

A matching M is a subset of edges of an undirected, finite graph $G = (V, E)$ with vertex set V and edge set E where every vertex is incident with at most one edge of M. The maximum cardinality matching problem asks for a maximum matching in G, i.e., for a matching with a maximum number of edges. The ground set E contains the edges of G, feasible sets are the matchings M. We want to formulate the maximum cardinality matching problem as a binary linear program. To this end we introduce for each edge $j \in E$ a variable x_j. Let $\delta(v)$ denote the set of edges incident with vertex v. Then we get the following obvious necessary inequalities:

$$\begin{aligned} \sum_{j \in \delta(v)} x_j &\leq 1 \text{ for all } v, \\ x &\geq 0. \end{aligned}$$

If we consider the graph K_3, i.e., the complete graph with three vertices and three edges (which form a triangle), then the vector $x = (1/2, 1/2, 1/2)$ fullfills the inequalities above, but does not correspond to a matching. Thus it is necessary to add additional constraints in the case of a non-bipartite graph. One can show that in the case of a bipartite graph the above mentioned constraints are sufficient for describing a matching. Let $\gamma(W)$ denote the subset of all edges with both endpoints in $W \subseteq V$. Edmonds (1965) introduced for the maximum cardinality matchings in non-bipartite graphs the additional constraints

$$\sum_{j \in \gamma(W)} x_j \leq 1/2(|W| - 1),$$

for all $W \subseteq V$ with $|W| \geq 3$, odd.

Theorem 3.1 *(Edmonds, 1965)*
The matching polytope is fully described by

$$\begin{aligned} x &\geq 0, \\ \sum_{j \in \delta(v)} x_j &\leq 1 \quad \textit{for all } v \in V, \\ \sum_{j \in \gamma(W)} x_j &\leq 1/2(|W|-1) \quad \textit{for all } W \subseteq V \textit{ with } |W| \geq 3, \quad \textit{odd.} \end{aligned}$$

Symmetric travelling salesman problems
As a second example we consider the symmetric travelling salesman problem (TSP). Let again a finite, undirected graph $G = (V,E)$ with vertex set V and edge set E be given. In order to describe the feasible sets (tours) by linear inequalities we introduce a binary variable $x(e)$ for every edge $e \in E$. Obviously the following inequalities must be fulfilled:

$$0 \leq x(e) \leq 1, \text{ for all } e \in E, \tag{2.1}$$

and

$$\sum_{e \in \delta(v)} x(e) = 2, \text{ for all } v \in V. \tag{2.2}$$

But these inequalities do not fully describe tours, since they may be incidence vectors of more than one cycle in G, so-called *subtours*. Therefore one requires also the so-called *subtour elimination constraints*

$$\sum_{e \in \delta(W)} x(e) \geq 2 \text{ for all } W \subseteq V, 2 \leq |W| \leq |V| - 2. \tag{2.3}$$

Now one can show

Theorem 3.2 *The integral points lying in the convex polyhedron (2.1)-(2.3) correspond exactly to tours.*

It should be noted that a linear program with constraints (2.1)-(2.3) can be solved in polynomial time, even if there are exponentially many inequalities of the form (2.3). The convex polytope described by (2.1)-(2.3) may, however, have fractional vertices which do not correspond to tours. Thus further inequalities must be added which cut off such fractional vertices. There are many classes of such additional inequalities known, e.g. comb inequalities, clique tree inequalities and many others. The interested reader is referred to e.g. Grötschel, Lovász and Schrijver (see Grötschel et al. (1988)). It should be noted that a complete characterization of the convex hull of all tours is not known in general.

Since the polytope described by (2.1)-(2.3) may have non-integral extreme points, the following *separation problem* plays an important role for solving the TSP: If the optimal solution for the linear program with the feasible set (2.1)-(2.3) is not integral, we have to add a so-called *cutting plane,* i.e., a linear constraint which is fulfilled by all tours, but which cuts off the current infeasible point. Usually such a cutting plane is determined by heuristics and is taken from the class of comb inequalities, clique tree inequalities or other facet defining families of linear inequalities for the TSP polytope.

4. (Generalized) Convexity and algorithms

In this section we will point out that convexity also plays an important role in algorithms for solving a convex or linear integer program. Let $f_i(x)$, $1 \leq i \leq m$, be quasiconvex functions defined on a region $D \subseteq \mathbb{R}^n$ and consider the *convex integer program*

$$\begin{aligned} &\textbf{minimize} && c'x && && (2.4)\\ &\textbf{subject to} && f_i(x) \leq 0 && 1 \leq i \leq m, && (2.5)\\ & && x \textbf{ integer.} && && (2.6) \end{aligned}$$

Branch and bound method

When we use a branch and bound method for solving (4)-(6), we first solve the underlying convex program without the constraint x being integral. If the solution x^* is integral, we are done. Otherwise, say, x_1^* is not integral. We create two new problems by adding either

$$x_1 \leq [x_1^*]$$

or

$$x_1 \geq [x_1^*] + 1.$$

Instead of solving these two subproblems we can - due to the *convexity of the level sets* - fix the variable x_1 to $[x_1^*]$ and $[x_1^*] + 1$, respectively. Therefore we solve a problem with $x_1 = [x_1^*]$ and a problem with $x_1 = [x_1^*] + 1$. Now assume that the solution of the first subproblem with the additional constraint $x_1 = [x_1^*]$ is still not integral. Then we must generate three new subproblems in the next branching step, namely two subproblems for fixing a new variable to an integer value and one subproblem with fixing x_1 to $[x_1^*] - 1$. For details, see e.g. Burkard (1972). Thus the convexity of the level sets helps to fix variables which accelerates the solution of the problem.

Cutting plane methods

Given problem (2.4)-(2.6), we first solve again the underlying convex program without the constraint x being integral. If the solution obtained in this way is not integral, we search for a valid inequality which cuts off this solution, but which does not cut off any feasible integral solution (separation problem). If no valid inequality can be found, we branch *(branch and cut method)*. This method uses essentially the fact that the intersection of two convex sets is again convex.

Subgradient optimization

For hard combinatorial optimization problems often a strong lower bound can be computed by a Lagrangean relaxation approach which uses the minimization of a non-smooth convex function. Held and Karp (1971) used such an approach very successfully for the symmetric travelling salesman problem, see also Held et al. (1974). We will illustrate this approach by considering the axial 3-dimensional assignment problem.

The axial 3-dimensional assignment problem can be formulated in the following way:

$$\begin{aligned}
\min \quad & \sum_{i=1}^{n}\sum_{j=1}^{n}\sum_{k=1}^{n} c_{ijk}x_{ijk} \\
\text{s.t.} \quad & \sum_{j=1}^{n}\sum_{k=1}^{n} x_{ijk} = 1, \quad i = 1,2,\ldots,n, \\
& \sum_{i=1}^{n}\sum_{k=1}^{n} x_{ijk} = 1, \quad j = 1,2,\ldots,n, \\
& \sum_{i=1}^{n}\sum_{j=1}^{n} x_{ijk} = 1, \quad k = 1,2,\ldots,n, \\
& x_{ijk} \in \{0,1\} \text{ for all } 1 \le i,j,k \le n\,.
\end{aligned}$$

Karp (1972) showed that this problem is $\mathcal{NP}$**-hard.** In order to compute strong lower bounds we take two blocks of the constraints into the objective function via Lagrangean multipliers:

$$L(\pi,\epsilon) :=$$

$$\min\left\{\sum_{i=1}^{n}\sum_{j=1}^{n}\sum_{k=1}^{n}(c_{ijk}+\pi_j+\epsilon_i)x_{ijk} - \sum_{j=1}^{n}\pi_j - \sum_{i=1}^{n}\epsilon_i\right\}$$

such that

$$\begin{aligned}
& \sum_{i=1}^{n}\sum_{j=1}^{n} x_{ijk} = 1, \quad k = 1,2,\ldots,n \\
& x_{ijk} \in \{0,1\}, \qquad 1 \le i,j,k \le n, \\
& \qquad \pi \in \mathbb{R}^n, \epsilon \in \mathbb{R}^n\,.
\end{aligned}$$

$L(\pi, \epsilon)$ is a concave function as minimum of affine-linear functions. For finding its maximum a subgradient method can be used: Start with $\pi^r := \epsilon^r := 0$, use a greedy algorithm for evaluating $L(\pi^r, \epsilon^r)$ and let x^r_{ijk} be the corresponding optimal solution. Define $v^r_{i_0} := |\{x^r_{i_0,j,k} | x_{i_0,j,k} = 1\}| - 1$ for all $i_0 = 1, 2, \ldots, n$ and $w^r_{j_0} := |\{x^r_{i,j_0,k} | x_{i,j_0,k} = 1\}| - 1$ for all $j_0 = 1, 2, \ldots, n$. If $v^r = w^r = (0, 0, \ldots, 0)$, then the maximum is reached. Otherwise π and ϵ are updated with a suitable step length λ,

$$\pi^{r+1} := \pi^r + \lambda_r w^r, \quad \epsilon^{r+1} := \epsilon^r + \lambda_r v^r$$

and the next iteration is started. For details see Burkard and Rudolf (1993).

Other techniques

In connection with the application of semidefinite programming to combinatorial optimization problems, various other techniques from convex optimization were applied to discrete optimization problems. One of the most interesting approaches is due to Brixius and Anstreicher (2001) and concerns quadratic assignment problems (QAPs). Quadratic assignment problems which are very important for the practice, but notoriously hard to solve, can be stated as trace minimization problems of the form

$$\min \operatorname{tr}\, (AXB + C)X^t,$$

where A, B and C are given $n \times n$ matrices and X is an $n \times n$ permutation matrix. First, one can relax the permutation matrix to an orthogonal matrix with row and column sum equal to 1. Then one can separate the linear and the quadratic term in the objective function. Brixius and Anstreicher interpret the relaxed problem in terms of semidefinite programming and evaluate a new bound which requires the solution of a convex quadratic program. This is performed via an interior point algorithm. The solution of the quadratic program allows to fix variables for the studied QAP and leads to very good computational results.

5. Convex configurations and combinatorial optimization problems

Many combinatorial optimization problems become easier to solve, if the input stems from convex sets. For example, the following fact about the **planar travelling salesman problem** (TSP), i.e., a TSP where the distances between the cities are given by (Euclidean) distances in the plane, is well known. Assume that the cities lie on the boundary of a convex set in the plane. Then an optimal solution is obtained by passing through the cities in clockwise or counterclockwise order on the boundary. The reason for this is that in an optimal Hamiltonian cycle in the Euclidean plane the edges of the cycle never cross due to the quadrilateral inequality. Due to convexity every other solution than the clockwise or anticlockwise tour would have some crossing edges. It can be tested in $O(n \log n)$ time whether n given points in the plane lie on the boundary of a convex set, see e.g. Preparata and Shamos (1988). Their cyclic order can be found within the same time. If a distance matrix for a planar TSP is given, it can be tested in $O(n^2)$ time whether this is a distance matrix of vertices of a convex polygon or not (see Hotje's procedure in Burkard (1990)). Thus the case of a planar TSP whose cities are vertices of a convex polygon can easily be recognized and solved even though the planar TSP is $\mathcal{NP}$**-hard** in general (see Papadimitriou (1977)).

The same arguments as above apply, if the distances between cities are measured in the l_1**-norm** and the cities are vertices of a rectilinearly convex set in the plane. A region R is called *rectilinearly convex* if every horizontal or vertical line intersects R in an interval.

The distance matrix $C = (c_{ij})$ of a planar TSP whose vertices lie on the boundary of a convex polygon has a special structure. The matrix fulfills the so-called *Kalmanson conditions*

$$c_{ij} + c_{kl} \leq c_{ik} + c_{jl} \quad \text{for all } 1 \leq i < j < k < l \leq n, \tag{2.7}$$

$$c_{il} + c_{jk} \leq c_{ik} + c_{jl} \quad \text{for all } 1 \leq i < j < k < l \leq n. \tag{2.8}$$

Kalmanson (1975) showed that a TSP whose distance matrix fulfills these Kalmanson conditions has the tour $< 1, 2, ..., n-1, n >$ as optimal solution, i.e. the travelling salesperson starts in city 1, goes then to city 2, and so on until she or he returns from city n to city 1. The definition of the Kalmanson property depends on a suited numbering of the rows and columns (i.e. of the cities) of the distance matrix. If after a renumbering of the rows and columns a matrix becomes a Kalmanson matrix, we speak of a *permuted Kalmanson matrix.* Permuted Kalmanson matrices can be recognised in $O(n^2)$ time by a method due to Christopher, Farach and Trick (see Christopher et al. (1996) and Burkard et al. (1998)). Permuted Kalmanson matrices are also interesting in connection with the so-called *master tour problem.* A *master tour* π for a set V of cities fulfills the following property: for every $V' \subseteq V$ an optimum travelling salesman tour for V' is obtained by removing from π the cities that are not in V'. Deĭneko, Rudolf and Woeginger (see Deĭneko et al. (1998)) showed that the master tour property holds if and only if the distance matrix is a permuted Kalmanson matrix.

Now let us turn to the **minimum spanning tree problem** (MST). Let a finite undirected and connected graph $G = (V, E)$ with vertex set V and edge set E be given. Every edge e has a positive length $c(e)$. (MST) asks for a spanning tree $T = (V, E_T)$, $E_T \subseteq E$, of G such that

$$\sum_{e \in E_T} c(e)$$

is minimum. If n points in the plane are given, the graph G is given by the complete complete graph K_n of these points and the edge lengths are given as (Euclidean) distances between the points. We have

Theorem 5.1 *A minimum spanning tree for n points in the plane can be computed in $O(n \log n)$ time. If the points lie on the boundary of a convex set and are given in cyclic order, the MST problem can be solved in $O(n)$ time.*

The idea behind this theorem is (see e.g. Mehlhorn (1984b)) that a minimum spanning tree of the given points contains only edges of the Delauney triangulation of these points. According to Aggarwal et al. (1989) the Delaunay triangulation of vertices of a convex polygon can be computed in $O(n)$ time. The Delaunay triangulation leads to a planar graph. Mehlhorn (1984a) showed that the MST in a planar graph can be solved in $O(n)$ time.

Similar results hold for the maximum spanning tree problem (see Monma et al. (1990)).

Now let us turn to the **Steiner tree problem** (STP) which has many applications in network design or VLSI design. The Steiner tree problem asks for the shortest connection of k given points, called *terminals* where it is allowed to introduce additional points, the so-called *Steiner points.* For example, if the terminals are the vertices of an equilateral triangle, then the center of gravity of the triangle is introduced as Steiner point. The connection of the Steiner point with each of the terminals yields the shortest Steiner tree of the given points. The length of a Steiner tree is again measured as sum of the lengths of all edges in the tree. The Steiner tree problem is $\mathcal{NP}$**-hard** in general (see Garey et al. (1977)). A Steiner tree problem is called *Euclidean,* if the terminals lie in the plane and all distances are measured in the Euclidean metric. For Euclidean Steiner tree problems, Provan (1988) showed the following result.

Theorem 5.2 *If the terminals of a Euclidean Steiner tree problem lie on the boundary of a convex set in the plane, then there exists a fully polynomial approximation scheme, i.e., there is an algorithm which constructs for any fixed $\epsilon > 0$ a Steiner tree T of length $l(T)$ such that*

$$l(T) \leq (1 + \epsilon) \cdot Opt$$

where Opt is the optimum value of the problem under consideration and where the running time of the algorithm is polynomial in n and $1/\epsilon$.

An even better result can be shown if the distances between vertices are measured in the l_1**-norm.** This problem plays a special role in VLSI design where the connections between points use only horizontal or vertical lines of a grid. Provan (1988) showed

Theorem 5.3 *If the n terminal nodes of a Steiner tree problem lie on the boundary of a rectilinearly convex set and the distances between vertices are measured in the l_1-norm, then the Steiner tree problem can be solved in $O(n^6)$ time.*

Now let us turn to **matching and assignment problems** in the plane. Let $2n$ points on the boundary of a convex set in the plane be given. We consider the complete graph K_{2n} whose vertices are these points and whose edge lengths are the Euclidean distances between the points. The weight of a matching M equals the sum of all edge lengths of M. Marcotte and Suri (1991) showed that a minimum weight matching in this K_{2n} can be found in $O(n \log n)$ time. Moreover, they showed that a maximum weight matching can be found in linear time.

Next we color n vertices of this K_{2n} red and n vertices blue and we allow edges only between vertices of different color. This gives rise to a matching problem in a bipartite graph *(assignment problem).* Marcotte and Suri (1991) showed also that the assignment problem defined above can be solved in $O(n \log n)$ time. Moreover, the verification of a minimum matching can be performed in $O(n \cdot \alpha(n))$ steps, where $\alpha(n)$ is the very slow growing inverse Ackermann function.

6. Conclusion

In the previous sections we outlined some of the important roles convexity plays in theory and practice of integer programming. But there are many other areas in discrete optimization, where (generalized) convexity is crucial. Let me just mention location problems, combinatorial optimization problems involving Monge arrays and submodular functions.

In **location theory** one wants to place one or more service centers such that the customers are served best. Classical location models lead to convex objective functions. The convexity of these functions is exploited in fast algorithms for solving these problems. For example, the simple form of Goldman's algorithm (see Goldman (1971)) for finding the 1-median in a tree is mainly due to the convexity of the corresponding objective function.

Secondly, I would like to mention **Monge arrays.** A real $m \times n$ matrix $C = (c_{ij})$ is called *Monge matrix,* if

$$c_{ij} + c_{rs} \leq c_{is} + c_{rj} \quad \text{for all } 1 \leq i < r \leq m, \; 1 \leq j < s \leq n. \tag{2.9}$$

Many combinatorial optimization problems turn out to be easier to solve, if the problems are related to a Monge matrix. For example, if the cost coefficients of a transportation problem fulfill the Monge property (2.9), then the transportation problem can be solved in a greedy way by the north west corner rule. Or, if the distances of a travelling salesman problem fulfill the Monge property, then the TSP can be solved in linear time. A survey on Monge properties and combinatorial optimization can be found in Burkard, Klinz and Rudolf (see Burkard et al. (1996)). Monge matrices are closely related to **submodular functions.** A set function $f : 2^V \to \mathbb{R}$ is called submodular, if

$$f(X) + f(Y) \geq f(X \cup Y) + f(X \cap Y) \quad \text{for all } X, Y \subseteq V.$$

Submodular functions exhibit many features similar to convex functions and they play among others an important role in combinatorial optimization problems involving matroids. For details, the reader is referred to the pioneering work of Murota (1998).

Acknowledgments

My thanks go to Bettina Klinz for various interesting discussions on the role of convexity in connection to the travelling salesman problem.

References

Aggarwal, A., Guibas, L.J., Saxe, J., and Shor, P.W. (1989), A linear algorithm for computing the Voronoi diagram of a convex polygon, *Discrete Comp. Geom.,* Vol. 4, 591–604.

Bank, B. and Mandel, R. (1988), (Mixed-) Integer solutions of quasiconvex polynomial inequalities. In: *Advances in Mathematical Optimization,* J. Guddat et al. (eds), Akademie Verlag, Berlin, pp. 20–34.

Berge, C. (1972), Balanced matrices, *Math. Programming,* Vol. 2, 19–31.

Brixius, N.W. and Anstreicher, K.W. (2001), Solving quadratic assignment problems using convex quadratic programming relaxations, dedicated to Professor Laurence C. W. Dixon on the occasion of his 65th birthday, *Optim. Methods Softw.,* Vol. 16, 49–68.

Burkard, R.E. (1972), *Methoden der ganzzahligen Optimierung.* (Springer, Vienna), 1972.

Burkard, R.E. (1990), Special cases of the travelling salesman problem and heuristics, *Acta Mathematicae Applicatae Sinica,* Vol. 6, 273–288.

Burkard, R.E., Deĭneko, V.G., van Dal, R., van der Veen, J.A.A., and Woeginger G.J. (1998), Well-solvable special cases of the travelling salesman problem: a survey, *SIAM Review,* Vol. 40, 496–546.

Burkard, R.E., Klinz, B., and Rudolf, R. (1996), Perspectives of Monge properties in optimization, *Discrete Appl. Mathematics* Vol. 70, 95–161.

Burkard, R.E. and Rudolf, R. (1993), Computational investigations on 3-dimensional axial assignment problems, *Belgian J. of Operations Research,* Vol. 32, 85–98.

Christopher, G., Farach, M., and Trick, M. (1996), The structure of circular decomposable metrics, in: *Algorithms – ESA '96, Lecture Notes in Comp. Sci.,* Vol. 1136, Springer, Berlin, pp.486–500.

Conforti, M., Cornuéjols, G., and Rao, M.R. (1999), Decomposition of balanced matrices, *J. Combinatorial Theory Ser. B,* Vol. 77, 292–406.

Cornuéjols, G. (2001), *Combinatorial Optimization: Packing and Covering,* SIAM, Philadelphia.

Deĭneko, V.G., Rudolf, R., and Woeginger, G.J. (1998), Sometimes traveling is easy: The master tour problem, *SIAM J. Discrete Math.,* Vol. 11, 81–83.

Edmonds, J. (1965), Maximum matching and a polyhedron with 0-1 vertices, *J. Res. Nat. Bur. Standards,* Vol 69b, 125–130.

Fulkerson, D.R., Hoffman, A.J., and Oppenheim, R. (1974), On balanced matrices, *Math. Programming Studies,* Vol. 1, 120–132.

Goldman, A.J. (1971), Optimal center location in simple networks, *Transportation Science,* Vol. 5, 212–221.

Garey, M.R., Graham, R.L., and Johnson, D.S. (1977), The complexity of computing Steiner minimal trees, *SIAM J. Appl. Math.,* Vol. 32, 835–859.

Grötschel, M., Lovász, L., and Schrijver, A. (1988), *Geometric Algorithms and Combinatorial Optimization,* Springer, Berlin.

Held, M. and Karp, R.M. (1971), The traveling-salesman problem and minimum spanning trees: Part II, *Math. Programming,* Vol. 1, 6–25.

Held, M., Wolfe, P., and Crowder, H.P. (1974), Validation of subgradient optimization, *Math. Programming,* Vol. 6, 62–88.

Hoffman, A. and Kruskal, J.B. (1956), Integral boundary points of convex polyhedra, in: *Linear Inequalities and Related Studies,* H. Kuhn and A. Tucker (eds.), Princeton University Press, Princeton, pp. 223–246.

Kalmanson, K. (1975), Edgeconvex circuits and the travelling salesman problem, *Canad. J. Math.*, Vol. 27, 1000–1010.

Karp, R.M. (1972), Reducibility among combinatorial problems, in: R.E. Miller and J.W. Thatcher (eds.), *Complexity of Computer Computations,* Plenum Press, New York, pp.85–103.

Marcotte, O. and Suri, S. (1991), Fast matching algorithms for points on a polygon, *SIAM J. Comput.*, Vol. 20, 405–422.

Mehlhorn, K. (1984a), *Data Structures and Algorithms 2: Graph Algorithms and NP-Completeness,* Springer, Berlin.

Mehlhorn, K. (1984b) *Data Structures and Algorithms 3: Multi-dimensional searching and Computational Geometry,* Springer, Berlin.

Minkowski, H. (1896), *Geometrie der Zahlen,* Teubner, Leipzig.

Monma, C., Paterson, M., Suri, S., and Yao, F. (1990), Computing Euclidean maximum spanning trees, *Algorithmica,* Vol. 5, 407–419.

Murota, K. (1998), Discrete convex analysis, *Mathematical Programming,* Vol. 83, 313–371.

Nemhauser, G.L. and Wolsey, L.A. (1988), *Integer and Combinatorial Optimization,* Wiley, New York.

Papadimitriou, C.H. (1977), The Euclidean TSP is NP-complete, *Theoret. Comp. Sci.,* Vol. 4, 237–244.

Preparata, F.P. and Shamos, M.I. (1988), *Computational Geometry: An Introduction,* Springer, Berlin.

Provan, J.S. (1988), Convexity and the Steiner tree problem, *Networks,* Vol. 18, 55–72.

Seymour, P.D. (1980), Decomposition of regular matroids, *J. Combinatorial Theory Ser. B,* Vol. 28, 305–359.

Chapter 3

LIPSCHITZIAN STABILITY OF PARAMETRIC CONSTRAINT SYSTEMS IN INFINITE DIMENSIONS

Boris S. Mordukhovich*
Dept of Mathematics
Wayne State University, USA

Abstract This paper mainly concerns applications of the generalized differentiation theory in variational analysis to robust Lipschitzian stability for various classes of parametric constraint systems in infinite dimensions including problems of nonlinear and nondifferentiable programming, implicit multifunctions, etc. The basic tools of our analysis involve coderivatives of set-valued mappings and associated limiting subgradients and normals for nonsmooth functions and sets. Using these tools, we establish new sufficient as well as necessary and sufficient conditions for robust Lipschitzian stability of parametric constraint systems with evaluating the exact Lipschitzian bounds. Most results are obtained for the class of Asplund spaces, which particularly includes all reflexive spaces, although some important characteristics are given in the general Banach space setting.

Keywords: Variational analysis, generalized differentiation, parametric constraint systems, Lipschitzian stability, coderivatives, Asplund spaces.

MSC2000: 49J52, 58C06, 90C31

---

* This research has been supported by the National Science Foundation under grants DMS-0072179 and DMS-00304989.
email :boris@math.wayne.edu

1. Introduction

The paper is mainly devoted to applications of modern tools of variational analysis and generalized differentiation to robust Lipschitzian stability of parametric constraint systems in infinite-dimensional spaces. We study a general class of set-valued mappings (multifunctions) $F\colon X \rightrightarrows Y$ given in the form

$$F(x) := \{y \in Y \mid g(x,y) \in \Theta, \quad (x,y) \in \Omega\}, \tag{3.1}$$

where $g\colon X \times Y \to Z$ is a single-valued mapping between Banach spaces, and where Θ and Ω are subsets of the spaces Z and $X \times Y$, respectively. Such set-valued mappings describe *constraint systems* depending on a parameter $x \in X$. One can view (3.1) as a natural generalization of the *feasible solution sets* to perturbed problems in nonlinear programming with inequality and equality constraints given by

$$F(x) := \Big\{y \in Y \Big| \begin{array}{ll} \varphi_i(x,y) \le 0, & i = 1,\dots,m; \\ \varphi_i(x,y) = 0, & i = m+1,\dots,m+r \end{array}\Big\}, \tag{3.2}$$

where φ_i are real-valued functions on $X \times Y$. Clearly (3.2) is a special case of (3.1) with $g = (\varphi_1,\dots,\varphi_{m+r})$, $\Omega = X \times Y$, $Z = \mathbb{R}^{m+r}$, and

$$\Theta := \Big\{(\alpha_1,\dots,\alpha_{m+r}) \Big| \begin{array}{ll} \alpha_i \le 0 \text{ for } i = 1,\dots,m & \text{and} \\ \alpha_i = 0 \text{ for } i = m+1,\dots,m+r \end{array}\Big\}. \tag{3.3}$$

Another special case of (3.1) with $\Theta = \{0\}$ and $\Omega = X \times Y$ is addressed by the classical implicit function theorem when the mapping

$$F(x) := \{y \in Y \mid g(x,y) = 0\} \tag{3.4}$$

is single-valued and smooth. In general we have *implicit multifunctions* in (3.4) and are interested in properties of their Lipschitz continuity. Some other important classes of systems that can be reduced to (3.1) include *parametric generalized equations,* in the sense of Robinson (1979),

$$0 \in f(x,y) + Q(x,y)$$

with $f\colon X \times Y \to Z$ and $Q\colon X \times Y \rightrightarrows Z$; see Mordukhovich (2002) for more details and references.

Our primary interest is *robust Lipschitzian stability* of parametric constraint systems (3.1) and their specifications. The main attention is paid to the concept of robust Lipschitzian behavior introduced by Aubin

(1984) under the name of "pseudo-Lipschitz" multifunctions. In our opinion, it would be better to use the term of *Lipschitz-like* multifunctions referring to this kind of Lipschitzian behavior, which is probably the most proper extension of the classical Lipschitz continuity to set-valued mappings (while "pseudo" means "false"; cf. Rockafellar and Wets (1998), where this property of multifunctions is called the *Aubin property* without specifying its Lipschitzian nature). It is well known that Aubin's Lipschitz-like property of an arbitrary set-valued mapping $F\colon X \rightrightarrows Y$ between Banach spaces is equivalent to *metric regularity* as well as to *linear openness* of its inverse $F^{-1}\colon Y \rightrightarrows X$. These properties play a fundamental role in nonlinear analysis, optimization, and their applications. Note that both Lipschitz-like and classical Lipschitz properties are *robust* (stable) with respect to perturbations of initial data, which is important for sensitivity analysis.

The main tools for our studying robust Lipschitzian stability in this paper involves *coderivatives* of set-valued mappings that give adequate extensions of the classical adjoint derivative operator, enjoy a comprehensive calculus, and play a crucial role in characterizations of Lipschitzian and related properties; see Mordukhovich (1997) and the references therein. Applications of coderivative analysis to various problems related to Lipschitzian stability of parametric constraint systems and generalized equations, mostly in finite dimensions, are given in Dontchev, Lewis and Rockafellar (2003), Dontchev and Rockafellar (1996), Henrion and Outrata (2001), Henrion and Römisch (1999), Jourani (2000), Klatte and Kummer (2002), Levy (2001), Levy and Mordukhovich (2002), Levy, Poliquin and Rockafellar (2000), Mordukhovich (1994a), Mordukhovich (1994b), Mordukhovich (2002), Mordukhovich and Outrata (2001), Mordukhovich and Shao (1997), Outrata (2000), Poliquin and Rockafellar (1998), Rockafellar and Wets (1998), Treiman (1999), Ye (2000), Ye and Zhu (2001) among other publications.

The main emphasis of this paper is a local coderivative analysis of Lipschitzian stability for constraint systems (3.1) and their specifications in infinite-dimensional (mostly Asplund) spaces. We base on the coderivative characterizations of the Lipschitz-like property for general multifunctions as in Mordukhovich (1997) using two kind of coderivatives–*normal* and *mixed*–that agree in finite dimensions. The mentioned characterizations involve also some *sequential normal compactness* (SNC) properties of multifunctions that are automatic in finite dimensions.

To apply the mentioned characterizations to the constraint systems (3.1) and their important specifications, we are going to use *coderivative calculus rules* available in Banach and Asplund spaces as well as the recently developed *SNC calculus* ensuring the preservation of the SNC and related properties under various operations. In this way we obtain efficient *sufficient* conditions, as well as *necessary and sufficient* conditions, for robust Lipschitzian stability of the parametric constraint systems under consideration with *upper estimating* (and also *exact computing* in some cases) the exact bounds of their Lipschitzian moduli.

The rest of the paper is organized as follows. Section 2 presents basic definitions and preliminary material needed in the sequel. In Section 3 we express (compute or upper estimate) coderivatives of general parametric constraint systems and their specifications in terms of initial data. These results are certainly of independent interest while playing a crucial role (along with the SNC calculus in infinite dimensions) for the study of robust Lipschitzian stability via the point-based coderivative criteria. The main results on Lipschitzian stability of constraint systems are established in Section 4.

Throughout the paper we use standard notation, with special symbols introduced where they are defined. Unless otherwise stated, all spaces considered are Banach whose norms are always denoted by $\|\cdot\|$. For any space X we consider its dual space X^* equipped with the weak* topology w^*, where $\langle\cdot,\cdot\rangle$ means the canonical pairing. For multifunctions $F\colon X \rightrightarrows X^*$ the expression

$$\operatorname*{Lim\,sup}_{x\to\bar{x}} F(x) := \Big\{x^* \in X^* \;\Big|\; \exists \text{ sequences } x_k \to \bar{x} \text{ and } x_k^* \xrightarrow{w^*} x^* \text{ with } x_k^* \in F(x_k) \text{ for all } k \in \mathbb{N}\Big\}$$

signifies the *sequential* Painlevé-Kuratowski upper/outer limit with respect to the norm topology in X and the weak* topology in X^*; $\mathbb{N} := \{1, 2, \ldots\}$. Recall that $F\colon X \rightrightarrows Y$ is *positively homogeneous* if $F(\alpha x) = \alpha F(x)$ for all $x \in X$ and $\alpha > 0$. The *norm* a positively homogeneous multifunction is defined by

$$\|F\| := \sup\big\{\|y\| \;\big|\; y \in F(x) \text{ and } \|x\| \le 1\big\}. \tag{3.5}$$

2. Basic Definitions and Preliminaries

Our primary interest in this paper is the following Lipschitzian property of multifunctions known also as pseudo-Lipschitzian or Aubin property. Given $F\colon X \rightrightarrows Y$ and $(\bar{x}, \bar{y}) \in \operatorname{gph} F$, we say that F is *Lipschitz-like* around $(\bar{x}, \bar{y})$ with modulus $\ell \ge 0$ if there are neighborhood U of $\bar{x}$ and V of $\bar{y}$ such that

$$F(x) \cap V \subset F(u) + \ell\|x - u\|\mathbb{B}_Y \quad \text{for all } x, u \in U, \tag{3.6}$$

where $\mathbb{B}_Y$ stands for the closed unit ball in Y. The infimum of all such moduli $\{\ell\}$ is called the *exact Lipschitzian bound* of F around $(\bar{x}, \bar{y})$ and is denoted by lip $F(\bar{x}, \bar{y})$.

If $V = Y$ is (3.6), the above Aubin's Lipschitz-like property reduces to the local Lipschitz continuity of F around $\bar{x}$ with respect to the Pompieu-Hausdorff distance on 2^Y, and for single-valued mappings $F = f\colon X \to Y$ it agrees with the classical local Lipschitz continuity. For general set-valued mappings F the (local) Lipschitz-like property can be viewed as a localization of Lipschitzian behavior not only relative to a point of the domain but also relative to a particular point of the image $\bar{y} \in F(\bar{x})$.

We are able to provide *complete dual characterizations* of the Lipschitz-like property (and hence the classical local Lipschitzian property) using appropriate constructions of generalized differentiation. To present them, we first recall the definitions of coderivatives for set-valued mappings, which are the basic constructions of our study. The reader may consult Mordukhovich (1997) for more references and discussions.

Given $F\colon X \rightrightarrows Y$ and $\varepsilon \geq 0$, define the ***ε-coderivative*** of F at $(\bar{x}, \bar{y}) \in \operatorname{gph} F$ as the set-mapping $\widehat{D}_\varepsilon^* F(\bar{x}, \bar{y})\colon Y^* \rightrightarrows X^*$ with the values

$$\widehat{D}_\varepsilon^* F(\bar{x}, \bar{y})(y^*) := \Big\{ x^* \in X^* \Big| \limsup_{(x,y) \overset{\operatorname{gph} F}{\to} (\bar{x},\bar{y})} \frac{\langle x^*, x - \bar{x}\rangle - \langle y^*, y - \bar{y}\rangle}{\|x - \bar{x}\| + \|y - \bar{y}\|} \leq \varepsilon \Big\},$$

where $x \overset{\Omega}{\to} \bar{x}$ means that $x \to \bar{x}$ with $x \in \Omega$. We put $\widehat{D}_\varepsilon^* F(\bar{x}, \bar{y})(y^*) = \emptyset$ for all $y^* \in Y^*$ and $\varepsilon \geq 0$ when $(\bar{x}, \bar{y}) \notin \operatorname{gph} F$, and denote $\widehat{D}^* F(\bar{x}, \bar{y}) := \widehat{D}_0^* F(\bar{x}, \bar{y})$. Then the *normal coderivative* of F at $(\bar{x}, \bar{y})$ is defined by

$$D_N^* F(\bar{x}, \bar{y})(\bar{y}^*) := \operatorname*{Lim\,sup}_{\substack{(x,y)\to(\bar{x},\bar{y}) \\ y^* \overset{w^*}{\to} \bar{y}^* \\ \varepsilon \downarrow 0}} \widehat{D}_\varepsilon^* F(x, y)(y^*), \tag{3.7}$$

i.e., $\bar{x}^* \in D_N^* F(\bar{x}, \bar{y})(\bar{y}^*)$ if and only if there are sequences $\varepsilon_k \downarrow 0$, $(x_k, y_k) \to (\bar{x}, \bar{y})$, and $(x_k^*, y_k^*) \overset{w^*}{\to} (\bar{x}^*, \bar{y}^*)$ with $(x_k, y_k) \in \operatorname{gph} F$ and $x_k^* \in \widehat{D}_{\varepsilon_k}^* F(x_k, y_k)(y_k^*)$. The *mixed coderivative* of F at $(\bar{x}, \bar{y})$ is

$$D_M^* F(\bar{x}, \bar{y})(\bar{y}^*) := \operatorname*{Lim\,sup}_{\substack{(x,y)\to(\bar{x},\bar{y}) \\ y^* \to \bar{y}^* \\ \varepsilon \downarrow 0}} \widehat{D}_\varepsilon^* F(x, y)(y^*), \tag{3.8}$$

i.e., $D_M^* F(\bar{x}, \bar{y})(\bar{y}^*)$ is the collection of such $\bar{x}^* \in X^*$ for which there are sequences $\varepsilon_k \downarrow 0$, $(x_k, y_k, y_k^*) \to (\bar{x}, \bar{y}, \bar{y}^*)$, and $x_k^* \overset{w^*}{\to} \bar{x}^*$ with $(x_k, y_k) \in \operatorname{gph} F$ and $x_k^* \in \widehat{D}_{\varepsilon_k}^* F(x_k, y_k)(y_k^*)$. One can equivalently put $\varepsilon = 0$ in (3.7) and (3.8) if F is closed-graph around $(\bar{x}, \bar{y})$ and if both X and Y are *Asplund,* i.e., such Banach spaces on which every convex continuous function is generically Fréchet differentiable (in particular, any reflexive spaces); see Phelps (1993) for more information on Asplund spaces.

It follows from the definitions that $D^*_M F(\bar{x},\bar{y})(y^*) \subset D^*_N F(\bar{x},\bar{y})(y^*)$ when the equality obviously holds if Y is finite-dimensional. Note that the above inclusion may be strict even for single-valued Lipschitzian mappings $f\colon I\!R \to Y$ with values in Hilbert spaces Y that are Fréchet differentiable at $\bar{x}$; see Example 2.9 in Mordukhovich and Shao (1998). We say that F is *coderivatively normal* at $(\bar{x},\bar{y})$ if

$$\|D^*_M F(\bar{x},\bar{y})\| = \|D^*_N F(\bar{x},\bar{y})\|, \tag{3.9}$$

where the norms of the coderivatives, as positively homogeneous multifunctions, are computed by (3.5). The mapping F is said to be *strongly coderivatively normal* at $(\bar{x},\bar{y})$ if

$$D^*_M F(\bar{x},\bar{y}) = D^*_N F(\bar{x},\bar{y}) := D^* F(\bar{x},\bar{y}). \tag{3.10}$$

Obviously (3.10) implies (3.9) but not vice versa, as shown by the mentioned example. Properties (3.9) and (3.10) hold if F is *graphically regular* at $(\bar{x},\bar{y})$ in the sense that

$$D^*_N F(\bar{x},\bar{y})(y^*) = \widehat{D}^* F(\bar{x},\bar{y})(y^*) \text{ for all } y^* \in Y^*.$$

The latter class includes set-valued mappings with convex graphs and also single-valued mappings strictly differentiable at $\bar{x}$ for which

$$\widehat{D}^* f(\bar{x})(y^*) = D^*_N f(\bar{x})(y^*) = D^*_M f(\bar{x})(y^*) = \{\nabla f(\bar{x})^* y^*\}.$$

Other sufficient conditions for properties (3.9) and (3.10) are presented and discusses in Mordukhovich (2002).

Next let us consider the subdifferential and normal cone constructions for functions and sets associated with the above coderivatives. Given an extended-real-valued function $\varphi\colon X \to \overline{I\!R} := [-\infty,\infty]$ finite at $\bar{x}$, we define its *subdifferential* at $\bar{x}$ by

$$\partial\varphi(\bar{x}) := D^* E_\varphi(\bar{x},\varphi(\bar{x}))(1) \text{ with } E_\varphi(x) := \{\nu \in I\!R \mid \nu \geq \varphi(x)\}, \tag{3.11}$$

where D^* stands for the common coderivative (3.10).

The *normal cone* to a set $\Omega \subset X$ at $\bar{x} \in \Omega$ can be defined as

$$N(\bar{x};\Omega) := D^*\Delta(\bar{x};\Omega)(1), \tag{3.12}$$

where $\Delta(x;\Omega) = 0$ if $x \in \Omega$ and $\Delta(x;\Omega) = \emptyset$ otherwise. The set Ω is *normally regular* at $\bar{x} \in \Omega$ if $\Delta(\cdot;\Omega)$ is graphically regular at $(\bar{x},0)$.

Intrinsic descriptions of $\partial\varphi$ and $N(\cdot;\Omega)$ with comprehensive theories for these objects can be found in Mordukhovich (1988) and Rockafellar and Wets (1998) in finite dimensions and in Mordukhovich and Shao (1996a) in infinite-dimensional (mostly Asplund) spaces.

Note the relationship

$$D_N^*F(\bar{x},\bar{y})(y^*) = \{x^* \in X^* \mid (x^*,-y^*) \in N((\bar{x},\bar{y});\operatorname{gph} F)\} \tag{3.13}$$

and the scalarization formulas

$$D_M^*f(\bar{x})(y^*) = \partial\langle y^*,f\rangle(\bar{x}), \quad D_N^*f(\bar{x})(y^*) = \partial\langle y^*,f\rangle(\bar{x}), \tag{3.14}$$

where the first formula holds in any Banach spaces, while the second one requires that X is Asplund and f is ***w^*-strictly** Lipschitzian* around $\bar{x}$ in the following sense: f is Lipschitz continuous around $\bar{x}$ and for every $v \in X$ and every sequences $x_k \to \bar{x}$, $t_k \downarrow 0$, and $y_k^* \xrightarrow{w^*} 0$ one has

$$\langle y_k^*, y_k\rangle \to 0 \text{ with } y_k := \frac{f(x_k+t_kv)-f(x_k)}{t_k} \quad \text{as } k \to \infty;$$

see Mordukhovich and Wang (2003b). The latter property always holds when f is compactly Lipschitzian in the sense of Thibault (1980).

The generalized differential constructions (3.7), (3.8), (3.11), and (3.12) enjoy fairly rich calculi in both finite-dimensional and infinite-dimensional settings; see Rockafellar and Wets (1998), Mordukhovich (1997), and Mordukhovich (2001) with the references therein. These calculi require natural qualification conditions and also the so-called "normal compactness" conditions needed only in infinite dimensions; see Borwein and Strojwas (1985), Ioffe (2000), Jourani and Thibault (1999), Mordukhovich and Shao (1997), and Penot (1998) for the genesis of such properties and various applications. The following two properties formulated in Mordukhovich and Shao (1996b) are of particular interest for applications in this paper.

A mapping $F\colon X \rightrightarrows Y$ is *sequentially normally compact* (SNC) at $(\bar{x},\bar{y}) \in \operatorname{gph} F$ if for any sequences $(\varepsilon_k, x_k, y_k, x_k^*, y_k^*) \in [0,\infty) \times (\operatorname{gph} F) \times X^* \times Y^*$ satisfying

$$\varepsilon_k \downarrow 0, \quad (x_k,y_k) \to (\bar{x},\bar{y}), \quad x_k^* \in \widehat{D}^*_{\varepsilon_k}F(x_k,y_k)(y_k^*) \tag{3.15}$$

one has $(x_k^*,y_k^*) \xrightarrow{w^*} (0,0) \Longrightarrow \|(x_k^*,y_k^*)\| \to 0$ as $k \to \infty$. A mapping F is *partially sequentially normally compact* (PSNC) at $(\bar{x},\bar{y})$ if for any above sequences satisfying (3.15) one has

$$\left[x_k^* \xrightarrow{w^*} 0 \text{ and } \|y_k^*\| \to 0\right] \Longrightarrow \|x_k^*\| \to 0 \quad \text{as } k \to \infty.$$

One may equivalently put $\varepsilon_k = 0$ in the above properties if both spaces X and Y are Asplund and the mapping F is closed-graph around $(\bar{x},\bar{y})$. Respectively, we say that a set $\Omega \subset X$ is *SNC* at $\bar{x} \in \Omega$ if the constant mapping $F(x) \equiv \Omega$ satisfies this property, and that a set $\Omega \subset X \times Y$ is *PSNC with respect to* X at $(\bar{x}, \bar{y}) \in \Omega$ if the mapping $F\colon X \rightrightarrows Y$ with $\operatorname{gph} F = \Omega$ is PSNC at this point.

Note that the SNC property of sets and mappings are closely related to the compactly epi-Lipschitzian property of Borwein and Strojwas (1985); see Ioffe (2000) and Fabian and Mordukhovich (2001) on recent results in this direction. For closed convex sets $\Omega \subset X$ the latter property holds if and only if the affine hull of Ω is a closed finite-codimensional subspace of X with $\operatorname{ri} \Omega \neq \emptyset$; cf. Borwein, Lucet and Mordukhovich (2000). On the other hand, every Lipschitz-like mapping $F\colon X \rightrightarrows Y$ between Banach spaces is PSNC at $(\bar{x}, \bar{y})$, and hence it is SNC at this point when $\dim Y < \infty$; see Theorem 4.1 in the next section. We refer the reader to the recent paper by Mordukhovich and Wang (2003a) for an extended calculus involving SNC and PSNC properties applied below.

3. Coderivatives of Constraint Systems

In this section we obtain results on computing and estimating coderivatives of the general constraint systems (3.1) and some of their specifications. They are used in the next section for deriving efficient conditions for robust Lipschitzian stability of these systems with respect to perturbation parameters. The next theorem provides *precise formulas* (equalities) for computing both coderivatives (3.7) and (3.8) in general Banach space and Asplund space settings.

Theorem 3.1 *Let $F\colon X \rightrightarrows Y$ be given in (3.1) with $g\colon X \times Y \to Z$, $\Theta \subset Z$, and $\Omega \subset X \times Y$. Take $(\bar{x}, \bar{y}) \in \operatorname{gph} F$ and put $\bar{z} := g(\bar{x}, \bar{y}) \in \Theta$. The following assertions hold:*

(i) *Assume that X, Y, Z are Banach spaces, that $\Omega = X \times Y$, and that g is strictly differentiable at $(\bar{x}, \bar{y})$ with the surjective derivative $\nabla g(\bar{x}, \bar{y})$. Then for all $y^* \in Y^*$ one has*

$$D_N^* F(\bar{x}, \bar{y})(y^*) = \big\{x^* \in X^* \big|\ (x^*, -y^*) \in \nabla g(\bar{x}, \bar{y})^* N(\bar{z}; \Theta)\big\}. \tag{3.16}$$

(ii) *Let X, Y, Z be Asplund, and let g be Lipschitz continuous around $(\bar{x}, \bar{y})$. Assume that*

$$\big(D_N^* g(\bar{x}, \bar{y}) \circ N(\bar{z}; \Theta)\big) \cap \big(-N((\bar{x}, \bar{y}); \Omega)\big) = \{0\}, \tag{3.17}$$

that either g is graphically regular at $(\bar{x}, \bar{y})$ with $\dim Z < \infty$ or g is strictly differentiable at $(\bar{x}, \bar{y})$, and that the sets Ω and Θ are locally closed around $(\bar{x}, \bar{y})$ and $\bar{z}$ and normally regular at these points, respec-

tively. Then one has

$$D^*F(\bar{x},\bar{y})(y^*) = \Big\{x^* \in X^* \Big| \begin{array}{l}(x^*,-y^*) \in D^*g(\bar{x},\bar{y}) \circ N(\bar{z};\Theta) \\ +N((\bar{x},\bar{y});\Omega)\Big\}, \quad y^* \in Y^*,\end{array} \tag{3.18}$$

for both coderivatives $D^* = D_N^*, D_M^*$ *provided that*

$$N(\bar{z};\Theta) \cap \ker D_N^* g(\bar{x},\bar{y}) = \{0\} \tag{3.19}$$

and that either Ω *is SNC at* $(\bar{x},\bar{y})$ *while* g^{-1} *is PSNC at* $(\bar{z},\bar{x},\bar{y})$, *or* Θ *is SNC at* $\bar{z}$. *Under the assumptions made* F *is graphically regular at* $(\bar{x},\bar{y})$, *and hence it is strongly coderivatively normal at this point.*

Proof. To prove (i), we observe that

$$\text{gph}\, F = g^{-1}(\Theta) \quad \text{when} \quad \Omega = X \times Y$$

for the mapping F in (3.1). Thus representation (3.16) follows directly from the exact formula for computing the normal cone (3.12) to inverse images established in Mordukhovich and Wang (2002) under the assumptions made in (i).

Now let us prove that, under the assumptions made in (ii), representation (3.18) holds for $D^* = D_N^*$ and also that F is graphically regular at $(\bar{x},\bar{y})$. Observe that in general one has

$$\text{gph}\, F = g^{-1}(\Theta) \cap \Omega \tag{3.20}$$

for the mapping F in (3.1). To prove (3.18) and the graphical regularity of F at $(\bar{x},\bar{y})$, we start with the case when Ω is SNC at $(\bar{x},\bar{y})$. Based on the results in Mordukhovich and Shao (1996a), we conclude that

$$N((\bar{x},\bar{y});\text{gph}\, F) = N((\bar{x},\bar{y});g^{-1}(\Theta)) + N((\bar{x},\bar{y});\Omega) \tag{3.21}$$

and the graph of F is normally regular at $(\bar{x},\bar{y})$ provided that

$$N((\bar{x},\bar{y});g^{-1}(\Theta)) \cap \big(-N((\bar{x},\bar{y});\Omega)\big) = \{0\}. \tag{3.22}$$

Specifying the general chain rule from Mordukhovich (1997) in this case, one has the equality

$$N((\bar{x},\bar{y});g^{-1}(\Theta)) = D_N^* g(\bar{x},\bar{y}) \circ N(\bar{z};\Theta)$$

provided that the qualification condition (3.19) holds and that either Θ is SNC at $\bar{z}$ or g^{-1} is PSNC at $(\bar{z},\bar{x},\bar{y})$. Substituting the latter equality into (3.21) and (3.22), we justify representation (3.18) for $D^* = D_N^*$ and the graphical regularity of F at $(\bar{x},\bar{y})$ under the assumptions made.

When Ω is not assumed to be SNC at $(\bar{x},\bar{y})$, we still get equality (3.21) and the graphical regularity of F at $(\bar{x},\bar{y})$ under condition (3.22) if the set $g^{-1}(\Theta)$ is SNC at $(\bar{x},\bar{y})$. Let us show that the latter holds under the assumptions imposed on g and Θ. To furnish this, we apply the SNC calculus rule from Theorem 3.8 in Mordukhovich and Wang (2003a) when the outer mapping is the indicator function $\delta(\cdot;\Theta)$. Then we conclude that $g^{-1}(\Theta)$ is SNC at $(\bar{x},\bar{y})$ if either g is SNC at $(\bar{x},\bar{y})$ or Θ is SNC at $\bar{z}$ under the qualification condition (3.19). Combining all the above, we complete the proof of the theorem. □

The next theorem gives *upper estimates* for the normal and mixed coderivatives of F under less restrictive assumptions on the initial data in comparison with Theorem 3.1(ii).

Theorem 3.2 *Let $g\colon X\times Y\to Z$ be a mapping between Asplund spaces continuous around $(\bar{x},\bar{y})\in \operatorname{gph} F$ for the constraint system F defined in (3.1), where $\Omega\subset X\times Y$ and $\Theta\subset Z$ are locally closed around $(\bar{x},\bar{y})$ and $\bar{z}=g(\bar{x},\bar{y})$, respectively. Assume the constraint qualifications (3.17), (3.19) and that one of the following conditions holds:*

(a) *Either Ω is SNC at $(\bar{x},\bar{y})$ and Θ is SNC at $\bar{z}$, or g is SNC at $(\bar{x},\bar{y})$.*

(b) *Ω is SNC at $(\bar{x},\bar{y})$ and g^{-1} is PSNC at $(\bar{z},\bar{x},\bar{y})$.*

(c) *g is PSNC at $(\bar{x},\bar{y})$ and Θ is SNC at $\bar{z}$.*

Then one has the inclusion

$$D^*F(\bar{x},\bar{y})(y^*)\subset\Big\{x^*\in X^*\Big|\ \ (x^*,-y^*)\in D_N^*g(\bar{x},\bar{y})\circ N(\bar{z};\Theta)+N((\bar{x},\bar{y});\Omega)\Big\},\quad y^*\in Y^*, \tag{3.23}$$

for both coderivatives $D^*=D_N^*,D_M^*$ of F at $(\bar{x},\bar{y})$.

Proof. It is sufficient to justify (3.23) for $D^*=D_N^*$. Applying the intersection rule from Corollary 4.5 in Mordukhovich and Shao (1996a) to the set in (3.20), we get the inclusion

$$N((\bar{x},\bar{y});\operatorname{gph} F)\subset N((\bar{x},\bar{y});g^{-1}(\Theta))+N((\bar{x},\bar{y});\Omega) \tag{3.24}$$

under the qualification condition (3.22) provided that either Ω is SNC at $(\bar{x},\bar{y})$ or $g^{-1}(\Theta)$ is SNC at $(\bar{x},\bar{y})$. Then we have

$$N((\bar{x},\bar{y});g^{-1}(\Theta))\subset D_N^*g(\bar{x},\bar{y})\circ N(\bar{z};\Theta) \tag{3.25}$$

from the mentioned chain rule in Mordukhovich (1997) under the qualification condition (3.19) if either g^{-1} is PSNC at $(\bar{z},\bar{x},\bar{y})$ or Θ is SNC at $\bar{z}$. By Corollary 3.8 from Mordukhovich and Wang (2003a) we know that

$g^{-1}(\Theta)$ is SNC at $(\bar{x}, \bar{y})$ if either g is SNC at $(\bar{x}, \bar{y})$, or Θ is SNC at $\bar{z}$ while g is PSNC at $(\bar{x}, \bar{y})$ (in particular, when g is locally Lipschitzian around this point). Combining all these conditions and substituting (3.25) into (3.22) and (3.24), we complete the proof of the theorem. □

Next we present some corollaries of the obtained results concerning the specific constraint systems (3.3) and (3.4) important in applications. We start with computing coderivatives of *implicit multifunction.*

Corollary 3.1 *Let $F\colon X \rightrightarrows Y$ given in (3.4), where $g\colon X \times Y \to Z$ with $g(\bar{x}, \bar{y}) = 0$. The following assertions hold for both coderivatives $D^* = D^*_N, D^*_M$:*

(i) *Assume that X, Y, Z are Banach spaces and that g is strictly differentiable at $(\bar{x}, \bar{y})$ with the surjective derivative $\nabla g(\bar{x}, \bar{y})$. Then F is strongly coderivatively normal at $(\bar{x}, \bar{y})$, and one has*

$$D^*F(\bar{x}, \bar{y})(y^*) = \{x^* \in X^* \mid (x^*, -y^*) = \nabla g(\bar{x}, \bar{y})^* z^* \text{ for some } z^* \in Z^*\}.$$

(ii) *Let X and Y be Asplund, and let $\dim Z < \infty$. Assume that g is Lipschitz continuous around $(\bar{x}, \bar{y})$, graphically regular at this point, and satisfies the condition*

$$\ker \partial\langle\cdot, g\rangle(\bar{x}, \bar{y}) = \{0\}.$$

Then F is graphically regular at $(\bar{x}, \bar{y})$, and one has

$$D^*F(\bar{x}, \bar{y})(y^*) = \{x^* \in X^* \mid (x^*, -y^*) \in \partial\langle z^*, g\rangle(\bar{x}, \bar{y}) \text{ for some } z^* \in Z^*\}.$$

(iii) *Let X, Y, Z be Asplund. Assume that g^{-1} is PSNC at $(\bar{z}, \bar{x}, \bar{y})$ and satisfies the qualification condition*

$$\ker D^*_N g(\bar{x}, \bar{y}) = \{0\}.$$

Then for all $y^* \in Y^*$ one has

$$D^*F(\bar{x}, \bar{y})(y^*) \subset \{x^* \in X^* \mid (x^*, -y^*) \in \operatorname{rge} D^*_N g(\bar{x}, \bar{y})\},$$

where rge *stands for the range of multifunctions.*

Proof. The coderivative representation in (i) for $D^* = D^*_N$ follows immediately from Theorem 3.1(i). It holds also for $D^* = D^*_M$, which can be obtained similarly to the proof of (3.16). Assertion (ii) is a direct consequence of Theorem 3.1(ii) and the coderivative scalarization (3.14). To prove (iii), we use Theorem 3.2 and observe that conditions (b) there are the most general among (a)–(c) ensuring inclusion (3.23) in the setting under consideration. □

Next let us consider consequences of Theorems 3.1 and 3.2 for parametric constraint systems given in form (3.2), which describe sets of *feasible solutions* to perturbed problems of mathematical programming in infinite-dimensional spaces. We present two results for such constraint systems. The first corollary concerns classical constraint systems in (smooth) *nonlinear programming* with equality and inequality constraints given by strictly differentiable functions. In this framework we obtain an *exact formula* for computing coderivatives of feasible solution maps under a parametric version of the *Mangasarian-Fromovitz constraint qualification.*

Corollary 3.2 *Let $F\colon X \rightrightarrows Y$ be a multifunction between Asplund spaces given in form (3.2), where all $\varphi_i\colon X \times Y \to I\!R$, $i = 1,\dots,m+r$, are strictly differentiable at $(\bar{x},\bar{y}) \in \operatorname{gph} F$. Denote*

$$\begin{aligned} \bar{z} &:= \big(\varphi_1(\bar{x},\bar{y}),\dots,\varphi_{m+r}(\bar{x},\bar{y})\big), \\ I(\bar{x},\bar{y}) &:= \big\{i = 1,\dots,m+r \big|\ \varphi_i(\bar{x},\bar{y}) = 0\big\} \end{aligned}$$

and assume that:

(a) $\nabla\varphi_{m+1}(\bar{x},\bar{y}),\dots,\nabla\varphi_{m+r}(\bar{x},\bar{y})$ *are linearly independent;*

(b) *there is $u \in X \times Y$ satisfying*

$$\begin{aligned} &\langle\nabla\varphi_i(\bar{x},\bar{y}),u\rangle < 0, \quad i \in \{1,\dots,m\} \cap I(\bar{x},\bar{y}), \\ &\langle\nabla\varphi_i(\bar{x},\bar{y}),u\rangle = 0, \quad i = m+1,\dots,m+r. \end{aligned}$$

Then F is graphically regular at $(\bar{x},\bar{y})$, and one has

$$\begin{aligned} D^*F(\bar{x},\bar{y})(y^*) = \Big\{x^* \in X^* \Big|\ (x^*,-y^*) = \sum_{i\in I(\bar{x},\bar{y})} \lambda_i\nabla\varphi_i(\bar{x},\bar{y}), \\ \lambda_i \geq 0 \ \text{ if } \ i \in \{1,\dots,m\} \cap I(\bar{x},\bar{y})\Big\} \end{aligned} \tag{3.26}$$

with arbitrary $\lambda_i \in I\!R$ for $i = m+1,\dots,m+r$.

Proof. It follows from Theorem 3.1(ii) with $g = (\varphi_1,\dots,\varphi_{m+r})$, $\Omega = X \times Y$, and Θ given in (3.3). The set Θ is convex (thus normally regular at every point), and one has

$$\begin{aligned} N(\bar{z};\Theta) = \big\{(\lambda_1,\dots,\lambda_{m+r}) \in I\!R^{m+r} \big|\ \lambda_i \geq 0, \\ \lambda_i\varphi_i(\bar{x},\bar{y}) = 0 \ \text{ if } \ i = 1,\dots,m\big\}. \end{aligned}$$

In this case the qualification condition (3.19) is equivalent to the fulfillment of (a) and (b) in the corollary, and (3.18) reduces to (3.26).
□

The following corollary of Theorem 3.2 gives upper estimates for both coderivatives of feasible solution maps in parametric problems of *non-differentiable programming* with equality and inequality constraints described by Lipschitz continuous functions on Asplund spaces.

Corollary 3.3 *Let* $F\colon X \rightrightarrows Y$ *be a multifunction between Asplund spaces given in (3.2), let* $(\bar{x},\bar{y}) \in \operatorname{gph} F$, *and let* $\bar{z}$ *and* $I(\bar{x},\bar{y})$ *be defined in Corollary 3.2. Assume that all* φ_i, $i = 1,\ldots,m+r$, *are Lipschitz continuous around* $(\bar{x},\bar{y})$ *and that*

$$\Big[\sum_{i\in I(\bar{x},\bar{y})} \lambda_i x_i^* = 0\Big] \Longrightarrow \Big[\lambda_i = 0,\ i \in I(\bar{x},\bar{y})\Big] \tag{3.27}$$

whenever $\lambda_i \geq 0$ *for* $i \in I(\bar{x},\bar{y})$, $x_i^* \in \partial\varphi_i(\bar{x},\bar{y})$ *for* $i \in \{1,\ldots,m\} \cap I(\bar{x},\bar{y})$, *and* $x_i^* \in \partial\varphi_i(\bar{x},\bar{y}) \cup \partial(-\varphi_i)(\bar{x},\bar{y})$ *for* $i = m+1,\ldots,m+r$. *Then one has the inclusion*

$$D^*F(\bar{x},\bar{y})(y^*) \subset \Big\{x^* \in X^* \Big|\ (x^*,-y^*) \in \sum_{i\in\{1,\ldots,m\}\cap I(\bar{x},\bar{y})} \lambda_i\partial\varphi_i(\bar{x},\bar{y})$$

$$+ \sum_{i=m+1}^{m+r} \lambda_i\Big(\partial\varphi_i(\bar{x},\bar{y}) \cup \partial(-\varphi_i)(\bar{x},\bar{y})\Big),\ \lambda_i \geq 0 \ \text{ as } \ i \in I(\bar{x},\bar{y})\Big\}$$

for both coderivatives $D^* = D^*_N, D^*_M$.

Proof. It follows from Theorem 3.2 under condition (c) with $g = (\varphi_1,\ldots,\varphi_{m+r})$ and Θ given in (3.3) due to the scalarization formula (3.14) for $D^*_N g(\bar{x},\bar{y})$ and the subdifferential sum rule from Theorem 4.1 in Mordukhovich and Shao (1996a). □

4. Robust Lipschitzian Stability

In this section we obtain sufficient conditions, as well as necessary and sufficient conditions, for the Lipschitz-like property of the parametric constraint systems (3.1) and their specifications (3.2) and (3.4). Our approach is based on the following *coderivative characterizations* of Lipschitzian behavior of multifunctions given in Mordukhovich (1997) (see also the references therein) combined with the *coderivative formulas* derived in the preceding section as well as with the *SNC calculus* developed in Mordukhovich and Wang (2003a).

Theorem 4.1 *Let* $F\colon X \rightrightarrows Y$ *be closed-graph around* $(\bar{x},\bar{y})$. *Consider the properties:*

(a) *F is Lipschitz-like around* $(\bar{x},\bar{y})$.

(b) *F is PSNC at $(\bar{x},\bar{y})$ and* $\|D^*_M F(\bar{x},\bar{y})\| < \infty$.

(c) *F is PSNC at $(\bar{x},\bar{y})$ and* $D^*_M F(\bar{x},\bar{y})(0) = \{0\}$.

Then (a)⇒(b)⇒(c) *while these properties are equivalent if both X and Y are Asplund. Moreover, one has the estimates*

$$\|D^*_M F(\bar{x},\bar{y})\| \le \operatorname{lip} F(\bar{x},\bar{y}) \le \|D^*_N F(\bar{x},\bar{y})\| \tag{3.28}$$

for the exact Lipschitzian bound of F around $(\bar{x},\bar{y})$, where the upper estimate holds if $\dim X < \infty$ *and Y is Asplund. Thus*

$$\operatorname{lip} F(\bar{x},\bar{y}) = \|D^*_M F(\bar{x},\bar{y})\| = \|D^*_N F(\bar{x},\bar{y})\| \tag{3.29}$$

if in addition F is coderivatively normal at $(\bar{x},\bar{y})$.

If both X and Y are finite-dimensional, then F is automatically PSNC and coderivatively normal at $(\bar{x},\bar{y})$, and we get the coderivative criterion for the Aubin Lipschitz-like property

$$D^* F(\bar{x},\bar{y})(0) = \{0\} \quad \text{with} \quad \operatorname{lip} F(\bar{x},\bar{y}) = \|D^* F(\bar{x},\bar{y})\|$$

from Mordukhovich (1993); see also Theorem 9.40 in Rockafellar and Wets (1998) with the references and commentaries therein.

First let us present *necessary and sufficient* conditions for robust Lipschitzian stability with *precise formulas* for computing the *exact Lipschitzian bound* of the general constraint systems (3.1) satisfying some *regularity* assumptions.

Theorem 4.2 *Let $F\colon X \rightrightarrows Y$ be a set-valued mapping between Asplund spaces defined by the constraint system (3.1), let $\bar{z} := g(\bar{x},\bar{y})$ with $(\bar{x},\bar{y}) \in \operatorname{gph} F$, and let Θ be locally closed around $\bar{z}$ and SNC at this point. The following assertions hold:*

(i) *Assume that Z is Banach, that $\Omega = X \times Y$, and that g is strictly differentiable at $(\bar{x},\bar{y})$ with the surjective derivatives $\nabla g(\bar{x},\bar{y})$. Then the condition*

$$(x^*,0) \in \nabla g(\bar{x},\bar{y})^* N(\bar{z};\Theta) \Longrightarrow x^* = 0 \tag{3.30}$$

is sufficient for the Lipschitz-like property of F around $(\bar{x},\bar{y})$ being necessary and sufficient for this property if F is strongly coderivatively normal at $(\bar{x},\bar{y})$ (in particular, when $\dim Y < \infty$*). If in addition* $\dim X < \infty$, *then one has*

$$\operatorname{lip} F(\bar{x},\bar{y}) = \sup\left\{\|x^*\| \,\middle|\, (x^*,-y^*) \in \nabla g(\bar{x},\bar{y})^* N(\bar{z};\Theta),\ \ \|y^*\| \le 1\right\}.$$

(ii) *Assume that Z is Asplund; that Θ is normally regular at $\bar{z}$; that Ω is locally closed around $(\bar{x},\bar{y})$, normally regular at $(\bar{x},\bar{y})$, and PSNC at* *this point with respect to X; and that* g *is either strictly differentiable at* $(\bar{x}, \bar{y})$ *or graphically regular at this point with* $\dim Z < \infty$. *Suppose also that both qualification conditions (3.17) and (3.19) are fulfilled. Then the implication*

$$(x^*, 0) \in D^*g(\bar{x}, \bar{y}) \circ N(\bar{z}; \Theta) + N((\bar{x}, \bar{y}); \Omega) \Longrightarrow x^* = 0 \tag{3.31}$$

is necessary and sufficient for the Lipschitz-like property of F *around* $(\bar{x}, \bar{y})$. *If in addition* $\dim X < \infty$, *then one has*

$$\begin{aligned} \operatorname{lip} F(\bar{x}, \bar{y}) = \sup \Big\{ \|x^*\| \,\Big|\, & (x^*, -y^*) \in D^*g(\bar{x}, \bar{y}) \circ N(\bar{z}; \Theta) \\ & + N((\bar{x}, \bar{y}); \Omega), \ \ \|y^*\| \le 1 \Big\}. \end{aligned}$$

Proof. We use characterization (c) and the exact bound formula (3.29) from Theorem 4.1 for the Lipschitz-like property of general closed-graph multifunctions between Asplund spaces. To justify (i), observe first that $\operatorname{gph} F = g^{-1}(\Theta)$; thus F is SNC at $(\bar{x}, \bar{y})$ under the assumptions made as proved in Mordukhovich and Wang (2002). Then using the coderivative formula (3.16), we get characterization (3.30) from the condition $D^*_M F(\bar{x}, \bar{y})(0) = \{0\}$ and the exact bound formula in (i) from (3.29).

To prove (ii), we represent gph F in the intersection form (3.20) and deduce from Corollary 3.5 in Mordukhovich and Wang (2003a) that F is PSNC at $(\bar{x}, \bar{y})$ if the qualification condition (3.22) is fulfilled and if Ω is PSNC at $(\bar{x}, \bar{y})$ with respect to X while $g^{-1}(\Theta)$ is SNC at this point. By Theorem 3.8 in Mordukhovich and Wang (2003a) the latter property holds if Θ is SNC at $\bar{z}$ under the qualification condition (3.19). Moreover, these assumptions ensure that the qualification conditions (3.17) and (3.19) imply (3.22) due to the inclusion for $N((\bar{x}, \bar{y}); g^{-1}(\Theta))$ following from Theorem 4.5 in Mordukhovich (1997). Involving the other assumptions in (ii), we get equality (3.18) for both normal and mixed coderivatives of F at $(\bar{x}, \bar{y})$ by Theorem 3.1(ii). Thus the condition $D^*F(\bar{x}, \bar{y})(0) = \{0\}$ is equivalent to (3.31), and the exact bound formula of the theorem reduces to (3.29) in Theorem 4.1. □

One can easily derive from Theorem 4.2, as $\Theta = \{0\}$ and $\Omega = X \times Y$, necessary and sufficient conditions for *Lipschitz-like implicit multifunctions* in (3.4) with computing their exact Lipschitzian bounds. Let us present a corollary of Theorem 4.2 characterizing robust Lipschitzian stability of the classical *feasible solution sets* in parametric *nonlinear programming* with strictly differentiable data.

Corollary 4.1 *Let* $F\colon X \rightrightarrows Y$ *be a constraint system given in (3.2), where X and Y are Asplund and where* $\varphi_i\colon X \times Y \to I\!R$ *are strictly* *differentiable at* $(\bar{x},\bar{y}) \in \operatorname{gph} F$ *for all* $i = 1,\ldots,m+r$. *Denote* $\bar{z}$ *and* $I(\bar{x},\bar{y})$ *as in Corollary 3.2 and assume that the parametric Mangasarian-Fromovitz constraint qualification* (a) *and* (b) *therein holds. Then the condition*

$$\Big[\sum_{i\in I(\bar{x},\bar{y})} \lambda_i \nabla_y \varphi_i(\bar{x},\bar{y}) = 0\Big] \Longrightarrow \Big[\sum_{i\in I(\bar{x},\bar{y})} \lambda_i \nabla_x \varphi_i(\bar{x},\bar{y}) = 0\Big]$$
$$\text{for any } \lambda_i \in \mathbb{R} \text{ with } \lambda_i \geq 0 \text{ if } i \in \{1,\ldots,m\} \cap I(\bar{x},\bar{y})$$

is necessary and sufficient for the Lipschitz-like property of F around $(\bar{x},\bar{y})$. *If in addition* $\dim X < \infty$, *then one has*

$$\operatorname{lip} F(\bar{x},\bar{y}) = \max\Big\{\Big\|\sum_{i\in I(\bar{x},\bar{y})} \lambda_i \nabla_x \varphi_i(\bar{x},\bar{y})\Big\| \text{ subject to } \lambda_i \in \mathbb{R},$$
$$\Big\|\sum_{i\in I(\bar{x},\bar{y})} \lambda_i \nabla_y \varphi_i(\bar{x},\bar{y})\Big\| \leq 1, \text{ and } \lambda_i \geq 0 \text{ if } i \in \{1,\ldots,m\} \cap I(\bar{x},\bar{y})\Big\}.$$

Proof. The necessary and sufficient condition of the corollary as well as the formula for the exact Lipschitzian bound with "sup" instead of "max" follow directly from Theorem 4.2 as $g = (\varphi_1,\ldots,\varphi_{m+r})$, $\Omega = X \times Y$, and Θ defined in (3.3). The only thing one needs to prove is that the maximum is attained in the formula for $\operatorname{lip} F(\bar{x},\bar{y})$. Assuming the contrary, we find sequences $\lambda_{ik} \in \mathbb{R}$, with $i \in I(\bar{x},\bar{y})$ and $k \in \mathbb{N}$, satisfying

$$\lambda_{ik} \geq 0 \text{ for } i \in \{1,\ldots,m\} \cap I(\bar{x},\bar{y}), \quad \lambda_k := \sum_{i\in I(\bar{x},\bar{y})} \big|\lambda_{ik}\big| \to \infty \text{ as } k \to \infty,$$
$$\lim_{k\to\infty} \Big\|\sum_{i\in I(\bar{x},\bar{y})} \lambda_{ik} \nabla_x \varphi_i(\bar{x},\bar{y})\Big\| = \ell, \quad \limsup_{k\to\infty} \Big\|\sum_{i\in I(\bar{x},\bar{y})} \lambda_{ik} \nabla_y \varphi_i(\bar{x},\bar{y})\Big\| \leq 1,$$

where $\ell := \operatorname{lip} F(\bar{x},\bar{y}) < \infty$. Consider the numbers

$$\widetilde{\lambda}_{ik} := \frac{\lambda_{ik}}{\lambda_k}, \quad i \in I(\bar{x},\bar{y}),\ k \in \mathbb{N}, \text{ with } \sum_{i\in I(\bar{x},\bar{y})} \big|\widetilde{\lambda}_{ik}\big| = 1,$$

and find subsequences (without relabeling) such that $\widetilde{\lambda}_{ik} \to \widetilde{\lambda}_i$ as $k \to \infty$ for $i \in I(\bar{x},\bar{y})$. Then $\widetilde{\lambda}_i$ are not equal to zero simultaneously for $i \in I(\bar{x},\bar{y})$, and one has $\widetilde{\lambda}_i \geq 0$ for $i \in \{1,\ldots,m\} \cap I(\bar{x},\bar{y})$,

$$\sum_{i\in I(\bar{x},\bar{y})} \widetilde{\lambda}_i \nabla_x \varphi_i(\bar{x},\bar{y}) = 0, \quad \sum_{i\in I(\bar{x},\bar{y})} \widetilde{\lambda}_i \nabla_y \varphi_i(\bar{x},\bar{y}) = 0.$$

The latter contradicts the assumed Mangasarian-Fromovitz constraint qualification. □

Next let us obtain *sufficient* conditions for robust Lipschitzian stability with *upper estimates* of the exact Lipschitzian bounds for *nonregular* constraint systems (3.1) and their specifications. For simplicity we consider only the case when g in (3.1) is w^***-strictly** *Lipschitzian* around the reference point.

Theorem 4.3 *Let $F\colon X \rightrightarrows Y$ be given in (3.1), where $g\colon X \times Y \to Z$ is a mapping between Asplund spaces that is assumed to be w^*-strictly Lipschitzian around $(\bar{x},\bar{y}) \in \operatorname{gph} F$, and where Ω and Θ are locally closed around $(\bar{x},\bar{y})$ and $\bar{z} := g(\bar{x},\bar{y})$, respectively. Then the condition*

$$\Big[(x^*,0) \in \partial\langle z^*,g\rangle(\bar{x},\bar{y}) + N((\bar{x},\bar{y});\Omega), \quad z^* \in N(\bar{z};\Theta)\Big] \Longrightarrow z^* = 0 \quad \text{and} \quad x^* = 0 \tag{3.32}$$

is sufficient for the Lipschitz-like property of F around $(\bar{x},\bar{y})$ provided that Ω is PSNC at $(\bar{x},\bar{y})$ with respect to X and that Θ is SNC at $\bar{z}$. If in addition $\dim X < \infty$*, then one has*

$$\operatorname{lip} F(\bar{x},\bar{y}) \le \sup\Big\{\|x^*\| \;\Big|\; (x^*,-y^*) \in \partial\langle z^*,g\rangle(\bar{x},\bar{y}) + N((\bar{x},\bar{y};\Omega), \; z^* \in N(\bar{z};\Theta), \quad \|y^*\| \le 1\Big\}. \tag{3.33}$$

Proof. To establish the Lipschitz-like property of the constraint system (3.1) and the exact bound estimate, we employ the point-based characterization (c) with the upper estimate (3.28) from Theorem 4.1. Following the proof of Theorem 4.2 and using the SNC calculus rules from Corollary 3.5 and Theorem 3.8 in Mordukhovich and Wang (2003a), we conclude that F is PSNC at $(\bar{x},\bar{y})$ under the assumed SNC/PSNC properties of Θ and Ω as well as the qualification conditions (3.17) and (3.19). Observe that these assumptions ensure the fulfillment of the coderivative inclusion (3.23) from Theorem 3.2. Thus $D^*_M F(\bar{x},\bar{y})(0) = \{0\}$ if

$$\Big[(x^*,0) \in D^*_N g(\bar{x},\bar{y}) \circ N(\bar{z};\Theta) + N((\bar{x},\bar{y});\Omega)\Big] \Longrightarrow x^* = 0. \tag{3.34}$$

This also ensures the upper estimate

$$\operatorname{lip} F(\bar{x},\bar{y}) \le \sup\Big\{\|x^*\| \;\Big|\; (x^*,-y^*) \in D^*_N g(\bar{x},\bar{y}) \circ N(\bar{z};\Theta) + N((\bar{x},\bar{y});\Omega), \quad \|y^*\| \le 1\Big\}$$

if in additions X is finite-dimensional. The latter implies (3.33) by the scalarization formula (3.14), since g is w^***-strictly** Lipschitzian around $(\bar{x},\bar{y})$. Furthermore, one can check that the mentioned scalarization ensures the equivalence between (3.32) and the simultaneous fulfillment of the qualification conditions (3.17), (3.19), and (3.34). □

We conclude the paper with two corollaries of Theorem 4.3 that give efficient conditions for robust Lipschitzian stability of two remarkable constraint systems: *implicit multifunctions* defined by *nonregular* mappings and *feasible solution* maps in problems of *nondifferentiable programming.*

Corollary 4.2 *Let $g: X\times Y \to Z$ be a mapping between Asplund spaces, and let $g(\bar{x},\bar{y}) = 0$. Assume that g is Lipschitz continuous around $(\bar{x},\bar{y})$ and that $\dim Z < \infty$. Then the condition*

$$(x^*,0) \in \partial\langle z^*,g\rangle(\bar{x},\bar{y}) \Longrightarrow z^* = 0$$

is sufficient for the Lipschitz-like property of the implicit multifunction (3.4) around $(\bar{x},\bar{y})$. If in addition $\dim X < \infty$, then

$$\operatorname{lip} F(\bar{x},\bar{y}) \le \sup\Big\{\|x^*\|\Big|\ (x^*,-y^*) \in \partial\langle z^*,g\rangle(\bar{x},\bar{y}),\ z^* \in Z^*,\ \|y^*\| \le 1\Big\}.$$

Proof. Follows from Theorem 4.3 with $\Theta = \{0\}$ and $\Omega = X \times Y$. □

Corollary 4.3 *Let $F: X \rightrightarrows Y$ be a multifunction between Asplund spaces given in (3.2), let $(\bar{x},\bar{y}) \in \operatorname{gph} F$, and let $\bar{z}$ and $I(\bar{x},\bar{y})$ be defined in Corollary 3.2. Assume that all φ_i, $i = 1,\ldots,m+r$, are Lipschitz continuous around $(\bar{x},\bar{y})$ and that the constraint qualification (3.27) holds. Then the condition*

$$\Big[(x^*,0) \in \sum_{i\in\{1,\ldots,m\}\cap I(\bar{x},\bar{y})} \lambda_i\partial\varphi_i(\bar{x},\bar{y}) + \sum_{i=m+1}^{m+r} \lambda_i\Big(\partial\varphi_i(\bar{x},\bar{y}) \cup \partial(-\varphi_i)(\bar{x},\bar{y})\Big),$$
$$\lambda_i \ge 0 \ \ \textit{for}\ \ i \in I(\bar{x},\bar{y})\Big] \Longrightarrow x^* = 0$$

is sufficient for the Lipschitz-like property of F around $(\bar{x},\bar{y})$. If in addition $\dim X < \infty$, then one has the upper estimate

$$\operatorname{lip} F(\bar{x},\bar{y}) \le \sup\Big\{\|x^*\|\Big|\ (x^*,-y^*) \in \sum_{i\in\{1,\ldots,m\}\cap I(\bar{x},\bar{y})} \lambda_i\partial\varphi_i(\bar{x},\bar{y})$$
$$+ \sum_{i=m+1}^{m+r} \lambda_i\Big(\partial\varphi_i(\bar{x},\bar{y}) \cup \partial(-\varphi_i)(\bar{x},\bar{y})\Big),\ \lambda_i \ge 0 \ \ \textit{for}\ \ i \in I(\bar{x},\bar{y}),\ \|y^*\| \le 1\Big\}.$$

Proof. Follows from Theorem 4.3 with $g = (\varphi_1,\ldots,\varphi_{m+r}): X \times Y \to \mathbb{R}^{m+r}$, $\Omega = X \times Y$, and Θ defined in (3.3). □

References

Aubin, J.-P. (1984), Lipschitz behavior of solutions to convex minimization problems, *Mathematics in Operations Research,* Vol. 9, pp. 87-111.

Borwein, J.M., Lucet, Y. and Mordukhovich, B.S. (2000), Compactly epi-Lipschitzian sets and functions in normed spaces, *J. of Convex Analysis,* Vol. 7, pp. 375–393.

Borwein, J.M. and Strojwas, H.M. (1985), Tangential approximations, *Nonlinear Analysis,* Vol. 9, pp. 1347–1366.

Dontchev, A.L., Lewis, A.S. and Rockafellar, R.T. (2003), The radius of metric regularity, *Transactions of the American Mathematical Society,* Vol. 355, pp. 493–517.

Dontchev, A.L. and Rockafellar, R.T. (1996), Characterization of strong regularity for variational inequalities over polyhedral convex sets, *SIAM J. on Optimization,* Vol. 7, pp. 1087–1105.

Henrion, R. and Outrata, J.V. (2001), A subdifferential condition for calmness of multifunctions, *J. of Mathematical Analysis and Applications,* Vol. 258, pp. 110–130.

Henrion, R. and Römisch, W. (1999), Metric regularity and quantitative stability in stochastic programming, *Mathematical Programming,* Vol. 84, pp. 55–88.

Jourani, A. (2000), Hoffman's error bound, local controllability, and sensitivity analysis, *SIAM J. on Control and Optimization,* Vol. 38, pp. 947–970.

Ioffe, A.D. (2000), Codirectional compactness, metric regularity and subdifferential calculus, in Théra, M. (ed.), *Experimental, Constructive, and Nonlinear Analysis,* CMS Conference Proceedings, Vol. 27, pp. 123–164, American Mathematical Society, Providence, Phode Island.

Jourani, A. and Thibault, L. (1999), Coderivatives of multivalued mappings, locally compact cones and metric regularity, *Nonlinear Analysis,* Vol. 35, pp. 925–945.

Fabian, M. and Mordukhovich, B.S. (2001), Sequential normal compactness versus topological normal compactness in variational analysis, to appear in *Nonlinear Analysis.*

Klatte, D. and Kummer, B. (2002), *Nonsmooth Equations and Optimization,* Kluwer, Dordrecht.

Levy, A.B. (2001), Solution stability from general principles, *SIAM J. on Control and Optimization,* Vol. 40, pp. 1–38.

Levy, A.B. and Mordukhovich, B.S. (2002), Coderivatives in parametric optimization, to appear in *Mathematical Programming.*

Levy, A.B., Poliquin, R. A. and Rockafellar, R.T. (2000), Stability of locally optimal solutions, *SIAM J. on Optimization,* Vol. 10, pp. 580–604.

Mordukhovich, B.S. (1988), *Approximation Methods in Problems of Optimization and Control,* Nauka, Moscow.

Mordukhovich, B.S. (1993), Complete characterizations of openness, metric regularity, and Lipschitzian properties of multifunctions, *Transactions of the American Mathematical Society,* Vol. 340, pp. 1–35.

Mordukhovich, B.S. (1994a), Lipschitzian stability theory of constraint systems and generalized equations, *Nonlinear Analysis,* Vol. 33, pp. 173–206.

Mordukhovich, B.S. (1994b), Stability theory for parametric generalized equations and variational inequalities via nonsmooth analysis, *Transactions of the American Mathematical Society,* Vol. 343, pp. 609–658.

Mordukhovich, B.S. (1997), Coderivatives of set-valued mappings: calculus and applications, *Nonlinear Analysis,* Vol. 30, pp. 3059–3070.

Mordukhovich, B.S. (2001), The extremal principle and its applications to optimization and economics, In *Optimization and Related Topics* (Rubinov, A. and Glover, B., eds.), Applied Optimization Volumes 47, pp. 323–370, Kluwer, Dordrecht.

Mordukhovich, B.S. (2002), Coderivative analysis of variational systems, to appear in *Journal of Global Optimization.*

Mordukhovich, B.S. and Outrata, J.V. (2001), On second-order subdifferentials and their applications, *SIAM J. on Optimization,* Vol. 12, pp.139–169.

Mordukhovich, B.S. and Shao, Y. (1996a), Nonsmooth sequential analysis in Asplund spaces, *Transactions of the American Mathematical Society,* Vol. 349, pp. 1235–1280.

Mordukhovich, B.S. and Shao, Y. (1996b), Nonconvex differential calculus for infinite-dimensional multifunctions, *Set- Valued Analysis ,* Vol. 4, pp. 205–236.

Mordukhovich, B.S. and Shao, Y. (1997), Stability of set-valued mappings in infinite-dimensions: point criteria and applications, *SIAM J. on Control and Optimization,* Vol. 35, pp. 285–314.

Mordukhovich, B.S. and Shao, Y. (1998), Mixed coderivatives of set-valued mappings in variational analysis, *J. of Applied Analysis,* Vol. 4, pp. 269–294.

Mordukhovich, B.S. and Wang, Y. (2003a), Calculus of sequential normal compactness in variational analysis, *J. of Mathematical Analysis and Applications,* Vol. 282, pp. 63–84.

Mordukhovich, B.S. and Wang, Y. (2003b), Differentiability and regularity of Lipschitzian mappings, *Proceedings of the American Mathematical Society,* Vol. 131, pp. 389–399.

Mordukhovich, B.S. and Wang, Y. (2002), Restrictive metric regularity and generalized differential calculus in Banach spaces, preprint.

Outrata, J.V. (2000), A generalized mathematical program with equilibrium constraints, *SIAM J. on Control and Optimization,* Vol. 38, pp. 1623–1638.

Penot, J.-P. (1998), Compactness properties, openness criteria and coderivatives, *Set-Valued Analysis,* Vol. 6, pp. 363–380.

Phelps, R.R. (1993), *Convex Functions, Monotone Operators and Differentiability,* 2nd edition, Springer, Berlin.

Poliquin, R.A. and Rockafellar, R.T. (1998), Tilt stability of a local minimum, *SIAM J. on Optimization,* Vol. 8, pp. 287–299.

Robinson, S.M. (1979), Generalized equations and their solutions, part I: basic theory, *Mathematical Programming Study,* Vol. 10, pp. 128–141.

Rockafellar, R.T. and Wets, R.J.-B. (1998), *Variational Analysis,* Springer, Berlin.

Thibault, L. (1980), Subdifferentials of compactly Lipschitzian vector-valued functions, *Ann. Mat. Pure Appl.,* Vol. 125, pp. 157–192.

Treiman, J.S. (1999), Lagrange multipliers for nonconvex generalized gradients with equality, inequality, and set constraints, *SIAM J. Control Optimization,* Vol. 37, pp. 1313–1329.

Ye, J.J. (2000), Constraint qualifications and necessary optimality conditions for optimization problems with variational inequality constraints, *SIAM J. on Optimization,* Vol. 10, pp. 943–962.

Ye, J.J. and Zhu, Q.J. (2001), Multiobjective optimization problems with variational inequality constraints, to appear in *Mathematical Programming.*

Chapter 4

MONOTONICITY IN THE FRAMEWORK OF GENERALIZED CONVEXITY

Hoang Tuy*
Institute of Mathematics, Vietnam

Abstract An increasing function $f : R^n \to R \cup \{+\infty\}$ is a function such that $f(x') \geq f(x)$ whenever $x' \geq x$ (component-wise). A downward set $G \subset R^n$ is a set such that $x \in G$ whenever $x' \geq x$ for some $x' \in G$. We present a geometric theory of monotonicity in which increasing functions relate to downward sets in the same way as convex functions relate to convex sets. By giving a central role to a separation property of downward sets similar to that of convex sets, a theory of monotonic optimization can be developed which parallels d.c. optimization in several respects.

Keywords: Monotonicity. Downward sets. Normal sets. Separation property. Polyblock. Increasing functions. Monotonic functions. Difference of monotonic functions (d.m. functions). Abstract convex analysis. Global optimization.

MSC2000: 26B09,49J52

1. Introduction

Convexity is essential to modern optimization theory. Since the set of d.c. functions (functions representable as difference of convex functions) is a lattice with respect to the operations of pointwise maximum and pointwise minimum, the d.c. structure underlies a wide variety of nonconvex problems. The study of these problems is the subject of the theory of d.c. optimization developed over the last three decades.

* This research has been supported in part by the VN National Program on Basic Research,
email: htuy@math.ac.vn

However, convexity or reverse convexity is not always the natural property to be expected from many nonlinear phenomena. Another property at least as pervasive in the real world as convexity and reverse convexity is monotonicity. A function $f(x)$ is said to be *increasing* if $f(x) \leq f(x')$ whenever $x \leq x'$; *decreasing* if $-f(x)$ is increasing; *monotonic* if it is either increasing or decreasing. Just as d.c. functions constitute the linear space generated by convex functions, d.m. functions, i.e. functions which can be represented as differences of two monotonic functions, form the linear space generated by increasing functions. Since any polynomial in $x \in R^n$ with positive coefficients is obviously increasing on R^n_+, it is easily seen that the linear space of d.m. functions on $[a, b] := \{x \in R^n | \ a \leq x \leq b\}$, $[a, b] \subset R^n_+$, is dense in $C_{[a,b]}$, the space of continuous functions on $[a, b]$, with supnorm topology.

In the last few years a theory of monotonic optimization (see e.g. Rubinov et al. (2001), Tuy (1999), Tuy (2000), Rubinov (2000)) has emerged with the aim to provide a general mathematical framework for the study of optimization problems described by means of monotonic and more generally, d.m. functions.

There is a striking analogy between several basic facts from monotonicity theory and convexity theory, so that monotonicity can be regarded as a kind of generalized convexity, or abstract convexity, using a term coined by Singer a few years ago (see Singer (1997)).

From the point of view of modern optimization theory, a fundamental property of convex sets is the separation property which, in its simplest form, states that any point lying outside a closed convex set can be separated from it by a halfspace. The geometric object analogue to a convex set is a *downward set* which is the lower level set of an increasing function. A separation property holds for downward sets which reminds the same property of convex sets, but with the difference that separation is performed by the complement of a cone congruent to the positive orthant, rather than by a halfspace.

An important role in convexity theory is played by polytopes which can be defined as convex hulls of finite sets. The analogue of a polytope is a *polyblock,* defined as the downward hull of a finite set, i.e. the smallest downward set containing the latter. As is well known, a consequence of the classical separation property of convex sets is that any compact convex set is the intersection of a family of enclosing polytopes. Likewise, from the separation property of downward sets it follows that any closed upper bounded downward set is the intersection of a family of enclosing polyblocks. Furthermore, just as the maximum of a convex function over a compact convex set is attained at one extreme point, the maximum of an increasing function over a closed upper bounded downward set is attained at one upper extreme point. This analogy allows the polyhedral outer approximation method for maximizing convex functions over compact convex sets to be extended, with suitable modifications, to a polyblock outer approximation method for maximizing increasing functions over closed upper bounded downward sets.

The intersection of R^n_+ with a downward set in R^n is a *normal set.* This concept was introduced more than twenty years ago in mathematical economics (see Makarov and Rubinov (1977)) to describe any set $G \subset R^n_+$ such that $x \in G$ whenever $0 \leq x \leq x'$, and $x' \in G$. In our earlier paper Tuy (1999) a systematic study of normal sets was presented with a view of application to the theory of monotonic inequalities and monotonic optimization. It turns out that almost all properties of normal sets remain essentially valid for downward sets, so that most properties established in Tuy (1999) could be transferred automatically to downward sets, mutatis mutandis.

In the present paper a geometric theory of monotonicity is developed which parallels d.c. optimization in several respects. Although for the foundation of this theory, just properties of normal sets are needed, it is more convenient to consider downward sets and to put the theory in a framework of generalized convexity. It should be noted in this connection that downward sets were first introduced and extensively studied in Martinez-Legaz et al. (2002). However, while these authors focussed on analytical properties pertinent to approximation, we shall be concerned more with geometric properties important for optimization.

The paper consists of 7 sections. After the Introduction, we will discuss in Section 2 basic approaches to monotonicity from the view point of abstract convexity. In Sections 3 and 4 we will present the essential properties of downward sets, increasing functions and d.m. functions, to be used for the foundation of monotonic optimization. In Section 5 devoted to the theory of monotonic optimization, we will review the concept of polyblock approximation and show how it can be applied to outer approximation or branch and bound methods for maximizing or minimizing increasing functions under monotonic constraints. In Section 6 this concept is extended to solve discrete monotonic optimization problems via a special operation called S-adjustment. Finally, Section 7 is devoted to the concepts of regularity, duality and reciprocity together with their applications to the study of nonregular problems.

2. Two approaches to abstract convexity

Whereas the fundamental role of convexity in modern optimization is well known, it is less obvious which key properties are responsible for much of this role.

Close scrutiny shows that the single property that lies at the foundation of almost all theoretical and algorithmic developments of convex and local optimization is the *separation* property of convex sets, namely:

Given a closed convex set $G \subset R^n$ and any point $x \in R^n \setminus G$ there exists a closed halfspace L in R^n such that

$$G \subset L, \quad \text{while } x \notin L. \tag{4.1}$$

It is this property that is used, in one or another of its equivalent formulations (such as the Hahn-Banach theorem), in such constructions as:

- Subdifferential of convex functions
- Linearization (approximation of convex functions by affine functions).
- Cutting plane (Outer Approximation methods)
- Optimality conditions (Kuhn-Tucker theorem, maximum principle, etc.)
- Duality
- Lagrange multipliers
  etc.

An equally important property is the approximation property which states that every closed convex function is the upper envelope of a family of affine functions. In fact, this property can be used to derive nearly all constructions listed above and serve as the foundation of most analytical developements in optimization theory.

It is natural that efforts to generalize the concept of convexity should focus on generalizing the above properties. If analytical aspects are emphasized (see e.g. Singer (1997), Rubinov (2000), Martinez-Legaz et al. (2002) and references therein), the concept of convex functions is generalized first, by defining an abstract convex function as a function which is the upper envelope of a subfamily of a given family H of elementary functions, devised to play a role analogous to that of affine functions in classical convex analysis. On the other hand, if the geometric and numerical point of view is predominant (Beckenbach and Bellman (1961),

Ben-Tal and Ben-Israel (1981), Tuy (1999)), the concept of separation is generalized first, by allowing a separation of a set from a point by something other than a halfspace. Thus, the primary concept in the former approach is that of abstract convex functions, wheras in the latter approach the concept of abstract convex sets characterized by a separation property is more central. Of course, if abstract convex sets relate to abstract convex functions in much the same way in the two approaches, then the results obtained will be essentially equivalent.

Aside from these two approaches (which are often used simultaneously), we should also mention a third approach with a primary concern about the economic meaning of the concept of convexity. In the latter approach, the defining property of convexity is generalized first, by allowing two points in the set to be connected by a more general path than a segment as in the definition of convex sets in the classical sense (see Hackman and Passy (1988) and references therein). However, to our knowledge little has been done so far regarding numerical-algorithmic or theoretical-analytical development in this direction.

3. Downward Sets

We begin with introducing some notations and concepts. For any two vectors $x', x \in R^n$ we write $x' \geq x$ and say that x' *dominates* x if $x'_i \geq x_i \ \forall i = 1, \ldots, n$. We write $x' > x$ and say that x' *strictly dominates* x if $x'_i > x_i \ \forall i = 1, \ldots, n$. Let $R^n_+ = \{x \in R^n | \ x \geq 0\}$ and $R^n_{++} = \{x \in R^n | \ x > 0\}$. For $a \in R^n$ denote

$$K_a = \{x \in R^n | \ x \geq a\}, \qquad K^\circ_a = \{x \in R^n | \ x > a\}.$$

Since K_a, K°_a are translates of the orthants R^n_+, R^n_{++} resp., it is convenient to refer to them as the *closed* and *open*, resp., *orthocones* vertexed at a. For $a \leq b$ the box (hyperrectangle) $[a, b]$ is defined to be the set of all x such that $a \leq x \leq b$. We also write $(a, b] := \{x | \ a < x \leq b\}$, $[a, b) := \{x | \ a \leq x < b\}$. As usual e is the vector of all ones and e^i the i-th unit vector of R^n. For any two vectors $x^1, x^2 \in R^n$ we write $y = x^1 \wedge x^2$ whenever $y_i = \min\{x^1_i, \ x^2_i, \}, i = 1, \ldots, n$ and $z = x^1 \vee x^2$ whenever $z_i = \max\{x^1_i, x^2_i\}, i = 1, \ldots, n$.

A set $G \subset R^n$ is called a *downward set,* or briefly, a *down set,* if for any two points $x, x' \in R^n : x' \in G$ whenever $x' \leq x \in G$. The emptyset and R^n are special down sets which we will refer to as *trivial down sets* in R^n. A nontrivial down set is thus neither empty nor the whole space. Many properties stated below are almost straightforward. Others are not new and can be found in Tuy (1999) or Martinez-Legaz et al. (2002). They are reviewed for completeness and for the convenience of the reader.

Proposition 3.1 *The intersection and the union of a family of down sets are down sets.*

Proof. Immediate. □

Proposition 3.2 *Every down set G is connected and has a nonempty interior.*

Proof. The first assertion is trivial because for any two points x, x' in a down set G, both segments joining $x \wedge x'$ to x and $x \wedge x'$ to x' belong to G. If $a \in G$, and $u > 0$ then $x = a - u$ is an interior point of G since $[x - 0.5u, x + 0.5u] \subset G$. □

For any set $D \subset R^n$ the whole space R^n is a down set containing D. The intersection of all down sets containing D, i.e. the smallest down set containing D, is called the *down hull* of D and denoted by $D^{\rceil}$. A set D is said to be *upper (lower,* resp.) *bounded* if there is $a \in R^n$ such that $x \le a$ ($x \ge a$, resp.) $\forall x \in D$.

Proposition 3.3 *The down hull of a set $D \subset R^n$ is the set $D^{\rceil} = (D - R^n_+) = \cup_{y \in D}\{x|\ x \le y\}$. If D is upper bounded then so is $D^{\rceil}$. If D is compact then $D^{\rceil}$ is closed and upper bounded.*

Proof. The set $D^{\rceil}$ is obviously down and any down set containing D obviously contains it. Therefore. $D^{\rceil}$ is the down hull of D. If $D \subset \{x|\ x \le a\}$ and $x \in D^{\rceil}$, i.e. $x \le y$ for some $y \in D$, then $x \le y \le a$, hence $D^{\rceil} \subset \{x|\ x \le a\}$. If D is compact and $x^k \in D^{\rceil}$, $x^k \to x^0$, then $x^k \le y^k \in D$, and by passing to subsequence if necessary, one can assume $y^k \to y^0 \in D$, hence $x^0 \le y^0 \in D$, i.e. $x^0 \in D^{\rceil}$. □

3.1 BOUNDARY AND EXTREME POINTS

A point $y \in R^n$ is called an *upper boundary point* of a set $G \subset R^n$ if $y \in \mathrm{cl}G$ while $K_y^\circ \subset R^n \setminus G$. The set of upper boundary points of G is called the *upper boundary* of G and is denoted by $\partial^+ G$. If G is closed then obviously $\partial^+ G \subset G$.

Proposition 3.4 *Let G be a closed nontrivial down set in R^n. For every $x \in R^n$ and $v \in R^n_{++}$ the line $\Gamma(x, v) := \{x - \alpha v|\ \alpha \in R\}$ meets the upper boundary of G at a unique point $\sigma_G(x, v)$ defined by*

$$\sigma_G(x, v) = x - \mu v, \quad \mu = \min\{\alpha|\ x - \alpha v \in G\}. \tag{4.2}$$

Proof. Since $\emptyset \ne G \ne R^n$ there are a point $a \in G$ and a point $b \in R^n \setminus G$. Then $x - \alpha_1 v < a$, hence $x - \alpha_1 v \in G$, for some $\alpha_1 > 0$ and

$x - \alpha_2 v > b$, hence $x - \alpha_2 v \notin G$, for some $\alpha_2 < 0$. Therefore, $\alpha_2 < \mu = \inf\{\alpha|\ x + \alpha v \in G\} < \alpha_1$. Since G is closed, clearly $y = x - \mu v \in G$, so $\mu = \min\{\alpha|\ x - \alpha v \in G\}$. If there were $z \in G \cap K_y^\circ$, then $[y, z] \subset G$ and since $y = x - \mu v$, we would have $x_i - \mu v_i < z_i\ \forall i$, hence there would exist $\alpha < \mu$ such that $x_i - \alpha v_i < z_i\ \forall i$, i.e. such that $x - \alpha v \in [y, z] \subset G$, contradicting (4.2). Therefore, $K_y^\circ \subset R^n \setminus G$, and so $y \in \partial^+ G$. For any $\alpha > \mu$ we have $x - \alpha v < x - \mu v = y \in G$, hence $[x - \alpha v, y] \subset G$, while for $\alpha < \mu$ we have $x - \alpha v > x - \mu v = y \in R^n \setminus G$, hence $[y, x - \alpha v] \subset R^n \setminus G$. Therefore, no point $x + \alpha v$ with $\alpha \neq \mu$ belongs to $\partial^+ G$, completing the proof of the Proposition. □

Corollary 3.1 *A closed nontrivial down set G has a nonempty upper boundary $\partial^+ G$ and is just equal to the down hull of this upper boundary. Furthermore, for any $v \in R^n_{++} : \partial^+ G = \{x \in R^n|\ \sigma_G(x, v) = 0\}$.*

Proof. For any $x \in G$ and any $v \in R^n_{++}$ the point $y = \sigma_G(x, v) \in \partial^+ G$ and satisfies $x \leq y$. Therefore, $\partial^+ G \neq \emptyset$ and $G \subset (\partial^+ G)^\rceil$. Conversely, if $x \in (\partial^+ G)^\rceil$ then $x \leq y$ for some $y \in \partial^+ G$, hence $x \in \partial^+ G \subset G$, i.e. $(\partial^+ G)^\rceil \subset G$. The last assertion of the Corollary is obvious. □

Let D be a subset of R^n. A point $v \in D$ is called an *upper extreme point* of D if $x \in D, x \geq v \Rightarrow x = v$. Clearly every upper extreme point v of a down set $G \subset R^n$ satisfies $K_v^\circ \subset R^n \setminus G$, hence is an upper boundary point of G. In other words, if $V = V(G)$ denotes the set of upper extreme points of G then $V \subset \partial^+ G$.

Proposition 3.5 *A closed upper bounded nontrivial down set $G \subset R^n$ has at least one upper extreme point and is equal to the down hull of the set V of its upper extreme points.*

Proof. In view of Corollary 3.1, $V^\rceil \subset (\partial^+ G)^\rceil = G$, so it suffices to show that $\partial^+ G \subset V^\rceil$. Let $y \in \partial^+ G$. Define $x^1 \in \text{argmax}\{x_1|\ x \in G,\ x \geq y\}$, and $x^i \in \text{argmax}\{x_i|\ x \in G,\ x \geq x^{i-1}\}$ for $i = 2, \ldots, n$. Then $v := x^n \in G$ and $v \geq x$ for all $x \in G$ satisfying $x \geq y$. Therefore, $x \in G, x \geq v \Rightarrow x = v$. This means that $y \leq v \in V$, hence $y \in V^\rceil$, as was to be proved. □

Proposition 3.6 *The set of upper extreme points of the down hull of a compact set $D \subset R^n$ is a subset of the set of upper extreme points of D.*

Proof. If $v \in D$ but v is not an upper extreme point, then there exists a point $x \in D$ satisfying $x \geq v, x \neq v$. Since $D \subset D^\rceil$ this implies that v is not an upper extreme point of $D^\rceil$. □

Remark 3.1 Upper extreme points play for down sets a role analogous to that of extreme points for convex sets. In fact, Propositions 3.5 and 3.6 are analogous to well known propositions in convex analysis, namely: a compact convex set is equal to the convex hull of the set of its extreme points (Krein-Milman's Theorem), and any extreme point of the convex hull of a compact set is an extreme point of this set.

Remark 3.2 Upper extreme points of a set $D \subset R^n$ are Pareto-maximal points of D, with respect to R^n_+ (considered as ordering cone), as defined in vector optimization (see e.g. D.T.Luc (1989)). Also upper boundary points of a set G are weak Pareto-maximal points, with respect to R^n_+. Therefore, properties of upper extreme and upper boundary points could also be derived from more general properties of Pareto-maximal and weak Pareto-maximal points with respect to R^n_+. Note, however, that a point v of a down set G is an upper extreme point if and only if it can be removed from G so as to leave a down set. This characterization of upper extreme points of a down set is analogous to the characterization of extreme points of convex sets as those whose removal from the set does not destroy its convexity. Furthermore, just as extreme points of a convex set are necessarily boundary points, upper extreme points of a down set are necessarily upper boundary points. This analogy motivates the terminology used here which stresses the geometric nature of the concepts independent from any optimization context and thus avoids likely confusion when considering, for instance, a vector optimization problem over a down set. Moreover, here and in the next subsections we focus on properties that are almost straigthforward though essential for a theory of monotonic optimization which parallels d.c. optimization, and do not attempt to formulate or prove strongest results. For instance, we only need Proposition 3.6 as stated, though it is almost obvious that conversely, any upper extreme point of a compact set D is also an upper extreme point of its down hull (an analogous fact does not hold for extreme points of convex sets).

3.2 POLYBLOCKS

The simplest nonempty down set is the down hull of a singleton $\{a\} \subset R^n$, i.e. the set $P_a := \{x \in R^n | \ x \leq a\}$. We call such a set a *block* of vertex a.

For every point $a \in R^n$, define $l_a(x) = \min_{i=1,\dots,n}(x_i - a_i)$. Clearly the orthocone K_a° can be defined as $K_a^\circ = \{x| \ l_a(x) > 0\}$. The set $L_a = \{x| \ l_a(x) \leq 0\}$, i.e. the complement to the orthocone K_a° is a down set referred to as a *hyperangle*. Denote $\mathcal{L} = \{l(x)| \ l(x) \equiv l_a(x)$ for some $a \in R^n\}$. We shall shortly see that functions $l(x) \in \mathcal{L}$ and hyperangles play in monotonic analysis essentially the same role as affine functions and hyperplanes in classical convex analysis.

By Proposition 3.1 the union of a family of blocks is a down set. Conversely it is obvious that

Proposition 3.7 *For any down set G we have*

$$G = \cup_{y \in G} P_y.$$

This motivates the concept of *polyblock,* which by definition is the union of finitely many blocks, i.e. the down hull of a finite set in R^n. More precisely, a set P is called a *polyblock* in R^n if $P = \cup_{z \in T} P_z$, where $T \subset R^n$ ($|T| < +\infty$). The set T is called the *vertex set* of the polyblock. A vertex $z \in T$ is said to be *improper* if it is dominated by some other $z' \in T$, i.e. if there is $z' \in T \setminus \{z\}$ such that $z \leq z'$. Of course a polyblock is fully determined by its *proper* vertices.

Proposition 3.8 *Any polyblock is down, closed and upper bounded. The union or intersection of finitely many polyblocks is a polyblock.*

Proof. The first assertion is immediate, since a finite set $T \subset R^n$ is bounded above by the point $a = \vee\{z \in T\}$ defined by $a_i = \max\{x_i |\ x \in T\}, i = 1, \ldots, n$. The union of finitely many polyblocks is obviously a polyblock. To see that the intersection of finitely many polyblocks is a polyblock it suffices to observe that $(\cup_i A_i) \cap (\cup_j B_j) = \cup_{i,j}(A_i \cap B_j)$ and $P_a \cap P_b = P_u$ with $u = a \wedge b$. □

A polyblock is the analogue of a polytope in convex analysis. In fact, just as a polytope is the convex hull of finitely many points in R^n, a polyblock is the down hull of finitely many points in R^n. It is well known that any convex compact set is the intersection of a nested family of polytopes and hence can be approximated, as closely as desired, by a polytope enclosing it. We next show that in an analogous manner, any closed, upper bounded, down set is the intersection of a nested family of polyblocks and can be approximated, as closely as desired, by a polyblock containing it.

Proposition 3.9 *Let $G \subset R^n$ be a closed nontrivial down set. For any $x \in R^n \setminus G$ there exists $y \in \partial^+ G$ such that the hyperangle L_y separates G strictly from x (i.e. contains G but not x).*

Proof. For any $v \in R^n_{++}$ let $y = \sigma_G(x, v)$. Then $y = x - \mu v \in \partial^+ G$ for some $\mu > 0$, so that $x \in K^\circ_y \subset R^n \setminus G$, i.e. $x \notin L_y$ and $G \subset L_y$. □

The hyperangle L_y is referred to as the *supporting hyperangle* of the down set G at y. Thus a closed down set has a supporting hyperangle at each upper boundary point.

Proposition 3.10 *If $\bar{x} < \bar{z} \leq b$ then $P = P_{\bar{z}} \setminus K^{\circ}_{\bar{x}}$ is a polyblock with vertices*

$$z^i = \bar{z} - (\bar{z}_i - \bar{x}_i)e^i \quad i = 1, \ldots, n.$$

Proof. Let $K^{\circ}_i = \{x \in R^n | \ \bar{x}_i < x_i\}$. Since $K^{\circ}_{\bar{x}} = \cap_{i=1}^n K^{\circ}_i$, we have $P = P_{\bar{z}} \setminus K^{\circ}_{\bar{x}} = \cup_{i=1}^n (P_{\bar{z}} \setminus K^{\circ}_i)$. But

$$P_{\bar{z}} \setminus K^{\circ}_i = \{x | \ x_i \leq \bar{x}_i, \ x_j \leq \bar{z}_j \ \forall j \neq i\} = P_{z^i}$$

where z^i denotes the vector such that $z^i_j = \bar{z}_j \ \forall j \neq i, \ z^i_i = \bar{x}_i$, i.e. $z^i = \bar{z} - (\bar{z}_i - \bar{x}_i)e^i$. □

Proposition 3.11 *Let G be a closed upper bounded set in R^n. Then the following assertions are equivalent:*

(i) *G is a down set;*

(ii) *For any point $z \in R^n \setminus G$ there exists a polyblock separating z from G (i.e. containing G but not z).*

(iii) *G is the intersection of a family of polyblocks.*

Proof. (i)⇒ (ii). If $z \in R^n \setminus G$ then by Proposition 3.9 there exists $y \in \partial^+ G$ such that $z \in K^{\circ}_y$ but $K^{\circ}_y \cap G = \emptyset$, i.e. $P_b \setminus K^{\circ}_y$ (which is a polyblock by Proposition 3.10) separates z from G.

(ii) ⇒ (iii) Let E be the intersection of all polyblocks containing G. Clearly $G \subset E$. If (ii) holds, then for any $z \in E \setminus G$ there is a polyblock containing G but not z, so $E \subset G$.

(iii) ⇒(i) Obvious because by Proposition 3.8 any polyblock is closed and down. □

A set G is said to be *robust* if any point of G is the limit of a sequence of interior points of G.

Proposition 3.12 *A nonempty closed down set G is robust.*

Proof. For any $x \in G$ and any $\lambda > 0$ the point $x - \lambda e$ belongs to the interior of G and so x is the limit point of a sequence of interior points of G. □

4. Increasing and d.m. functions

A function $f : R^n \to R$ is said to be *increasing* if $f(x) \leq f(x')$ whenever $x \leq x'$; it is said to be increasing on a box $[a, b] \subset R^n$ if $f(x) \leq f(x')$ whenever $a \leq x \leq x' \leq b$. Functions increasing in this sense abound in economics, engineering, and many other fields. Outstanding examples of increasing functions on R^n_+ are production functions, cost functions and utility functions in Mathematical Economics, polynomials (in particular quadratic functions) with nonnegative coefficients, posynomials $\sum_{j=1}^m c_j \prod_{i=1}^n (x_i)^{a_{ij}}$ $(c_j \geq 0,\ a_{ij} \geq 0)$ in engineering design problems, etc. Other non trivial examples are functions of the form $f(x) = \sup\{g(u)|\ u \in D(x)\}$, where $g : R^n_+ \to R$ is a continuous function and $D : R^n_+ \to 2^{R^n_+}$ is a compact-valued multimapping such that $D(x') \supset D(x)$ for $x' \geq x$.

Proposition 4.1 (i) *If f_1, f_2 are increasing functions then for any nonnegative numbers λ_1, λ_2 the function $\lambda_1 f_1 + \lambda_2 f_2$ is increasing.*

(ii) *The pointwise supremum of a bounded above family $(f_\alpha)_{\alpha \in A}$ of increasing functions and the pointwise infimum of a bounded below family $(f_\alpha)_{\alpha \in A}$ of increasing functions are increasing.*

Proof. Immediate. □

It is well known that the maximum of a quasiconvex function over a compact set is equal to its maximum over the convex hull of this set and is attained at one extreme point. Analogously:

Proposition 4.2 *The maximum of an increasing function $f(x)$ over a compact set D is equal to its maximum over the down hull of D and is attained at at least one upper extreme point.*

Proof. Let $\bar{x}$ be a maximizer of $f(x)$ on $G = D^\rceil$. Since by Proposition 3.5 *G* is equal to the down hull of the set *V* of its upper extreme points, there exists $v \in V$ such that $\bar{x} \leq v$. Then $f(v) \geq f(\bar{x})$, hence v is also a maximizer of $f(x)$ on *G*. But by Proposition 3.6, v is also an upper extreme point of *D,* hence it is also a maximizer of $f(x)$ on *D*. □

Just as convex sets are essentially lower level sets of quasiconvex functions, down sets are essentially lower level sets of increasing functions, as shown by the next proposition.

Proposition 4.3 *For any increasing function $g(x)$ on R^n the level set $G = \{x \in R^n|\ g(x) \leq 0\}$ is a down set, closed if $g(x)$ is lower semicontinuous. Conversely, for any nontrivial, closed down set $G \subset R^n$ there exists a lower semicontinuous, strictly increasing function $g : R^n \to R$ such that $G = \{x \in R^n|\ g(x) \leq 0\}$.* ($g$ is said to be strictly increasing if it is increasing and $g(x') < g(x)$ whenever $x' < x$)

Proof. We need only prove the second assertion. For a $v \in R^n_{++}$ let $g(x) = \min\{\lambda \in R|\ x - \lambda v \in G\}$ (so $g(x) = \mu$, where μ is defined according to (4.2)). If $x' \leq x$ then $x' - \lambda v \leq x - \lambda v$, hence $x' - \lambda v \in G$ whenever $x - \lambda v \in G$. This proves that $g(x') \leq g(x)$, i.e., $g(x)$ is increasing. Furthermore, if $x' < x$ then $x' \leq x - \mu v$ for some $\mu > 0$ and since $x - \lambda v \in G$ if and only if $(x - \mu v) - (\lambda - \mu)v \in G$, it follows that $g(x') \leq g(x - \mu v) = g(x) - \mu < g(x)$, so g is strictly increasing. That $G = \{x| \; g(x) \leq 0\}$ is obvious from the definition of $g(x)$, so it only remains to prove that $g(x)$ is lower semicontinuous. Let $\{x^k\} \subset R^n$ be a sequence such that $x^k \to x^0$ and $g(x^k) \leq \alpha \; \forall k$. Since $x^k - \lambda_k v \in G$ for $\lambda_k = g(x^k)$ it follows from $g(x^k) \leq \alpha$ that $x^k - \alpha v \in G \; \forall k$, hence $x^0 - \alpha v \in G$ in view of the closedness of the set G. Therefore $g(x^0) \leq \alpha$, proving that the set $\{x \in R^n + | \; g(x) \leq \alpha\}$ is closed, and hence, that $g(x)$ is lower semi-continuous. □

Note that if $G = \{x| \; g(x) \leq 0\}$, where $g(x)$ is a continuous increasing function, then, obviously, $\partial^+ G \subset \{y \in R^n_+| \; g(y) = 0\}$, but the converse may not be true.

Many functions encountered in different fields of pure and applied mathematics are not monotonic, but can be represented as differences of monotonic functions. A function $f(x)$ for which there exist two increasing functions $f_1(x), f_2(x)$ satisfying $f(x) = f_1(x) - f_2(x)$ is called a *d.m. function.* The set of all d.m. functions on a given hyperrectangle $[a, b]$ forms a linear space, denoted by $DM[a, b]$, which is the linear space generated by increasing functions on $[a, b]$. The following properties have been established in Tuy (1999) or Tuy (2000):

Proposition 4.4 (i) $DM[a, b]$ *is a lattice with respect to the operations*

$$(f_1 \vee f_2)(x) = \max\{f_1(x), f_2(x)\}, \qquad (f_1 \wedge f_2)(x) = \min\{f_1(x), f_2(x)\}.$$

(ii) $DM[a, b]$ *is dense in the space* $C[a, b]$ *of continuous functions on* $[a, b]$ *endowed with the usual supnorm.*

A d.m. constraint is a constraint of the form $g(x) \leq 0$ where g is a d.m. function.

Proposition 4.5 *Any optimization problem which consists in maximizing or minimizing a d.m. function under d.m. constraints can be reduced to the canonical form:*

$$\max\{f(x)| \; g(x) \leq 1 \leq h(x), \; x \in R^n_+\}, \tag{4.3}$$

where f, g, h *are increasing functions.*

In the next section we shall discuss methods for solving this problem which will be referred to as the *basic monotonic optimization problem.*

5. The basic monotonic optimization problem

By defining $G = \{x|\ g(x) \leq 1\},\ H = \{x|\ h(x) < 1\}$ the basic monotonic optimization problem (4.3) is : given a closed down set G, an open down set H in R^n, and an increasing function $f(x)$, find

$$\max\{f(x)|\ x \in G \setminus H\}.$$

Assuming that there exists a box $[a,b]$ such that

$$G \setminus H \subset [a,b] \subset R^n_+ \tag{4.4}$$

we can rewrite the problem as

$$\max\{f(x)|\ x \in [a,b],\ x \in G \setminus H\}. \tag{BMO}$$

A feasible solution $\bar{x}$ of (BMO) which is an upper extreme point of the feasible set is called an *upper basic solution.* Such a point must belong to $\partial^+ G$.

Proposition 5.1 *If* (BMO) *is feasible, at least an optimal solution of it is an upper basic solution.*

Proof. This follows from Proposition 4.2. □

Thus, a global maximizer of $f(x)$ must be sought among the upper extreme points of the set $(G \setminus H) \cap [a,b]$.

Remark 5.1 A minimization problem such as

$$\min\{f(x)|\ x \in [a,b],\ x \in G \setminus H\}.$$

can be converted to an equivalent maximization problem. To be specific, by setting $x = a+b-y$, $\tilde{f}(y) = -f(a+b-y)$, $\tilde{G} = a+b-G$, $\tilde{H} = a+b-H$, this problem is easily seen to be equivalent to the following (BMO):

$$\max\{\tilde{f}(y)|\ y \in [a,b],\ y \in \tilde{G} \setminus \tilde{H}\}.$$

Therefore, in the sequel, we will restrict attention to the problem (BMO).

Based on the polyblock approximation of down sets and the upper basic solution property (Proposition 5.1) several methods have been developed for solving (BMO).

5.1 Outer Approximation

We only briefly describe the basic ideas of the POA (Polyblock Outer Approximation) method for solving (BMO). For a detailed discussion of this method and its implementation the reader is referred to Tuy (2000), Tuy and Luc (2000) , Hoai Phuong and Tuy (2003), Hoai Phuong and Tuy (2002), and also Tuy et al. (2002).

At a general iteration of the procedure a set W_k is available such that $W_k \cap H = \emptyset$ and the polyblock $P_k := W_k^{\rceil}$ has a nonempty intersection with the optimal solution set Ω of the problem when $\Omega \neq \emptyset$ (for example $W_1 = \{b\}$). Also, a number $\gamma_k \in R \cup \{+\infty\}$ is known which, if finite, is the objective function value of the best feasible solution $\bar{x}^k$ so far available, such that $f(z) > \gamma_k \; \forall z \in W_k$.

Let $v^k \in \mathbf{argmax}\{f(z)| \; z \in W_k\}$ and let $y^k = \sigma_G(v^k, b)$ be the intersection of $\partial^+ G$ with the halfline $\{v^k - \lambda b| \; \lambda \geq 0\}$. If $\|y^k - v^k\| \leq \varepsilon$, where $\varepsilon \geq 0$ is the tolerance, then v^k yields an **ε-approximate** optimal solution. Otherwise, separate v^k from G by a hyperangle $R^n \setminus \{z| \; z > y^k\}$, determining with P_k a polyblock P_{k+1}. Let γ_{k+1} be the updated current best objective function value. Compute the proper vertex V_{k+1} of P_{k+1}, let $W_{k+1} = \{z \in V_k \setminus H| \; f(z) > \gamma_{k+1}\}$. If $W_{k+1} = \emptyset$, then γ_{k+1} is the global optimal value (so the problem is infeasible if $\gamma_{k+1} = +\infty$ and the associated feasible solution $\bar{x}^{k+1}$ is an optimal solution if $\gamma_{k+1} < +\infty$). If $W_{k+1} \neq \emptyset$, then go to the next iteration.

It can be proved (see e.g. Tuy (2000)) under assumption (4.4) that for $\varepsilon > 0$ the above procedure is finite, whereas for $\varepsilon = 0$ the algorithm generates an infinite sequence $\{y^k\}$ converging to an optimal solution.

The implementation of this method requires efficient procedures for two operations:

1) Given a point $v \notin G$ compute $y = \sigma_G(v, b)$, the intersection point of $\partial^+ G$ with the halfline $\{v - \lambda b| \; \lambda \geq 0\}$. In many cases this subproblem reduces to solving a simple equation or a linear program. In the most general case, it can always be solved by a binary search, using the downwardness of the set G.

2) Given a polyblock P with proper vertex set V, a point $v \in V$ and a point $x \in [a, v)$ such that $G \setminus H \subset P \setminus (x, b]$, determine a new polyblock P' satisfying $P \setminus (x, b] \subset P' \subset P \setminus \{v\}$.

A simple procedure was first proposed in Tuy (2000) and Tuy (1999) for computing the proper vertex set of a polyblock P' satisfying $P \subset P' \subset P \setminus \{v\}$. However, the polyblock P' obtained that way is generally larger than $P \setminus (x, b]$. Since the smallest polyblock $P' \supset P \setminus (x, b]$ is $P \setminus (x, b]$, it is more efficient to use $P' = P \setminus (x, b]$, but then the following slightly more involved procedure is needed to derive the proper vertex set of $P \setminus (x, b]$ from that of P (see Tuy et al. (2002)).

For any two z, y define $J(z, y) = \{j | \ z_j > y_j\}$ and if $a \leq x \leq z \leq b$ then define $z^i = z + (x_i - z_i)e^i$, $i = 1, \ldots, n$, so that $z^i := z \wedge u^i$ for $u^i = b + (x - b)e^i$.

Proposition 5.2 *Let P be a polyblock with proper vertex set $V \subset [a, b]$, let $x \in [a, b]$, be such that $V_* := \{z \in V | \ z > x\} \neq \emptyset$. Then the polyblock $P \setminus (x, b]$ has vertex set*

$$V' = (V \setminus V_*) \cup \{z^i = z \wedge u^i | \ z \in V_*, \ i \in \{1, \ldots, n\}\}, \tag{4.5}$$

and its proper vertex set is obtained from V' by removing every z^i for which there exists $y \in V_*^+ := \{z \in V | \ z \geq x\}$ such that $J(z, y) = \{i\}$.

Proof. Since $[a, z] \cap (x, b] = \emptyset$ for every $z \in V \setminus V_*$, it follows that $P \setminus (x, b] = P_1 \cup P_2$, where P_1 is the polyblock with vertex $V \setminus V_*$ and $P_2 = (\cup_{z \in V_*} [a, z]) \setminus (x, b] = \cup_{z \in V_*} ([a, z] \setminus (x, b])$. Noting that $[a, b] \setminus (x, b]$ is a polyblock with vertices $u^i = b + (x - b)e^i$, $i = 1, \ldots, n$, we can then write $[a, z] \setminus (x, b] = [a, z] \cap ([a, b] \setminus (x, b]) = [a, z] \cap (\cup_{i=1,\ldots,n} [a, u^i]) = \cup_{i=1,\ldots,n} [a, z] \cap [a, u^i] = \cup_{i=1,\ldots,n} [a, z \wedge u^i]$, hence $P_2 = \cup\{[a, z \wedge u^i] | \ z \in V_*, i = 1, \ldots, n\}$, which shows that the vertex set of $P \setminus (x, b]$ is the set V' given by (4.5).

It remains to show that every $y \in V \setminus V_*$ is proper, while a z^i with $z \in V_*$ is improper if and only if $J(z, y) = \{i\}$ for some $y \in V_*^+$.

Since every $y \in V \setminus V_*$ is proper in V, while $z^i \leq z \in V$ for every z^i it is clear that every $y \in V \setminus V_*$ is proper. Therefore, any improper element must be some z^i such that $z^i \leq y$ for some $y \in V'$. Two cases are possible: either $y \in V$ or $y \in V' \setminus V$. In the former case $(y \in V)$ since obviously $x \leq z^i$ we must have $x \leq y$, i.e. $y \in V_*^+$; furthermore, $z_j = z_j^i \leq y_j \ \forall j \neq i$, hence, since $z \not\leq y$ it follows that $z_i > y_i$, i.e. $J(z, y) = \{i\}$. In the latter case $(y \in V' \setminus V)$, $z^i \leq y^l$ for some $(y, l) \neq (z, i), y \in V_*, \ i, l \in \{1, \ldots, n\}$. We cannot have $y = z, l \neq i$ for then the relation $z^i \leq z^l$ would imply $z_l = z_l^i \leq z_l^l = x_l$, conflicting with $z \in V_*$. So $y \neq z$ and $z_j^i \leq y_j^l \ \forall j = 1, \ldots, n$. Remembering that

$$z_j^i = \begin{cases} x_i \leq z_i & \text{for } j = i \\ z_j & \text{for } j \neq i \end{cases} \qquad y_j^l = \begin{cases} x_l \leq y_l & \text{for } j = l \\ y_j & \text{for } j \neq l \end{cases}$$

we infer that if $l = i$, then $z_j = z_j^i \leq y_j^i \quad \forall j \neq i$, and since $z \not\leq y$ it follows that $z_i > y_i$, and hence $J(z, y) = \{i\}$. On the other hand, if $l \neq i$ then from $z_j^i \leq y_j^l \ \ \forall j = 1, \ldots, n$, we have $z_j \leq y_j \ \ \forall j \notin \{i, l\}$, while for $j = l : z_l^i \leq y_l^l$, i.e. $z_l \leq x_l \leq y_l$. Hence, $z_j \leq y_j \ \forall j \neq i$, and again since $z \not\leq y$, we derive $z_i > y_i$ and $J(z, y) = \{i\}$. Thus any improper z^i must satisfy $J(z, y) = \{i\}$ for some $y \in V_*$. Conversely, if $J(z, y) = \{i\}$ for some $y \in V_*^+$ then $z_j \leq y_j\ \forall j \neq i$, hence $z^i \leq y^i$, i.e. z^i is improper. This completes the proof of the Proposition. □

Preliminary computational experience has shown that the above POA method, even in its original version (see e.g. Rubinov et al. (2001) and Tuy and Luc (2000)), works well on problems of relatively small dimension. Fortunately, a variety of highly nonconvex large scale problems can be converted into monotonic optimization problems of much reduced dimension. This class includes, for example, problems of the form

$$\max\{\varphi(u(x))|\quad x \in D\} \tag{4.6}$$

where $D \subset R^n_+$ is a nonempty compact convex set, $\varphi : R^m_+ \to R$ is an increasing function, $u(x) = (u_1(x), \ldots, u_m(x))$, $u_i : D \to R_+$ being nonnegative-valued continuous functions on D. For these problems, the monotonic approach has proved to be quite efficient, especially when existing methods cannot be used or encounter difficulties due to high nonconvexity (see e.g. Hoai Phuong and Tuy (2003)).

5.2 BRANCH AND BOUND

As an outer approximation, the POA method suffers from drawbacks inherent to this kind of procedures and is generally slow on high dimension. For dealing with large scale problems whose dimension cannot be significantly reduced by monotonicity, branch and bound procedures are usually more efficient. The POA method then furnishes a tool for computing good bounds.

A branch and bound is characterized by two basic operations:

1) branching: the space is partitioned into rectangles (rectangular algorithm) or cones vertexed at 0 and having each exactly n edges ;

2) bounding: for each partition set M (rectangle, or cone, according to the subdivision used) compute an upper bound $\beta(M)$ for the objective function value $f(x)$ over all feasible points $x \in M$, i.e. a number $\beta(M)$ satisfying

$$\beta(M) \geq \max\{f(x)|\ x \in M \cap (G \setminus H)\}.$$

Bounding over a rectangle. After reducing the size of the rectangle, whenever possible (by replacing it with a smaller rectangle still containing all feasible solutions in it), let $M = [p, q]$. If either of the following conditions fails: $p \in G \setminus H$, $q \notin H$, then $\beta(M) = -\infty$ (no feasible solution better than the current incumbent exists in M) and M is discarded from further consideration. Otherwise, apply a number of iterations of the POA procedure for computing $\max\{f(x)|\ x \in [p, q] \cap (G \setminus H)$ and let $\beta(M)$ be the incumbent value in the last iteration.

In many cases, a tighter upper bound $\beta(M)$ can also be obtained by combining monotonicity with convexity, as discussed in Tuy et al. (2002). For instance, if the normal set G happens to be also convex, while $H = \emptyset$, or if a convex approximation of G (and/or H) is readily available, then good upper bounds may often be obtained by combining polyblock with polyhedral outer approximation.

A key subproblem in solving (BMO) is to transcend a given incumbent solution $\bar{x}$, i.e. to find a better feasible solution than $\bar{x}$ if there is one. Setting $\gamma = f(\bar{x})$, $C(\gamma) = \{x \in [a, b] | \ f(x) \leq \gamma\}$, this reduces to recognizing whether $G \subset C(\gamma)$. Denote by E^* the polar set of E. Since $G \subset C(\gamma) \Rightarrow [C(\gamma)]^* \subset G^*$, the optimal value of the problem

$$\max\{\gamma | \ [C(\gamma)]^* \subset G^*\} \tag{4.7}$$

yields an upper bound for the optimal value of (BMO). But, as can easily be proved, *the polar of a normal set is a normal set,* so the problem (4.7) only involves closed convex normal sets. By exploiting this copresence of monotonicity and convexity it is often possible to obtain quite efficient bounds.

Bounding over a cone. To exploit the propriety that the optimum is attained on the upper boundary of G (Proposition 5.1), it appears that conical partition should be more appropriate than rectangular partition. Let $M = \text{cone}(u^1, \ldots, u^n)$, where $u^1, \ldots, u^n$ are vertices of an $(n-1)$-**subsimplex** of the unit simplex $[e^1, \ldots, e^n]$ in R^n. For each $i = 1, \ldots, n$ let v^i be the intersection of the ray through u^i with $\partial^+ G$ and define

$$p = \wedge_{i=1}^n v^i, \quad q = \vee_{i=1}^n v^i$$

Then

$$\partial^+ G \cap M \subset [p, q].$$

Since $p \leq v^i \in G$ it follows that $p \in G$. Furthermore, if $p \in H$ then $[p, q] \subset H$ so $[p, q]$ contains no point of $G \setminus H$ and M can be discarded from consideration. Assuming $p \notin H$, we thus have $p \in G \setminus H$, i.e. p is feasible and we can compute an upper bound for $f(x)$ over $[p, q] \cap (G \setminus H)$ by performing a number of iterations of the POA procedure on $[p, q]$.

It can be easily seen that with this bounding method the search is concentrated on the upper boundary of $G \setminus H$. For this reason the bound computed in conical partition is expected to be tighter than the bound computed in rectangular partition. Consequently, the convergence of a conical algorithm will generally be faster.

6. Discrete Monotonic Optimization

Consider the *discrete monotonic optimization problem*

$$\max\{f(x)|\ g(x) \leq 1 \leq h(x),\ x \in S\}.$$

where $f, g, h : R^n_+ \to R$ are increasing functions and S is a given finite subset of R^n_+. Defining as previously $G = \{x|\ g(x) \leq 1\}$, $H = \{x|\ h(x) < 1\}$ and assuming that $G \cap S \subset [a, b]$ with $[a, b] \subset R^n_+$, we can rewrite this problem as

$$\max\{f(x)|\ x \in [a, b],\ x \in G \cap S \setminus H\} \qquad \text{(DMO)}$$

Let $D = (G \cap S)^{\rceil}$. Clearly D is a polyblock with vertex set $G \cap S$.

Proposition 6.1 *Problem (DMO) is equivalent to*

$$\max\{f(x)|\ x \in D \setminus H\} \qquad (4.8)$$

Proof. This follows from Proposition 5.1 and the fact that an upper basic solution of (4.8) must be an upper extreme point of D, hence must belong to $G \cap S$. □

Solving problem (DMO) is thus reduced to solving (4.8) which is a monotonic optimization problem without explicit discrete constraint. The difficulty, now, is how to handle the polyblock D which is defined only implicitly as the normal hull of $G \cap S$. In Tuy, Minoux and Hoai-Phuong (2004) the following method was proposed for overcoming this difficulty and solving (DMO).

Without loss of generality we can assume that

$$a_i = \min\{x_i|\ x \in S\} \quad i = 1, \ldots, n \qquad (4.9)$$

Now define an operation $\lfloor\ \rfloor_S$ by setting, for any $v \in (a, b]$,

$$\lfloor v \rfloor_S = \tilde{v}, \text{ with } \quad \tilde{v}_i = \begin{cases} \max_{y \in S}\{y_i|\ y_i < v_i\} & (i = 1, \ldots, s) \\ v_i & i = s+1, \ldots, n \end{cases} \qquad (4.10)$$

In the frequently encountered special case when $S = S_1 \times \cdots \times S_s$, and every S_i is a finite set of real numbers we have

$$\tilde{v}_i = \begin{cases} \max\{\xi|\ \xi \in S_i,\ \xi < v_i\} & i = 1, \ldots, s \\ v_i & i = s+1, \ldots, n. \end{cases} \qquad (4.11)$$

so $\tilde{v} \in S$. (For example, if each S_i is the set of integers, then $\tilde{v}_i$ is the largest integer still less than v_i).

Clearly $\tilde{v} = \lfloor v \rfloor_S$ is uniquely defined for every $v \in (a, b]$. We shall refer to $\lfloor v \rfloor$ as the *S-adjustment* of v.

Proposition 6.2 *If* $[x, b] \cap (G \cap S) = \emptyset$ *and* $\tilde{v} = \lfloor x \rfloor_S$ *then* $(\tilde{v}, b] \cap G \cap S = \emptyset$.

Proof. Suppose there is $y \in (\tilde{v}, b] \cap G \cap S$. Since $y \in (\tilde{v}, b]$ we have $y_i > \tilde{v}_i$ for every $i = 1, \ldots, s$. On the other hand, since $y \in G \cap S$ while $[x, b] \cap G \cap S = \emptyset$, there is at least one $i \in \{1, \ldots, s\}$ such that $y_i < x_i$. From the definition of $\tilde{v}$ it then follows that $\tilde{v}_i \geq y_i$, a contradiction. □

Proposition 6.3 *Let P be a polyblock containing* $D \setminus H$, *let* v *be a proper vertex of P such that* $v \in H \setminus D$, *let x be the intersection of* $\partial^+ G$ *with the ray through* v *and*

$$\tilde{v} = \begin{cases} x & \text{if } x \in S \\ \lfloor x \rfloor_S & \text{if } x \notin S. \end{cases} \tag{4.12}$$

Then $(\tilde{v}, b] \cap D = \emptyset$, *i.e. the cone* $\{x | \ x \geq \tilde{v}\}$ *separates* v *from* D.

Proof. If $x \in S$ so that $\tilde{v} = x$ then $(\tilde{v}, b] \cap G = \emptyset$, hence $(\tilde{v}, b] \cap D = \emptyset$. If $x \notin S$, then, since $[x, b] \cap G \cap S = \emptyset$, by Proposition 6.2 $(\tilde{v}, b] \cap G \cap S = \emptyset$, i.e. $G \cap S \subset [a, b] \setminus (\tilde{v}, b]$, hence $D \subset [a, b] \setminus (\tilde{v}, b]$, i.e. $(\tilde{v}, b] \cap D = \emptyset$. □

With the S-adjustment an outer approximation method can be developed for solving (DMO) which works essentially in the same way as the outer approximation method for solving (BMO), except that, instead of using the separation property in Proposition 3.9, we now use the separation property in Proposition 6.3 to separate an unfit solution v from $D = (G \cap S)^{\rceil}$. For details we refer the reader to the above mentioned paper of Tuy, Minoux and Hoai-Phuong.

7. Regularity, Duality and Reciprocity

Consider the monotonic optimization problem (A) depicted in Fig. 1, where the feasible set is composed of the shaded area plus the isolated point x^*, and the optimum is attained at x^*. Clearly if the constraint $g(x) \leq \alpha$ is replaced by $g(x) \leq \alpha'$ with $\alpha' < \alpha$, then x^* will become infeasible and the optimal solution will move to some point around x'^*, far away from x^*. Thus, a slight error of the data may cause a significant error for the optimal value, and solving the problem by the previous algorithms may be a difficult task. In this section we shall show how the difficulty can be overcome by using the concepts of duality and reciprocity to be defined shortly.

7.1 DUALITY BETWEEN OBJECTIVE AND CONSTRAINT

Recall that cost functions, utility functions are typical examples of increasing functions, whereas a production set (set of technologically

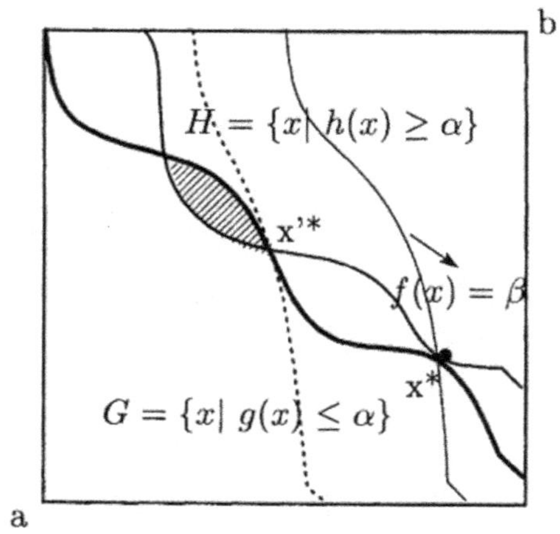

Fig. 1: Nonregular problem

feasible production programs) is naturally a normal set. Therefore, by interpreting $f(x)$ as a utility, $g(x)$ a cost and $X \subset R^n_+$ a production set, the optimization problem

$$\sup\{f(x)|\ x \in X,\ g(x) \leq \alpha\} \tag{A}$$

is to find the maximum utility of a production program with cost no more than α. We call *dual* of (A) the problem

$$\inf\{g(x)|\ x \in X,\ f(x) \geq \beta\} \tag{B}$$

i.e. to find the minimum cost of a production program with utility no less than β. Clearly when $X = R^n_+$, and f, g are increasing functions both (A) and (B) are monotonic optimization problems. However, the results below are valid for arbitrary nonempty set $X \subset R^n$ and for arbitrary functions $f, g : X \to R$.

If $\alpha \geq \inf(\text{B})$ then $\beta \leq \sup(\text{A})$ because an optimal solution x^* of (B) will satisfy $g(x^*) = \inf(\text{B}) \leq \alpha$, hence x^* will also be feasible to (A), which implies that $\beta \leq f(x^*) \leq \sup(\text{A})$. However $\alpha = \inf(\text{B})$ does not necessarily imply $\beta = \sup(\text{A})$, as can be shown by easily constructed examples. The question arises under which conditions:

$$\alpha = \inf(\text{B}) \Rightarrow \beta = \sup(\text{A}). \tag{4.13}$$

We say that problem (A) is *regular* if

$$\sup(\text{A}) = \sup\{f(x)|\ x \in X,\ g(x) < \alpha\}.$$

Analogously, problem (B) is regular if

$$\inf(\text{B}) = \inf\{g(x)|\ x \in X,\ f(x) > \beta\}.$$

Proposition 7.1 (Duality principle, Tuy (1987)) (i) *If (A) is regular then* $\alpha = \inf(B) \Rightarrow \beta = \sup(A)$.

(ii) *If (B) is regular then* $\beta = \sup(A) \Rightarrow \alpha = \inf(B)$.

Proof. By symmetry it suffices to prove (i). Suppose $\alpha = \inf(B)$. Then, as we have observed, $\beta \leq \sup(A)$. But by regularity of (A):

$$\beta \leq \sup(A) = \sup\{f(x) | \, x \in X, \; g(x) < \alpha\},$$

so if $\beta < \sup(A)$ then there is $x \in X$ satisfying $g(x) < \alpha, f(x) > \beta$. Since x is then feasible to (B), we must have $g(x) \geq \inf(B) = \alpha$, a contradiction. Therefore, $\beta = \sup(A)$. □

Corollary 7.1 *If both problems (A) and (B) are regular then*

$$\alpha = \inf(B) \Leftrightarrow \beta = \sup(A). \quad \square$$

From an heuristic point of view, if $f(.)$ is a utility and $g(.)$ a cost then the regularity condition means that a slight change of the minimal cost should not cause drastic change of the utility received. Under this condition, it is natural that, as asserted in Proposition 7.1, if it costs at least α to achieve a utility no less than β then a utility at most β can be achieved at a cost no more than α.

7.2 OPTIMALITY CONDITION

A consequence of Proposition 7.1 is the following

Proposition 7.2 (Optimality criterion, Tuy (1987)) *Let* x^* *be a feasible solution of problem (A). If problem (B) is solvable and regular for* $\alpha = g(x^*)$, *then a necessary condition for* x^* *to be optimal for problem (A) is that*

$$\inf\{g(x) | \, x \in X, \; f(x) \geq f(x^*)\} = \alpha. \tag{4.14}$$

This condition is also sufficient, provided problem (A) is solvable and regular .

Proof. If x^* is optimal to (A) then $f(x^*) = \sup(A)$, whence (4.14), by Proposition 7.1, (ii). Conversely, if (4.14) holds, i.e. $\alpha = \inf(B)$ for $\beta = f(x^*)$, then $f(x^*) = \sup(A)$ by Proposition 7.1, (i). □

Proposition 7.3 *Suppose problem (B) is solvable and regular. For a given value* β_1 *let*

$$\alpha_1 = \min\{g(x) | \, x \in X, \; f(x) \geq \beta_1\}. \tag{4.15}$$

Then

(i) $\alpha_1 > \alpha \Rightarrow \max(\mathrm{A}) < \beta_1$;

(ii) $\alpha_1 < \alpha \Rightarrow \max(\mathrm{A}) > \beta_1$;

(iii) $\alpha_1 = \alpha \Rightarrow \begin{cases} \max(\mathrm{A}) = \beta_1 & \text{if (A) is regular,} \\ \max(\mathrm{A}) \geq \beta_1 & \text{otherwise.} \end{cases}$

Proof. Observe that by Proposition 7.2

$$\alpha = \min\{g(x) |\ x \in X,\ f(x) \geq \sup(\mathrm{A})\} \tag{4.16}$$

Therefore, (i) and (ii) follow from (4.15) and (4.16). Suppose now that $\alpha_1 = \alpha$. Then by (4.15) $\alpha = \min\{g(x) |\ x \in X,\ f(x) \geq \beta_1\}$, so if problem (A) is regular then, by Proposition 7.2, $\beta_1 = \max\{f(x) |\ x \in X, g(x) \leq \alpha\} = \max(\mathrm{A})$. In any case, let x^1 be an optimal solution of (4.15) . If $\beta_1 > \sup(\mathrm{A})$, then $f(x^1) \geq \beta_1 > \sup(\mathrm{A})$, so x^1 is infeasible to problem (A), i.e. $g(x^1) > \alpha$, conflicting with x^1 being an optimal solution of (4.15) while $\alpha_1 = \alpha$. Therefore, $\sup(\mathrm{A}) \geq \beta_1$. □

Application. Suppose that an original problem (A) is difficult to solve directly, while problem (B) is easy or can be solved efficiently. Then, provided problem (B) is solvable and regular, by solving a sequence of problems 4.15 where β_1 is iteratively adjusted according to Proposition 7.3 we can eventually determine max (A) with any desired accuracy.

In particular, this method can be used to solve any nonregular problem (A), whose dual (B) is regular (as it happens with the problem depicted in Fig. 1),

7.3 RECIPROCITY

A concept closedly related to the above duality concept is that of reciprocity introduced by Tikhonov (1980), as early as in 1980, for the study of ill-posed problems.

Two problems (A), (B) are said to be *reciprocal* if they have the same set of optimal solutions.

Observe that an obvious sufficient condition for reciprocity is

$$\alpha = \sup(\mathrm{B}), \quad \beta = \inf(\mathrm{A}).$$

Indeed, if these equalities hold then any optimal solution x^* of (A) is feasible to (B) and satisfies $f(x^*) \geq \alpha$, hence is optimal to (B). Similarly any optimal solution x^* to (B) is optimal to (A). A consequence of Proposition 7.1 is then

Proposition 7.4 (Reciprocity principle) (i) *If problem (A) is regular, while problem (B) is solvable and* $\alpha = \inf(B)$, *then* $\beta = \sup(A)$ *and the two problems are reciprocal.*

(ii) *If problem (B) is regular, while problem (A) is solvable and* $\beta = \sup(A)$, *then* $\alpha = \inf(B)$ *and the two problems are reciprocal.*

A special case of Proposition 7.4 is the following result, first established in Tikhonov (1980) but using a much more elaborate argument:

Corollary 7.2 (Tikhonov (1980), Fundamental Theorem) *Let* $\psi : R^n \to R$ *be a continuous function such that* $\delta_0 := \inf\{\psi(x)|\ x \in R^n\} \leq \delta < +\infty$. *If* $\eta = \inf\{\|x\| \mid \psi(x) \leq \delta\}$ *then the following two problems are reciprocal:*

$$\inf\{\|x\| \mid x \in R^n,\ \psi(x) \leq \delta\} \tag{4.17}$$

$$\inf\{\psi(x) \mid x \in R^n,\ \|x\| \leq \eta\}. \tag{4.18}$$

Proof. In fact, problem (4.18) is regular, and problem (4.17) is solvable ($\eta = \min\{\|x\| \mid \psi(x) \leq \delta\}$), so Proposition 7.4 applies, with $X = R^n,\ g(x) = \|x\|,\ f(x) = -\psi(x),\ \alpha = -\delta,\ \beta = \eta$. □

A detailed discussion of the relation of global optimality conditions to reciprocity conditions, together with an analysis of erroneous results that have appeared in the recent literature on this subject can be found in Tuy (2003).

References

E.F. Beckenbach and R. Bellman, *Inequalities,* Springer-Verlag 1961.

A. Ben-Tal and A. Ben-Israel, F-convex functions: Properties and applications, in : *Generalized concavity in optimization and economics,* eds. S. Schaible and W.T. Ziemba, Academic Press, New York 1981.

Z. First, S.T. Hackman and U. Passy, Local-global properties of bifunctions, *Journal of Optimization Theory and Applications* 73 (1992) 279-297.

S.T. Hackman and U. Passy, Projectively-convex sets and functions, *Journal of Mathematical Economics* 17 (1988) 55-68.

N.T. Hoai Phuong and H. Tuy, A Monotonicity Based Approach to Nonconvex Quadratic Minimization, *Vietnam Journal of Mathematics* 30:4 (2002) 373-393.

N.T. Hoai Phuong and H. Tuy, A unified approach to generalized fractional programming, *Journal of Global Optimization,* 26 (2003) 229-259.

R. Horst and H. Tuy, *Global Optimization (Deterministic Approaches),* third edition, Springer-Verlag, 1996.

H. Konno and T. Kuno, Generalized multiplicative and fractional programming, *Annals of Operations Research,* **25** (1990) 147-162.

H. Konno, Y. Yajima and T. Matsui, Parametric simplex algorithms for solving a special class of nonconvex minimization problems, *Journal of Global Optimization,* **1** (1991) 65-81.

H. Konno, P.T. Thach and H. Tuy, *Optimization on Low Rank Nonconvex Structures,* Kluwer Academic Publishers, 1997.

D.T. Luc, *Theory of Vector Optimization,* Lecture Notes in Economics and Mathematical Systems 319, Springer-Verlag, 1989.

V.L. Makarov and A.M. Rubinov, *Mathematical Theory of Economic Dynamic and Equilibria,* Springer-Verlag, 1977.

J.-E. Martinez-Legaz, A.M. Rubinov and I. Singer, Downward sets and their separation and approximation properties, *Journal of Global Optimization,* **23** (2002) 111-137.

P. Papalambros and H.L. Li, Notes on the operational utility of monotonicity in optimization, *ASME Journal of Mechanisms, Transmissions, and Automation in Design,* **105** (1983) 174-180.

P. Papalambros and D.J. Wilde, *Principles of Optimal Design - Modeling and Computation,* Cambridge University Press, 1986

U. Passy, Global solutions of mathematical programs with intrinsically concave functions, in M. Avriel (ed.), *Advances in Geometric Programming,* Plenum Press, 1980.

A. Rubinov, *Abstract Convexity and Global Optimization* Kluwer Academic Publishers, 2000.

A. Rubinov, H. Tuy and H. Mays, Algorithm for a monotonic global optimization problem, *Optimization,* 49 (2001), 205-221.

I. Singer, *Abstract convex analysis,* Wiley-Interscience Publication, New York, 1997.

A. N. Tikhonov, On a reciprocity principle, *Soviet Mathematics Doklady,* vol.22, pp. 100-103, 1980.

H. Tuy, Convex programs with an additional reverse convex constraint, *Journal of Optimization Theory and Applications* 52 (1987) 463-486

H. Tuy, D.C. Optimization: Theory, Methods and Algorithms, in R. Horst and P.M. Pardalos (eds.), *Handbook on Global Optimization,* Kluwer Academic Publishers, 1995, pp. 149-216.

H. Tuy, *Convex Analysis and Global Optimization,* Kluwer Academic Publishers, 1998.

H. Tuy, Normal sets, polyblocks and monotonic optimization, *Vietnam Journal of Mathematics* 27:4 (1999) 277-300.

H. Tuy, Monotonic optimization: Problems and solution approaches, *SIAM J. Optimization* 11:2 (2000), 464-494.

H. Tuy and Le Tu Luc, A new approach to optimization under monotonic constraint, *Journal of Global Optimization,* 18 (2000) 1-15.

H. Tuy and F. Al-Khayyal, Monotonic Optimization revisited, Preprint, Institute of Mathematics, Hanoi, 2003.

H. Tuy, On global optimality conditions and cutting plane algorithms, *Journal of Optimization Theory and Applications,* Vol. 118 (2003), No. 1, 201-216.

H. Tuy, M. Minoux and N.T. Hoai-Phuong: Discrete monotonic optimization with application to a discrete location problem, Preprint, Institute of Mathematics, Hanoi, 2004.

II

CONTRIBUTED PAPERS

Chapter 5

ON THE CONTRACTION AND NONEXPANSIVENESS PROPERTIES OF THE MARGINAL MAPPINGS IN GENERALIZED VARIATIONAL INEQUALITIES INVOLVING CO-COERCIVE OPERATORS

Pham Ngoc Anh
Posts and Telecommunications Institute of Technology, Vietnam

Le Dung Muu*
Hanoi Institute of Mathematics, Vietnam

Van Hien Nguyen
Department of Mathematics University of Namur (FUNDP), Belgium

Jean-Jacques Strodiot
Department of Mathematics University of Namur (FUNDP), Belgium

Abstract We investigate the contraction and nonexpansiveness properties of the marginal mappings for gap functions in generalized variational inequalities dealing with strongly monotone and co-coercive operators in a real

* This work was completed during the visit of the second author at the Department of Mathematics, University of Namur (FUNDP), Namur, Belgium
E-mail: ldmuu@math.ac.vn

Hilbert space. We show that one can choose regularization operators such that the solution of a strongly monotone variational inequality can be obtained as the fixed point of a certain contractive mapping. Moreover a solution of a co-coercive variational inequality can be computed by finding a fixed point of a certain nonexpansive mapping. The results give a further analysis for some methods based on the auxiliary problem principle. They also lead to new algorithms for solving generalized variational inequalities involving co-coercive operators. By the Banach contraction mapping principle the convergence rate can be easily established.

Keywords: Generalized variational inequality, co-coercivity, contractive and nonexpansive mapping, Banach iterative method.

MSC2000: 90C29

1. Introduction

Let H be a real Hilbert space, C be a nonempty closed convex subset of H, $F : C \to H$ be a monotone mapping and φ be a closed proper convex function on H. We consider the following generalized variational inequality:

Find $x^* \in C$ such that

$$\langle F(x^*), x - x^* \rangle + \varphi(x) - \varphi(x^*) \geq 0 \;\; \forall\, x \in C, \tag{5.1}$$

where $\langle .,. \rangle$ denotes the inner product in H. The norm associated with this inner product will be denoted by $||.||$.

This generalized variational inequality problem was introduced by Browder (1966) and studied by a number of authors (see e.g. Hue (2004); Konnov (2001); Muu (1986); Noor (2001); Patriksson (1997); Patriksson (1999); Verma (2001); Zhu (1996)). Among various iterative methods for solving variational inequalities the gap function method is widely used (see e.g. Auslender (1976); Fukushima (1992); Marcotte (1995); Noor (1993); Patriksson (1997); Patriksson (1999); Taji (1996); Taji (1993); Zhu (1994); Wu (1992) and the references cited therein). The first gap function was given by Auslender (1976) for the variational inequality problem (5.1) where the function φ is absent. This gap function, in general, is not differentiable even F is. The first differentiable gap function has been introduced by Fukushima (1992). Extended differentiable gap functions have been studied in Zhu (1994). The gap function approach has been used to monitor the convergence of iterative sequences to a solution of a variational inequality problem and to develop descent algorithms for solving variational inequalities (see e.g.

Fukushima (1992); Konnov (2001); Noor (1993); Marcotte (1995); Patriksson (1997); Patriksson (1999); Zhu (1995); Zhu (1996)). For a good survey of solution methods for variational inequality problems, the reader is refered to Pang and Harker (1990).

In this paper, we will use the fixed point approach (see e.g. Gol'stein (1989); Noor (1993); Patriksson (1999)) to the variational inequality problem (5.1) by using a gap function which is an extension of the projection gap function introduced in Fukushima (1992). Actually, for solving the variational inequality problem (5.1), instead of considering the problem of minimizing the gap function over C, we consider the problem of finding fixed points of the marginal mapping given as the solution of the mathematical programming problems of evaluating the associated gap function. By choosing suitable regularization operators we show that the marginal mapping is contractive on C when either F is strongly monotone or φ is strongly convex. We weaken the strong monotonicity and strong convexity by co-coercivity and show that the marginal mapping is nonexpansive. These results allow that a solution of the variational inequality problem (5.1) can be obtained by the Banach contraction iterative procedure or its modifications. This fixed point approach gives a new analysis for some existing algorithms based on the auxiliary problem principle (see e.g. Cohen (1988); Konnov (2001); Hue (2004); Marcotte (1995); Zhu (1994); Zhu (1996)). It also yields new algorithms for solving generalized variational inequalities involving co-coercive operators. By the Banach contraction principle, the convergence is straightforward and bound errors are easy to obtain. From a computational view point for solving variational inequalities, this is essential for those methods where the strongly monotone variational inequalities appear as subproblems. Actually, in our algorithms the subproblems, at each iteration k, are strongly convex mathematical programs of the form

$$\min\left\{\frac{1}{2}(x-x^k)^T G(x-x^k) + \langle F(x^k), x-x^k\rangle + \varphi(x) | x \in C\right\},$$

or when φ is differentiable, the objective function of the subproblems is quadratic of the form

$$\min\left\{\frac{1}{2}(x-x^k)^T G(x-x^k) + \langle F(x^k) + \nabla\varphi(x^k), x-x^k\rangle \;| x \in C\right\},$$

where G is a suitable self-adjoint positive bounded operator from H into itself.

The paper will be organized as follows. In the next section we recall and prove some results on co-coercivity and the projection gap functions. In Section 3 we show how to choose the regularization operator such that the marginal mappings defined by these gap functions are contractive when, in the variational inequality problem (5.1), either F is strongly monotone or φ is strongly convex. The Section 4 studies the nonexpansiveness of the marginal mapping when the cost mapping is co-coercive. In the last section we describe the algorithms and discuss some algorithmic aspects.

2. Preliminaries on the Projection Gap Function

Note that when φ is differentiable on some open set containing C, then, since φ is lower semicontinuous proper convex, the variational inequality (5.1) is equivalent to the following one (see e.g. Patriksson (1997) Proposition 2.3):

Find $x^* \in C$ such that

$$\langle F(x^*) + \nabla\varphi(x^*), x - x^* \rangle \geq 0 \ \ \forall\, x \in C. \tag{5.2}$$

For the problem (5.1) we consider the following gap function:

$$g(x) = -\min\left\{ \langle F(x), y - x\rangle + \frac{1}{2}\langle y - x, G(y-x)\rangle + \varphi(y) - \varphi(x) \mid y \in C \right\} \tag{5.3}$$

where G is a self-adjoint positive linear bounded operator from H into itself. In the case when φ is differentiable we can use the formulation (1.2) to obtain the projection gap function

$$g_1(x) = -\min\left\{ \langle F(x) + \nabla\varphi(x), y - x\rangle + \frac{1}{2}\langle y - x, G(y-x)\rangle \mid y \in C \right\}. \tag{5.4}$$

Note that the objective function in the problem of evaluating $g_1(x)$ is always strongly convex quadratic.

Since C is closed convex and the objective functions are strongly convex, the mathematical programming problems (5.3) and (5.4) are always solvable for any $x \in C$. Let $h(x)$ and $h_1(x)$ denote the unique solution of problems (5.3) and (5.4), respectively. Both h and h_1 are marginal mappings onto C.

Observe that when φ is a constant function, these two mappings, $h(x)$ and $h_1(x)$, coincide and become the marginal mapping for the projection gap function introduced in Fukushima (1992). Thus, in this case $h(x) = h_1(x)$ for all $x \in C$. In general, $h \not\equiv h_1$. However both h and h_1 have a common property that a point x^* is a solution to the variational inequality problem (5.1) if and only if $h(x^*) = h_1(x^*) = x^*$. The following lemma is a consequence of Proposition 2.7 in Patriksson (1997). Below we give a direct proof.

Lemma 2.1 *Suppose that the variational inequality problem (5.1) has a solution. Then a point* x^* *is a solution of problem (5.1) if and only if* x^* *is a fixed point of* h. *The same claim is also true for* h_1.

Proof. Let x^* be a solution of (5.1) and $h(x^*)$ be the unique solution of the problem evaluating $g(x^*)$. Then

$$\langle F(x^*), h(x^*) - x^* \rangle + \varphi(h(x^*)) - \varphi(x^*) \geq 0. \tag{5.5}$$

Since $h(x^*)$ is the solution of the convex problem of evaluating $g(x^*)$, there exists a $z^* \in \partial\varphi(h(x^*))$ such that

$$\langle F(x^*) + G(h(x^*) - x^*) + z^*, y - h(x^*) \rangle \geq 0 \quad \forall\, y \in C. \tag{5.6}$$

Replacing $y = x^*$ in this inequality we get

$$\langle F(x^*) + G(h(x^*) - x^*) + z^*, x^* - h(x^*) \rangle \geq 0. \tag{5.7}$$

Adding these two inequalities (5.5) and (5.7) we obtain

$$\langle G(h(x^*) - x^*), x^* - h(x^*) \rangle + \langle z^*, x^* - h(x^*) \rangle + \varphi(h(x^*)) - \varphi(x^*) \geq 0. \tag{5.8}$$

Since $z^* \in \partial\varphi(h(x^*))$, we have

$$\langle z^*, x^* - h(x^*) \rangle \leq \varphi(x^*) - \varphi(h((x^*)).$$

Thus

$$\langle z^*, x^* - h(x^*) \rangle - \varphi(x^*) + \varphi(h((x^*)) \leq 0. \tag{5.9}$$

From inequalities (5.8) and (5.9), it follows that

$$\langle G(h(x^*) - x^*), x^* - h(x^*) \rangle \geq 0.$$

Hence $h(x^*) = x^*$, since G is self-adjoint and positive.

Conversely, suppose now $h(x^*) = x^*$. Then, by (5.6) we have

$$\langle F(x^*) + z^*, y - x^* \rangle \geq 0 \;\; \forall\, y \in C.$$

Since $z^* \in \partial\varphi(h(x^*))$,

$$\langle z^*, y - x^* \rangle \leq \varphi(y) - \varphi(x^*) \;\; \forall\, y \in C.$$

Adding the last two inequalities we have

$$\langle F(x^*), y - x^* \rangle + \varphi(y) - \varphi(x^*) \geq 0 \;\; \forall\, y \in C,$$

which means that x^* is solution of problem (5.1). The proof for h_1 can be done by the same way using the formulation (5.2) as a particular case of (5.1).

□

We recall that

Definition 2.1 *A multivalued mapping* $\Phi : C \to H$ *is said to be monotone on C if*

$$\langle z - z', x - x' \rangle \geq 0 \;\; \forall\, x, x' \in C, \;\; z \in \Phi(x), \;\; z' \in \Phi(x').$$

Φ *is called strongly monotone with modulus* $\beta > 0$ *(briefly* β*-strongly monotone) if*

$$\langle z - z', x - x' \rangle \geq \beta \|x - x'\|^2 \quad \forall\, x, x' \in C, \;\; z \in \Phi(x), \;\; z' \in \Phi(x').$$

A mapping $\Phi : C \to H$ *is said to be Lipschitz continuous on C with modulus* $\delta > 0$*, if*

$$\|\Phi(x) - \Phi(x')\| \leq \delta \|x - x'\| \;\; \forall\, x, x' \in C. \tag{5.10}$$

If (5.10) is satisfied with $\delta < 1$*, then the mapping* Φ *is said to be contractive on C; it is said to be nonexpansive on C if* $\delta = 1$*.*

The mapping Φ *is said to be co-coercive with modulus* $\delta > 0$*, shortly* δ*-co-coercive, on C if*

$$\langle \Phi(x) - \Phi(x'), x - x' \rangle \geq \delta \|\Phi(x) - \Phi(x')\|^2 \;\; \forall\, x, x' \in C.$$

The number δ *is called co-coercivity modulus.*

A real-valued function f *is said to be* δ*-co-coercive on C if its gradient* ∇f *is* δ*-co-coercive on C, i.e.,*

$$\langle \nabla f(x) - \nabla f(x'), x - x' \rangle \geq \delta \|\nabla f(x) - \nabla f(x')\|^2 \;\; \forall\, x, x' \in C.$$

The co-coercivity was introduced in Gol'stein (1989) and used by Browder and Petryshyn in Browder (1967) in the context of computing fixed points. Recently it has been used to establish the convergence of some methods based on the auxiliary problem principle (see e.g. Cohen (1988); Marcotte (1995); Salmon (2000); Zhu (1995); Verma (2001)).

It is easy to see that a co-coercive mapping is also Lipschitzian and that any Lipschitzian and strongly monotone mapping is co-coercive. A co-coercive mapping is not necessarily strongly monotone, the constant mapping is an example. Co-coercivity of Lipschitz gradientmaps was established in Gol'stein (1989) where it is proven (Chapter 1, Lemma 6.7, see also Zhu (1995)) that a function f is convex and its gradient is Lipschitz continuous on C with Lipschitz constant L if and only if ∇f

is co-coercive (with constant $\frac{1}{L}$). From this result it follows that any affine mapping $Qx + q$ with Q being a symmetric positive semidefinite matrix is co-coercive. More properties about co-coercive mappings can be found in Anh (2002); Marcotte (1995); Gol'stein (1989); Zhu (1994); Zhu (1995); Zhu (1996).

3. A Contraction Fixed Point Approach

In what follows we suppose that the regularization operator $G = \alpha I$ with $\alpha > 0$ and I being the identity operator. First let us consider the mapping h. For this case we do not require the convex function φ to be differentiable.

The next lemma gives a relationship between $h(x)$ and $F(x)$, which will be very useful for our purpose.

Lemma 3.1 *Let $h(x)$ denote the unique solution of the convex optimization problem (5.3). Then*

$$\|h(x) - h(x')\|^2 \leq \left\| x - x' - \frac{1}{\alpha}(F(x) - F(x')) \right\|^2 .$$

Proof. Since G is positive definite and φ is convex on C, problem (5.3) is strongly convex. Thus $h(x)$ is uniquely defined as the solution of the unconstrained problem

$$\min\left\{ \frac{1}{2}\langle y - x, G(y - x)\rangle + \langle F(x), y - x\rangle + \varphi(y) - \varphi(x) + \delta_C(y) | \ y \in H \right\}$$

where $\delta_C(y)$ stands for the indicator function of C. Noting that the subdifferential of the indicator function of C is just the outward normal cone of C, we have

$$0 \in G(h(x) - x) + F(x) + N_C(h(x)) + \partial\varphi(h(x)),$$

which implies that there exist $z_1 \in N_C(h(x))$ and $z_2 \in \partial\varphi(h(x))$ such that

$$G(h(x) - x) + F(x) + z_1 + z_2 = 0,$$

where $N_C(h(x))$ denotes the outward normal cone of C at $h(x)$. Since $G = \alpha I$, it follows that

$$h(x) = x - \frac{1}{\alpha}F(x) - \frac{1}{\alpha}z_1 - \frac{1}{\alpha}z_2. \tag{5.11}$$

By the same way

$$h(x') = x' - \frac{1}{\alpha}F(x') - \frac{1}{\alpha}z_1' - \frac{1}{\alpha}z_2'. \tag{5.12}$$

From (5.11) and (5.12) we can write

$$\|h(x) - h(x')\|^2 = \langle h(x) - h(x'), h(x) - h(x')\rangle$$

$$= \left\langle x - x' - \frac{1}{\alpha}(F(x) - F(x')) - \frac{1}{\alpha}(z_1 - z_1') - \frac{1}{\alpha}(z_2 - z_2'), h(x) - h(x')\right\rangle.$$

Since the subdifferential of a convex function is monotone, we have

$$\langle z_1 - z_1', h(x) - h(x')\rangle \geq 0 \ \ \forall\, z_1 \in N_C(h(x)),\ \ z_1' \in N_C(h(x')),$$

$$\langle z_2 - z_2', h(x) - h(x')\rangle \geq 0 \ \ \forall\, z_2 \in \partial\varphi_C(h(x)),\ \ z_2' \in \partial\varphi_C(h(x')). \tag{5.13}$$

Thus from (5.13) we obtain

$$\|h(x) - h(x')\|^2 \leq \left\langle x - x' - \frac{1}{\alpha}(F(x) - F(x')), h(x) - h(x')\right\rangle$$

$$\leq \left\| x - x' - \frac{1}{\alpha}(F(x) - F(x'))\right\| \, \|h(x) - h(x')\|.$$

Hence

$$\|h(x) - h(x')\| \leq \left\| x - x' - \frac{1}{\alpha}(F(x) - F(x'))\right\|.$$

□

Let us first consider the variational inequality problem (5.1) where either F is strongly monotone or φ is strongly convex on C. In this case, the mapping h defined by the unique solution of problem (5.3) is contractive on C as the following theorem states.

Theorem 3.1 *(i) If F is β-strongly monotone and L-Lipschitz continuous on C, then h is contractive on C with modulus $\delta := \sqrt{1 - \frac{2\beta}{\alpha} + \frac{L^2}{\alpha^2}}$ whenever $\alpha > \frac{L^2}{2\beta}$.*

(ii) If φ is ρ-strongly convex, then h is contractive on C with modulus

$$\delta := \frac{\sqrt{L^2 + \alpha^2}}{\alpha + \rho}$$

whenever $\alpha > \frac{L^2 - \rho^2}{2\rho}$.

Proof. (i) Suppose first that F is β-**strongly** monotone and L-Lipschitz continuous on C. From

$$\left\| x - x' - \frac{1}{\alpha}(F(x) - F(x')) \right\|^2$$
$$= \|x - x'\|^2 - \frac{2}{\alpha}\langle x - x', F(x) - F(x')\rangle + \frac{1}{\alpha^2}\|F(x) - F(x')\|^2,$$

by Lemma 3.1, it follows

$$\|h(x) - h(x')\|^2 \leq \|x - x'\|^2 - \frac{2}{\alpha}\langle x - x', F(x) - F(x')\rangle + \frac{1}{\alpha^2}\|F(x) - F(x')\|^2.$$

Since F is β-**strongly** monotone and L-Lipschitz continuous on C, we have

$$\langle x - x', F(x) - F(x')\rangle \geq \beta\|x - x'\|^2$$

and

$$\|F(x) - F(x')\|^2 \leq L^2\|x - x'\|^2.$$

Thus

$$\|h(x) - h(x\prime)\|^2 \leq \|x - x'\|^2 - \frac{2\beta}{\alpha}\|x - x'\|^2 + \frac{L^2}{\alpha^2}\|x - x'\|^2$$
$$= \left(1 - \frac{2\beta}{\alpha} + \frac{L^2}{\alpha^2}\right)\|x - x'\|^2.$$

Hence

$$\|h(x) - h(x')\| \leq \sqrt{1 - \frac{2\beta}{\alpha} + \frac{L^2}{\alpha^2}}\,\|x - x'\|.$$

Clearly, if $\alpha > \frac{L^2}{2\beta}$, then $\delta := \sqrt{1 - \frac{2\beta}{\alpha} + \frac{L^2}{\alpha^2}} \in (0, 1)$. Hence h is contractive on C with modulus δ.

(ii) Now assume that φ is ρ-**strongly** convex on C. From (5.11) and (5.12) in the proof of Lemma 3.1 it follows that

$$\|h(x) - h(x')\|^2 \leq \langle x - x' - \frac{1}{\alpha}(F(x) - F(x')), h(x) - h(x')\rangle$$
$$-\frac{1}{\alpha}\langle z_2 - z_2', h(x) - h(x')\rangle \tag{5.14}$$

where

$$z_2 \in \partial\varphi(h(x)),\ z_2' \in \partial\varphi(h(x')).$$

By ρ-**strong** convexity of φ, we have

$$\langle z_2 - z_2', h(x) - h(x')\rangle \geq \rho\|h(x) - h(x')\|^2.$$

Then from (5.14), it follows that

$$\|h(x) - h(x')\|^2 \leq \left\langle x - x' - \frac{1}{\alpha}(F(x) - F(x')), h(x) - h(x') \right\rangle$$

$$-\frac{\rho}{\alpha}\|h(x) - h(x')\|^2 \Longleftrightarrow$$

$$\left(1 + \frac{\rho}{\alpha}\right)\|h(x) - h(x')\|^2 \leq \left\langle x - x' - \frac{1}{\alpha}(F(x) - F(x')), h(x) - h(x') \right\rangle$$

$$\leq \left\| x - x' - \frac{1}{\alpha}(F(x) - F(x') \right\| \|h(x) - h(x')\| \Longleftrightarrow$$

$$\left(1 + \frac{\rho}{\alpha}\right)^2 \|h(x) - h(x')\|^2 \leq \left\| x - x' - \frac{1}{\alpha}(F(x) - F(x')) \right\|^2$$

$$= \|x - x'\|^2 + \frac{1}{\alpha^2}\|F(x) - F(x')\|^2 - \frac{2}{\alpha}\langle x - x', F(x) - F(x')\rangle. \quad (5.15)$$

Since F is Lipschitz continuous on C with constant $L > 0$, and monotone on C, we have

$$\|F(x) - F(x')\| \leq L\|x - x'\| \ \ \forall\, x, x' \in C$$

$$\langle F(x) - F(x'), x - x'\rangle \geq 0 \ \ \forall\, x, x' \in C.$$

Combining with (5.15) yields

$$\left(1 + \frac{\rho}{\alpha}\right)^2 \|h(x) - h(x')\|^2 \leq \|x - x'\|^2 + \frac{L^2}{\alpha^2}\|x - x'\|^2$$

$$= \left(1 + \frac{L^2}{\alpha^2}\right)\|x - x'\|^2.$$

Hence

$$\|h(x) - h(x')\| \leq \frac{\sqrt{L^2 + \alpha^2}}{\alpha + \rho}\|x - x'\|.$$

Clearly, $0 < \delta := \frac{\sqrt{L^2+\alpha^2}}{\alpha+\rho} < 1$ whenever $\alpha > \frac{L^2-\rho^2}{2\rho}$. □

Now we suppose that φ is differentiable on some open set containing C. As we have mentioned in the preceding section, the objective function of the problem (5.4) for evaluating $g_1(x)$ is always quadratic, whereas the objective function of the problem (5.3), in general, is not quadratic. Note that the use of the marginal mapping h_1 can be considered as a way for iteratively approximating the convex function φ by its minorant affine function. In the case where the mapping F is strongly monotone on C, φ is a constant function and H is a finite-dimensional Euclidean space, it has been proved (see Anh (2002)) that h_1 is contractive on C when $G = \alpha I$ with a suitable $\alpha > 0$. The Corollary 3.1 below is an extension of this result to the case where φ may be any differentiable convex function and H is a real Hilbert space.

Let $\phi := F + \nabla\varphi$. Then, by (5.4), $h_1(x)$ is the unique solution of the strongly convex quadratic programming problem

$$\min\left\{\langle\phi(x), y - x\rangle + \frac{\alpha}{2}||y - x)||^2 \; |y \in C\right\}$$

which can be also written as

$$\min\left\{||y - (x - \frac{1}{\alpha}\phi(x)||^2 \; |y \in C\right\}.$$

Hence $h_1(x) = Pr_C\left(x - \frac{1}{\alpha}\phi(x)\right)$, where Pr_C denotes the projection operator onto C. It is well known that this projection operator is nonexpansive, i.e.,

$$\|Pr_C(x) - Pr_C(x')\| \leq \|x - x'\| \;\; \forall\, x, x' \in H.$$

Corollary 3.1 *Suppose that either F is strongly monotone or φ is strongly convex, and that $F + \varphi$ is L-Lipschitz continuous on C. Then one can choose a regularization parameter α such that h_1 is contractive on C.*

Namely,

(i) If F is β-strongly monotone on C, then h_1 is contractive on C whenever $\alpha > \frac{L^2}{2\beta}$.

(ii) If φ is ρ-strongly convex on C, then h_1 is contractive on C whenever $\alpha > \frac{L^2}{2\rho}$.

This result is a consequence of Theorem 3.1. Below is a direct proof which is very simple.

Proof. (i) Suppose first that F is strongly monotone. For simplicity of notation, we will write h_1 for $h_1(x)$, h_1' for $h_1(x')$, and the same conventions also hold for F, $\nabla\varphi$ and ϕ.

Using the nonexpansiveness property of the projection we have

$$\begin{aligned}\|h_1 - h_1'\|^2 &\leq \left\|x - \frac{1}{\alpha}(F + \nabla\varphi) - (x' - \frac{1}{\alpha}(F' + \nabla\varphi'))\right\|^2 \\ &= \|x - x'\|^2 - \frac{2}{\alpha}\langle x - x', F - F'\rangle - \frac{2}{\alpha}\langle x - x', \nabla\varphi - \nabla\varphi'\rangle \\ &\quad + \frac{1}{\alpha^2}\|\phi - \phi'\|^2. \qquad (5.16)\end{aligned}$$

Since F is **β-strongly** monotone and $F + \nabla\varphi$ is L-Lipschitz continuous, we have

$$\langle x - x', F - F' \rangle \geq \beta \|x - x'\|^2$$

and

$$\|(F + \nabla\varphi) - (F' + \nabla\varphi')\|^2 \leq L^2 \|x - x'\|^2.$$

Thus, by monotonicity of $\nabla\varphi$, it follows from (5.16) that

$$\begin{aligned}\|h_1 - h_1'\|^2 &\leq \|x - x'\|^2 + \frac{L^2}{\alpha^2}\|x - x'\|^2 - \frac{2\beta}{\alpha}\|x - x'\|^2 \\ &= (1 + \frac{L^2}{\alpha^2} - \frac{2\beta}{\alpha^2})\|x - x'\|^2.\end{aligned}$$

Hence h_1 is contractive whenever $\alpha > \frac{L^2}{2\beta}$.

(ii) Now suppose that φ is **ρ-strongly** convex on C. Then for any $x, x' \in C$, we have

$$\varphi(x) \geq \varphi(x') + \langle \nabla\varphi(x'), x - x' \rangle + \frac{\rho}{2}\|x - x'\|^2,$$

$$\varphi(x') \geq \varphi(x) + \langle \nabla\varphi(x), x' - x \rangle + \frac{\rho}{2}\|x - x'\|^2.$$

Adding these two inequalities, we see that $\nabla\varphi$ is **ρ-strongly** monotone on C. The claim thus follows from part (i). □

4. Nonexpansiveness Fixed-Point Formulation

In this section we weaken the strongly monotonicity assumption of F in Theorem 5.1 by co-coercivity. The variational inequality (5.1) then may have many solutions. So it is not expected that there exists some $\alpha > 0$ such that the mapping h remains contractive. However, it will be nonexpansive as the following theorem states.

Theorem 4.1 *Suppose that F is δ-co-coercive on C with $\alpha \geq \frac{1}{2\delta}$. Then h is nonexpansive mapping on C, i.e.*

$$\|h(x) - h(x')\| \leq \|x - x'\| \quad \forall\, x, x' \in C.$$

Proof. For any x and x' one has

$$\begin{aligned}&\left\| x - x' - \frac{1}{\alpha}(F(x) - F(x')) \right\|^2 \\ &\quad = \|x - x'\|^2 - \frac{2}{\alpha}\langle x - x', F(x) - F(x') \rangle + \frac{1}{\alpha^2}\|F(x) - F(x')\|^2.\end{aligned}$$

On the other hand, since F is **δ-co-coercive** on C with modulus δ, and $\alpha \geq \frac{1}{2\delta}$, we have

$$\langle x - x', F(x) - F(x')\rangle \geq \frac{1}{2\alpha}\|F(x) - F(x')\|^2 \quad \forall\, x, x' \in C.$$

Thus

$$\left\| x - x' - \frac{1}{\alpha}(F(x) - F(x'))\right\|^2 \leq \|x - x'\|^2$$

or

$$\left\| x - x' - \frac{1}{\alpha}(F(x) - F(x'))\right\| \leq \|x - x'\|. \tag{5.17}$$

In view of Lemma 3.1, it follows from (5.17) that

$$\|h(x) - h(x')\| \leq \|x - x'\| \quad \forall\, x, x' \in C.$$

□

As before when φ is differentiable we have the following result which is a consequence of Theorem 4.1.

Corollary 4.1 *Suppose that the mapping $\phi := F + \nabla\varphi$ is Δ-co-coercive on C. Then the mapping h_1 is nonexpansive on C whenever $\alpha > \frac{1}{2\Delta}$.*

This corollary can be proven by using Theorem 4.1. Below is a direct proof using the nonexpansiveness of the projection.

Proof. Using again the nonexpansiveness of the projection we have

$$\|h_1 - h_1'\|^2 \leq \left\| x - \frac{1}{\alpha}\phi - (x' - \frac{1}{\alpha}\phi')\right\|^2$$

$$= \left\| x - x' - \frac{1}{\alpha}(\phi' - \phi)\right\|^2$$

$$= \|x - x'\|^2 - \frac{2}{\alpha}\langle x - x', \phi - \phi'\rangle + \frac{1}{\alpha^2}\|\phi - \phi'\|^2.$$

Since ϕ is **Δ-co-coercive,**

$$\langle x - x', \phi - \phi'\rangle \geq \Delta\|\phi - \phi'\|^2.$$

Thus

$$\|h_1 - h_1'\|^2 \leq \|x - x'\|^2 - \frac{2\Delta}{\alpha}\|\phi - \phi'\|^2 + \frac{1}{\alpha^2}\|\phi - \phi'\|^2$$

$$= \|x - x'\|^2 + \left(\frac{1}{\alpha^2} - \frac{2\Delta}{\alpha}\right) \|\phi - \phi'\|^2,$$

which implies

$$\|h_1 - h_1'\| \leq \|x - x'\|$$

whenever $\alpha > \frac{1}{2\Delta}$. □

Note that the gradient of a convex function is L-co-coercive on C if and only if it is $\frac{1}{L}$-**Lipschitz** continuous on C (see e.g. Zhu (1995)), by applying Theorem 4.1 with $F \equiv 0$, we have the following corollary.

Corollary 4.2 *Suppose that φ is convex and its gradient $\nabla\varphi$ is L-Lipschitz continuous on C. Let $h_1(x)$ be the marginal mapping of the strongly convex quadratic programming problem*

$$\min\{ \langle \nabla\varphi(x), y\rangle + \alpha\|y - x\|^2 \}$$

with $\alpha > \frac{L}{2}$. Then h_1 is nonexpansive on C, and $x^* = h_1(x^*)$ if and only if x^* is a solution of the convex program

$$\min\{\varphi(x) \;|x \in C\}.$$

Remark 4.1 *When $F \equiv 0$, the variational inequality problem (5.1) becomes the convex programming problem*

$$\min\{\varphi(x)|x \in C\}. \tag{5.18}$$

Since the constant mapping is co-coercive with any modulus $\delta > 0$, by Theorem 4.1, we have the following corollary from which it turns out that h is just the proximal mapping for convex programming problem (5.18) (see Rockafellar (1976)).

Corollary 4.3 *For each x, let h be the unique solution of the strongly convex program*

$$\min\{\varphi(y) + \alpha\|y - x\|^2 \;|y \in C\}.$$

Then for any $\alpha > 0$, the mapping h is nonexpansive on C, and x^* is a solution to (5.18) if and only if it is a fixed point of h.

5. On Solution Methods

The results in the preceding sections lead to algorithms for solving the generalized variational inequality problem (5.1) by the Banach contraction mapping principle or its modifications.

By Theorem 3.1, when either F is strongly monotone or φ is strongly convex on C, one can choose a suitable regularization parameter such that the mapping h is contractive on C. The same result is true for h_1 when φ is differentiable. In this case, by the Banach contraction principle the unique fixed point of h and of h_1, thereby the unique solution of the variational inequality (5.1), can be approximated by iterative procedures

$$x^{k+1} = h(x^k), \ k = 0, 1, \ldots$$

or

$$x^{k+1} = h_1(x^k), \ k = 0, 1, \ldots$$

where x^0 can be any starting point in C.

According to the definition of h and h_1, evaluating $h(x^k)$ and $h_1(x^k)$ amounts to solving the strongly convex programs (5.3) and (5.4), respectively.

The algorithms then can be described in detail as follows.

Algorithm 5.1 *(strongly monotone case)*
Choose a tolerance $\epsilon \geq 0$.
If F *is* β*-strongly monotone, choose* $\alpha > \frac{L^2}{2\beta}$.
If φ *is* ρ*-strongly convex, choose* $\alpha > \frac{L^2-\rho^2}{2\beta}$, *where* L *is the Lipschitz constant of* F.
Select $x^0 \in C$.
Iteration k $(k = 0, 1, \ldots)$*:*
Solve the strongly convex program

$$\min\{\frac{1}{2}\alpha||x - x^k||^2 + \langle F(x^k), x - x^k\rangle + \varphi(x) | x \in C\} \tag{5.19}$$

to obtain its unique solution x^{k+1}.
If $||x^{k+1} - x^k|| \leq \epsilon$, *then terminate:* x^k *is an* ϵ*-solution to the variational inequality problem (5.1).*
Otherwise, if $||x^{k+1} - x^k|| > \epsilon$, *then increase* k *by* 1 *and go to iteration* k.

By Theorem 3.1 and the Banach contraction principle, if the algorithm does not stop after a finite number of iterations, then the sequence $\{x^k\}$ generated by the algorithm strongly converges to the unique solution x^*

of the variational inequality problem (5.1). Moreover, at each iteration k we have the following convergence estimation:

$$\|x^{k+1} - x^*\| \leq \frac{\delta^{k+1}}{1-\delta}\|x^0 - x^1\|,$$

where $0 < \delta < 1$ is the contraction modulus of h. According to Theorem 3.1, $\delta := \sqrt{1 - \frac{2\beta}{\alpha} + \frac{L^2}{\alpha^2}}$ when F is β-**strongly** monotone, and $\delta := \frac{\sqrt{L^2+\alpha^2}}{\alpha+\rho}$ when φ is ρ-**strongly** convex.

From

$$\|x^{k+1} - x^*\| \leq \frac{\delta^{k+1}}{1-\delta}\|x^0 - x^1\|$$

with $0 < \delta < 1$, it follows that

$$\limsup_k \frac{\|x^k + 1 - x^*\|}{\|x^k - x^*\|} = \delta.$$

Thus the sequence $\{x^k\}$ generated by Algorithm 5.1 converges Q-linearly to x^*. Note that when F and φ have Lipschitz continuous gradients and F is strongly monotone on C, the Q-linear rate of convergence has been obtained in Patriksson (1999) (Theorem 6.9) for a generalized algorithmic scheme called CA algorithms.

Remark 5.1 *The above algorithm belongs to the well known general algorithmic scheme (see e.g. Cohen (1988); Konnov (2001); Patriksson (1999); Zhu (1994); Zhu (1996)) based on the so-called auxiliary problem principle. When φ is absent, this algorithm becomes the projection procedure.*

The main point here is the choice of the regularization parameter α such that the marginal mapping h is contractive.

In the case φ is differentiable and its gradient is easy to compute, it is suggested to use the marginal mapping h_1, since the objective function of the strongly convex program for evaluating h_1 is quadratic.

The algorithm for this case can be described similarly as before in the following manner.

Algorithm 5.2 *(strongly monotone and differentiable case)*
Choose a tolerance $\epsilon \geq 0$.
*If F is β-**strongly** monotone, choose $\alpha > \frac{L^2}{2\beta}$.*
*If φ is ρ-**strongly** convex, choose $\alpha > \frac{L^2-\rho^2}{2\beta}$, where L is the Lipschitz constant of F.*

Select $x^0 \in C$.
Iteration k ($k = 0, 1, \ldots$):
Solve the strongly convex quadratic program

$$\min\{\frac{1}{2}\alpha||x - x^k||^2 + \langle F(x^k) + \nabla\varphi(x^k), x - x^k\rangle | x \in C\}, \tag{5.20}$$

to obtain its unique solution x^{k+1}.
If $\|x^{k+1} - x^k\| \leq \epsilon$, then stop: x^k is an **ϵ-solution** *to problem (5.1).*
Otherwise, if $\|x^{k+1} - x^k\| > \epsilon$, then increase k by 1 and go to iteration k.

As before, the sequence $\{x^k\}$ generated by this algorithm also converges strongly to the unique solution of the variational inequality problem (5.1) and, as before, the geometric convergence can be easily obtained.

Remark 5.2 *Algorithm 5.2 belongs to the well known projection method for the problem (5.1). The use of subprograms (5.20) gives a way to approximate the convex function φ by its gradients at iteration points. Note that in this algorithm the feasible domain of the subproblems at each iteration is the same as the feasible set of the original problem. From a computational point of view this is important, since in some practical problems such as traffic equilibrium models, the feasible set C having specific structure.*

Now we turn to the nonexpansiveness case. For computing fixed points of nonexpansive mappings h and h_1, thereby a solution of problem (5.1), we shall use the following results.

Lemma 5.1 *(Browder (1967) Theorem 8, see also Goebel (1990) Corollary 9) Let X be a Banach space, K be a closed, convex subset of X, and $T : K \to K$ be a nonexpansive mapping for which T(K) is compact (weakly compact, resp.). Then for each $\lambda \in (0, 1)$, the iterates of the mapping $T_\lambda := (1 - \lambda)I + \lambda T$ converges (weakly converges, resp.) to a fixed point of T.*

In the next lemma the nonexpansive mapping is replaced by the contractive mappings

$$T_k = \frac{1}{k}x^0 + \left(1 - \frac{1}{k}\right)T, \; k = 1, 2, \ldots$$

where $x^0 \in K$. Let x^k be the unique fixed point of T_k. It has been shown in Aubin (1984) that if K is a closed and bounded subset in a Hilbert space, then the sequence of points x^k $(k = 0, 1, \ldots)$ has a weak limit point which is a fixed point of T in K.

As usual, for a mapping $T : K \to K$, the mapping T^k denotes the **k-times** composition mapping of T. In general, it is not true that the sequence of points $T^k(x^0)$ $(k = 0, 1 \ldots)$ tends to a fixed point of the nonexpansive mapping T. However, if we use the Cesàro means, then a fixed point of a nonexpansive mapping can be approximated as stated by the following lemma.

Lemma 5.2 *(Aubin (1984) Theorem 7, page 253). Let T be a nonexpansive mapping from a closed convex bounded subset K of a Hilbert space to itself. For any initial point $x^0 \in K$, the sequence of elements $y^{k+1} := \frac{1}{k+1} \sum_{j=1}^{k+1} T^j(x^0)$ $(k = 0, 1, \ldots)$ converges weakly to a fixed point of T.*

Under the assumptions of Theorem 4.1 and Corollary 4.1, the mappings h and h_1 are nonexpansive on C. In order to apply Lemma 5.1 we select any $\lambda \in (0, 1)$ and any $x^0 \in C$. From x^0 we construct the sequence $\{x^k\}$ $(k = 1, 2, \ldots)$ by setting

$$x^{k+1} := T_\lambda(x^k) := \lambda x^k + (1 - \lambda)T(x^k), \tag{5.21}$$

where we take $T = h$ (for Theorem 4.1) and $T = h_1$ (for Corollary 4.1). When $T = h$, to compute $T(x^k)$ we have to solve subproblem (5.19) with $\alpha > 0$ being chosen as in Theorem 4.1. When $T = h_1$ to compute $T(x^k)$ we have to solve subproblem (5.20) with $\alpha > 0$ being chosen as in Corollary 4.1.

Proposition 5.1 *Under the assumption of Theorem 4.1 (Corollary 4.1, resp.), the sequence $\{x^k\}$ $(k = 1, 2, \ldots)$ generated by (5.21) with $T \equiv h$ ($T \equiv h_1$, resp.) converges weakly to a solution of the variational inequality problem (5.1).*

Proof. Let x^* be any solution of the variational inequality (1.1). Since h is nonexpansive and $T \equiv h$, $h(x^*) = x^*$, by (5.21) we have

$$\begin{aligned} x^{k+1} - x^* &= \lambda x^k + (1 - \lambda)h(x^k) - x^* \\ &= \lambda x^k - \lambda x^* + (1 - \lambda)h(x^k) - h(x^*) + \lambda h(x^*). \end{aligned}$$

Thus

$$\|x^{k+1} - x^*\| \le \lambda\|x^k - x^*\| + (1 - \lambda)\|h(x^k) - h(x^*)\|$$

$$\leq \lambda\|x^k - x^*\| + (1-\lambda)\|x^k - x^*\| = \|x^k - x^*\|$$

from which it follows that

$$\|x^{k+1} - x^*\| \leq \|x^k - x^*\| \leq \|x^0 - x^*\| \;\forall\; k.$$

Hence $x^k \in B_0(x^*)$ for all k, where $B_0(x^*)$ stands for the closed ball centered at x^* with the radius $\|x^0 - x^*\|$. Applying Lemma 5.1 with $T = h$ and $K = B_0(x^*)$ we see that the sequence of points x^k ($k = 0, 1, ...$) weakly converges to a fixed point of h.

The same argument is true for h_1. □

Remark 5.3 *If, in addition,* C *is compact, then* $h(C)$ *and* $h_1(C)$ *are compact. By Lemma 5.1, the sequence* $\{x^k\}$ $(k = 1, 2, \ldots)$ *generated by (5.21) with* $T \equiv h$ *or* $T \equiv h_1$ *strongly converges to a solution of the variational inequality problem (5.1).*

Remark 5.4 *In Browder (1967) Browder and Petryshyn presented an iterative procedure for computing a fixed point of pseudocontractive mappings. A mapping* $T : C \to H$ *is said to be pseudocontractive or (pseudononexpansive) on* C *with modulus* ξ *if*

$$\|T(x) - T(y)\|^2 \leq \|x - y\|^2 + \xi\|(I - T)(x) - (I - T)(y)\|^2,$$

where $0 \leq \xi < 1$ *and* I *is the identity mapping.*

Clearly, nonexpansive mappings are always pseudononexpansive with any modulus $\xi \geq 0$.

It has been shown (Browder (1967) Theorem 12) that if T is pseudononexpansive with modulus $0 < \xi < 1$ on a closed convex set C in a real Hilbert space, then, for any $1 - \xi < \lambda < 1$ and $x^0 \in C$, the sequence x^k ($k = 0, 1, \ldots$) defined by

$$x^{k+1} := \lambda x^k + (1 - \lambda)T(x^k)$$

converges weakly to a fixed point of T.

Clearly, for a nonexpansive mapping, the sequence defined by (5.21) coincides with the sequence defined by Browder and Petryshyn.

Remark 5.5 *Note that, by (5.21),*

$$x^{k+1} = T(x^k) + \lambda(x^k - T(x^k)).$$

So this procedure can be considered as a line search on the line segment $(x^k,\ T(x^k))$ *where* λ *plays the role of the stepsize.*

Remark 5.6 *From $\|x^k - x^*\| \leq \|x^0 - x^*\|$, it is expected that the procedure (5.21) converges quickly to a solution of the variational inequality problem (5.1), provided that the initial point x^0 is near to some solution of (5.1).*

By applying Lemma 5.2 we may have another method for solving the variational inequality problem (5.1). Note that the Cesàro means in Lemma 5.2 can be rewritten as

$$y^{k+1} = \frac{1}{k+1}\sum_{j=0}^{k} T^j(x^0) + \frac{1}{k+1}T^{k+1}(x^0) \quad (k = 0, 1, \ldots)$$

or

$$\begin{aligned} y^{k+1} &= \frac{1}{k+1}\sum_{j=0}^{k} T^j(x^0) + \frac{1}{k+1}T^{k+1}(x^0) \\ &= \frac{k}{k+1}y^k + \frac{1}{k+1}T^{k+1}(x^0) \quad (k = 0, 1, \ldots). \end{aligned}$$

Let $x^{k+1} := T^{k+1}(x^0)$ $(k = 0, 1, \ldots)$, then we can write

$$y^{k+1} = \frac{k}{k+1}y^k + \frac{1}{k+1}x^{k+1} \quad (k = 0, 1, \ldots). \tag{5.22}$$

So to compute y^{k+1} we have to compute only $x^{k+1} = T(x^k)$, since other iteration points have been computed at the previous iterations. As before, when $T \equiv h$ ($T \equiv h_1$, resp.) the subproblem for computing $T(x^k)$ is (5.19)((5.20), resp.). The weak convergence of the sequence $\{y^k\}$ generated by (5.22) to a solution of problem (5.1) is ensured by Lemma 5.2 with the same argument as in the proof of Proposition 5.1 (the boundedness of the sequence $\{x^k\}$ follows from the nonexpansiveness of T).

6. Conclusion

We have used the contraction mapping fixed point principle for solving monotone variational inequalities. We have shown how to choose regularization parameters such that the marginal mappings determining the projection gap functions to be contractive under the strong monotonicity, and to be nonexpansive under the co-coercivity. The result leads to the Banach iterative method and its modifications for solving generalized variational inequalities involving strongly monotone and co-coercive operators.

References

Anh, P.N. and Muu, L.D. (2002), The Banach Iterative Procedure for Solving Monotone Variational Inequality. *Hanoi Institute of Mathematics,* Preprint No. 05.

Aubin, J.P. and Ekeland, I. (1984), *Applied Nonlinear Analysis,* Wiley , New York.

Auslender, A. (1976), *Optimisation: Méthodes Numériques,* Masson, Paris.

Browder, F.E. (1966) On the Unification of the Calculus of Variations and the Theory of Monotone Nonlinear Operators in Banach Spaces, *Proc. Nat. Acad. Sci. USA,* Vol. 56, pp. 419-425.

Browder, F.E. and Petryshyn, W.V. (1967), Construction of Fixed Points of Nonlinear Mappings in Hilbert Space, *J. on Mathematical Analysis and Applications,* Vol. 20, pp. 197-228.

Clarke, F.H. (1983), *Optimization and Nonsmooth Analysis,* Wiley, New-Yowk.

Cohen, G. (1988), Auxiliary Problem Principle Extended to Variational Inequalities, *J. of Optimization Theory and Applications,* Vol. 59, pp. 325-333.

Fukushima, M. (1992), Equivalent Differentiable Optimization Problems and Descent Methods for Asymmetric Variational Inequality Problems, *Mathematical Programming,* Vol. 53, pp. 99-110.

Goebel, K. and Kirk, W.A. (1990), *Topics in Metric Fixed Point Theory,* Cambridge University Press, Cambridge.

Golshtein E.G. and Tretyakov N.V. (1996), *Modified Lagrangians and Monotone Maps in Optimization,* Wiley, New York.

Harker, P.T. and Pang, J.S. (1990), Finite-Dimensional Variational Inequality and Nonlinear Complementarity Problems: a Survey of Theory, Algorithms, and Applications, *Mathematical Programming,* Vol. 48, pp. 161-220.

Hue, T.T., Strodiot, J.J. and Nguyen, V.H. (2004), Convergence of the Approximate Auxiliary Problem Method for Solving Generalized Variational Inequalities, *J. of Optimization Theory and Applications,* Vol. 121, pp. 119-145.

Kinderlehrer, D. and Stampacchia, G. (1980), An Introduction to Variational Inequalities and Their Applications, Academic Press, New York.

Konnov, I. (2001), *Combined Relaxation Methods for Variational Inequalities,* Springer, Berlin.

Konnov, I. and Kum S. (2001), Descent Methods for Mixed Variational Inequalities in a Hilbert Space, *Nonlinear Analysis: Theory, Methods and Applications,* Vol. 47, pp. 561-572.

Luo, Z. and Tseng, P. (1991), A Decomposition Property of a Class of Square Matrices, *Applied Mathematics,* Vol. 4, pp. 67-69.

Marcotte, P. (1995), A New Algorithm for Solving Variational Inequalities, *Mathematical Programming,* Vol. 33, pp. 339-351.

Marcotte, P. and Wu, J.H. (1995), On the Convergence of Projection Methods: Application to the Decomposition of Affine Variational Inequalities, *J. of Optimization Theory and Applications,* Vol. 85, pp. 347-362.

Muu, L.D. and Khang, D.B. (1983), Asymptotic Regularity and the Strongly Convergence of the Proximal Point Algorithm, *Acta Mathematica Vietnamica,* Vol. 8, pp. 3-11.

Muu. L.D. (1986), An Augmented Penalty Function Method for Solving a Class of Variational Inequalities, *USSR Computational Mathematics and Mathematical Physics,* Vol. 12, pp. 1788-1796.

Noor M.A. (1993) General Algorithm for Variational Inequalities, *J. Math. Japonica,* Vol. 38, pp. 47-53.

Noor M.A. (2001) Iterative Schemes for Quasimonotone Mixed Variational Inequalities, *Optimization,* Vol. 50, pp. 29-44.

Patriksson M, (1997) Merit Functions and Descent Algorithms for a Class of Variational Inequality Problems. *Optimization,* Vol. 41, pp. 37-55.

Patriksson M, (1999), *Nonlinear Programming and Variational Inequality Problems,* Kluwer, Dordrecht.

Rockafellar, R.T. (1976), Monotone Operators and the Proximal Point Algorithm, *SIAM J. on Control,* Vol. 14, pp. 877-899.

Rockafellar, R.T. (1979), *Convex Analysis,* Princeton Press, New Jersey.

Salmon, G., Nguyen, V.H. and Strodiot, J.J. (2000), Coupling the Auxiliary Problem Principle and Epiconvergence Theory to Solve General Variational Inequalities, *J. of Optimization Theory and Applications,* Vol. 104, pp. 629-657.

Taji, K. and Fukushima, M. (1996), A New Merit Function and a Successive Quadratic Programming Algorithm for Variational Inequality Problems, *SIAM J. on Optimization,* Vol. 6, pp. 704-713.

Taji, K., Fukushima, M. and Ibaraki (1993), A Global Convergent Newton Method for Solving Monotone Variational Inequality Problems, *Mathematical Programming,* Vol. 58, pp. 369-383.

Tseng, P. (1990), Further Applications of Splitting Algorithm to Decomposition Variational Inequalities and Convex Programming, *Mathematical Programming,* Vol. 48, pp. 249-264.

Verma, R.U. (2001), Generalized Auxiliary Problem Principle and Solvability of a Class of Nonlinear Variational Inequalities Involving Co-coercive and Co-Lipschitzian Mappings, *J. of Inequalities in Pure and Applied Mathematics,* Vol. 2, pp. 1-9.

Wu, J.H., Florian, M. and Marcotte, P. (1992), A General Descent Framework for the Monotone Variational Inequality Problem, *Mathematical Programming,* Vol. 53, pp. 99-110.

Zhu, D. and Marcotte, P. (1994), An Extended Descent Framework for Variational Inequalities, *J. of Optimization Theory and Applications,* Vol. 80, pp. 349-366.

Zhu, D. and Marcotte, P. (1995), A New Class of Generalized Monotonicity, *J. of Optimization Theory and Applications,* Vol. 87, pp. 457-471.

Zhu, D. and Marcotte, P. (1996), Co-coercivity and its Role in the Convergence of Iterative Schemes for Solving Variational Inequalities, *SIAM J. on Optimization,* Vol. 6, pp. 714-726.

Chapter 6

A PROJECTION-TYPE ALGORITHM FOR PSEUDOMONOTONE NONLIPSCHITZIAN MULTIVALUED VARIATIONAL INEQUALITIES

T. Q. Bao
Department of Mathematics
Wayne State University, U.S.A.

P. Q. Khanh
Department of Mathematics
International University,
Vietnam National University of Hochiminh City

Abstract We propose a projection-type algorithm for variational inequalities involving multifunction. The algorithm requires two projections on the constraint set only in a part of iterations (one third of the subcases). For the other iterations, only one projection is used. A global convergence is proved under the weak assumption that the multifunction of the problem is pseudomonotone at a solution, closed, lower hemicontinuous, and bounded on each bounded subset (it is not necessarily continuous). Some numerical test problems are implemented by using MATLAB with encouraging effectiveness.

Keywords: Variational inequalities, multifunctions, projections, pseudo monotonicity, closedness, lower hemicontinuity, boundedness.

MSC2000: 65K10, 90C25

1. Introduction

We consider the multivalued variational inequality problem: finding $x^* \in K$ such that there is $t^* \in Tx^*$ satisfying

$$(VI) \qquad \langle t^*, x - x^* \rangle \geq 0, \quad \forall x \in K,$$

where K is a closed convex set in R^n, $T : R^n \to 2^{R^n}$ is a multifunction, and $\langle .,. \rangle$ denotes the usual inner product in R^n.

For the single-valued case of *(VI)*, where T is a (single-valued) mapping, there are many numerical methods: projection, the Weiner - Hopf equations, proximal point, descent, decomposition and auxiliary principle. Among these methods, projection algorithms appeared first and are experiencing an explosive development due to their natural arguments, global convergence and simplicity of implementation. The first works were by Goldstein (1964), Levitin et al (1966), and Sibony (1970), where the authors proposed an extension of the projected gradient algorithm for convex minimization problems based on the iteration:

$$x^{k+1} = P_K(x^k - \rho T x^k), \tag{6.1}$$

where $\rho > 0$ is a parameter and $P_K(.)$ stands for the projection on K. If T is strongly monotone with modulus $\alpha > 0$, i.e. $\langle Ty - Tx, y - x \rangle \geq \alpha \|y - x\|^2$ for all $x, y \in R^n$, and Lipschitz with constant L, then the classical projection algorithm (6.1) globally converges to a solution for any $\rho \in (0, \frac{2\alpha}{L^2})$.

The extragradient algorithm, proposed by Korpelevich (1976), using two projections per iteration:

$$x^{k+1} = P_K(x^k - \rho T P_K(x^k - \rho T x^k)) \tag{6.2}$$

goes an important step for improving the classical algorithms (6.1). It requires T to be monotone and Lipschitz for its global convergence. Until now many projection-type algorithms have been developed to reduce assumptions which guarantee the convergence, and to improve the effectiveness of convergence rate, computational costs and implementation. Noor, see e.g. the recent Noor (1999); Noor (2003a); Noor (2003c) and the references therein, motivated by various fixed point presentations, which may be equivalent to single-valued variational inequalities, and Weiner-Hopf equations, proposed many variants of projection algorithms using two or more projections at each iteration. The stepsize ρ was designed to be ρ_k depending on iterations in these and the most other papers on projection methods, e.g. He (1997) - He et al (2002), Iusem et al (1997), Noor et al (1999) - Noor et al (2003), Solodov et al (1999) - Zhao (1999). Moreover, the two stepsizes in (6.2) may be different as ρ_k for the first projection and as α_k for the second one.In this case, the first projection is called the predictor step and the second one is the corrector step.

It is observed that the proximal point algorithm introduced by Martinet (Martinet (1970); Martinet (1972)) and generalized by Rockafellar (Rockafellar (1976a); Rockafellar (1976b)) has also been combined with projection algorithms. The proximal point algorithm for (VI) is

$$x^{k+1} = (I + \alpha_k(T + N_K))^{-1}x^k, \tag{6.3}$$

where I is the identity mapping, since (VI) is equivalent to finding $x^* \in R^n$ such that

$$0 \in T(x^*) + N_K(x^*),$$

where $N_K(x^*)$ stands for the normal cone to K at x^*. (6.3) can be checked to be equivalent to

$$x^{k+1} = P_K(x^k - \alpha_k T x^{k+1}). \tag{6.4}$$

So the proximal point algorithm may be viewed as an implicit projection algorithm (since x^{k+1} occurs on both sides of (6.4)). Moreover, the proximal point algorithm was also combined with projection algorithms to solve mixed variational inequalities, see e.g. Noor's recent papers (Noor (2001c); Noor (2003b)) and references therein. In this context, the combined algorithms are closely related to splitting algorithms and forward-backward algorithms.

Rather few algorithms have been developed for multivalued variational inequality problems. We observe such algorithms, which are of the projection type, only in Alber (1983), Iusem et al (2000), and Noor (2001a). For the convergence of the algorithms, in Alber (1983) and Noor (2001a), T should be Lipschitz and uniformly monotone, i.e. $\langle u - v, x - y\rangle \geq \varphi(\|x - y\|)$ for all $x, y \in R^n$ and $u \in Tx, v \in Ty$, where φ is a monotone increasing function with $\varphi(0) = 0$, or partially relaxed strongly monotone, i.e. $\langle u - v, z - y\rangle \geq -\alpha\|x - z\|^2$ for all $x, y, z \in R^n$ and $u \in Tx, v \in Ty$ and for some $\alpha > 0$. Note that these two monotonicity properties are only slightly weaker than strong monotonicity. Moreover, in Noor (2001a), three projections are needed at each iteration and $u^k \in Tx^k$ cannot be chosen arbitrarily at each iteration k. In Iusem et al (2000), T is assumed to be maximal monotone and $u^k \in Tx^k$ is obtained by solving a minimization problem on $T^{\epsilon_k}x^k$, where T^{ϵ_k} is an ϵ_k-**enlargement** of T. So u^k may be not implementable because performing a projection on K is equivalent to solving a quadratic minimization problem on K; it may be computationally expensive if K is not simple.

Motivated by these arguments, the aim of the present work is to develop an implementable algorithm for multivalued variational inequalities so that:

- it is implementable, in particular, $u^k \in Tx^k$ can be taken arbitrarily;
- it needs as less as possible projections per iteration;
- it globally converges under rather weak assumptions. In particular, Lipschitz continuity should be avoided since this is strict, and if satisfied, it is difficult to determine Lipschitz constants in general; even for affine mappings.

The paper is organized as follows. The remaining part of this Section contains some preliminaries. In Section 2, we present the proposed algorithm. A global convergence is established in Section 3. Finally, Section 4 provides some numerical examples.

A multifunction T: $R^n \to 2^{R^n}$ is said to be pseudomonotone at $x \in R^n$ if $\forall y \in R^n$, from $\langle u, y-x\rangle \geq 0$ for some $u \in Tx$ it follows that $\langle v, y-x\rangle \geq 0$ for all $v \in Ty$. T is called pseudomonotone if it is pseudomonotone at all $x \in R^n$. In the sequel, all properties defined at a point will be extended for all points in the same way. T is termed upper semicontinuous (usc for short) at x if for all neighborhood V of Tx, there is a neighborhood U of x such that $TU \subseteq V$. T is said to be closed at x if for any sequence $x^k \to x$ and $u^k \in Tx^k$ such that $u^k \to u$, one has $u \in Tx$. T is called to be lower semicontinuous (lsc) at x if $\forall x^k \to x, \forall u \in Tx, \exists u^k \in Tx^k$ such that $u^k \to u$. T is said lower hemicontinuous (lhc) at x if $\forall \alpha_k \to 0^+$, $\forall y \in R^n$,$\forall u \in Tx$, $\exists u^k \in T(x+\alpha_k(x-y))$ such that $u^k \to u$. Note that lower hemiconnuity is weaker than lower semicontinuity and closedness is different from upper semicontinuity.

In the sequel we need the following well known and basic facts.

Lemma 1.1 *A point $x^* \in R^n$ is a solution to (VI) if and only if there is $t^* \in Tx^*$ such that $x^* - P_K(x^* - \rho t^*) = 0$, where $\rho > 0$ is arbitrary.*

Lemma 1.2 *Let K be a closed and convex subset in R^n and $z \in R^n$. The following hold:*

(i) $x = P_K(z)$ if and only if $\langle x-z, y-x\rangle \geq 0$, $\forall y \in K$;

(ii) for any $y \in K$,

$$\|P_K(z) - y\|^2 \leq \|z-y\|^2 - \|z - P_K(z)\|^2.$$

2. The proposed algorithm

To explain what has suggested the algorithm, let us consider the single-valued case of (VI). If x^* is a solution, it holds that

$$x^* = P_K(x^* - \rho Tx^*) + \rho(Tx^* - TP_K(x^* - \rho Tx^*)),$$

for any $\rho > 0$. The converse may be untrue. However, this fixed-point formulation suggests the following iterations for the multivalued case. Given x^k, take arbitrarily $t^k \in Tx^k$, compute $\bar{x}^k = P_K(x^k - \rho t^k)$, take arbitrarily $\bar{t}^k \in T\bar{x}^k$, and finally set

$$x^{k+1} = \bar{x}^k + \rho(t^k - \bar{t}^k). \tag{6.5}$$

We checked the convergence to a solution but we failed. The reason may be the arbitrariness in taking $\bar{t}^k$. Setting instead

$$\bar{t}^k = P_{T\bar{x}^k} t^k, \tag{6.6}$$

we have proved that if T is pseudomonotone at a solution x^* of (VI), Lipschitz with constant L, and bounded on bounded sets, then the sequence generated by (6.5) from any starting point $x^0 \in R^n$ with $\bar{t}^k$ in (6.6) converges to a solution of (VI) for any $\rho > 0$ such that $\rho L < 1$.

Observe that the Lipschitz condition is strict and performing $P_{T\bar{x}^k} t^k$ can be quite difficult or even impossible in practice. We introduce a parameter γ_k to control the convergence and replace projection (6.6) by any $\bar{t}^k \in T\bar{x}^k$:

$$x^{k+1} = x^k + \gamma_k(\rho(t^k - \bar{t}^k) - (x^k - \bar{x}^k)). \tag{6.7}$$

(Observe that if $\gamma_k = 1$, then (6.7) collapses to (6.5).) To control the convergence, it is reasonable to choose γ_k so that the difference of the distances from x^{k+1} and from x^k to the solution set $S*$ is largest. Considering the global convergence of (6.7), unfortunately, we see that the assumptions remain the same as above (see the "first possibility" of Algorithm 2.1 and the related propositions), including the Lipschitz condition. To overcome the obstacle, we have to apply a linesearch on the interval $[x^k, \bar{x}^k]$ to get any point $\hat{t}^k \in T(x^k - \lambda^{m_k}(x^k - \bar{x}^k))$ which satisfies the Lipschitz condition

$$\|t^k - \hat{t}^k\| \leq L\|x^k - \bar{x}^k\|.$$

Then, with $\bar{t}^k$ replaced by $\hat{t}^k$, we choose the optimal γ_k as above (see the "second possibility" of Algorithm 2.1) to omit the assumed Lipschitz condition. However, we can establish the convergence for this case only if $x^k \in K$. So we are reluctant to use the second projection $P_K(x^k)$ in this case.

Before we state the resulting algorithm, observe that we have kept ρ constant and performed the linesearch, i.e. chosen an approximate m_k only outside projections. Note that a linesearch by choosing an appropriate ρ_k in $P_K(x^k - \rho_k t^k)$ may lead to many projections. We also note that problem (VI) for T and problem (VI) for ρT are equivalent. Therefore, we can fix $\rho = 1$. In the sequel, we denote $x - \bar{x} \equiv x - P_K(x-t)$ for a chosen $t \in Tx$ by $r(x)$ since it is the projection residue.

Algorithm 2.1 *We require two exogenous parameters λ and L taken* in (0,1).

1. **Initialization.** $x^0 \in R^n$.

2. **Iteration.** *Given x^k. If $r(x^k) = 0$, then stop x^k is a solution of (VI). Otherwise, take arbitrarily $t^k \in Tx^k$ and $\bar{t}^k \in T\bar{x}^k =: TP_K(x^k - t^k)$. Partition the iteration into three possibilities.*

First possibility. If

$$\left\langle t^k - \bar{t}^k, r(x^k) \right\rangle \leq L\|r(x^k)\|^2, \tag{6.8}$$

then take

$$\gamma_k = \frac{\|r(x^k)\|^2 - \langle t^k - \bar{t}^k, r(x^k)\rangle}{\|t^k - \bar{t}^k - r(x^k)\|^2},$$

$$x^{k+1} = x^k + \gamma_k(t^k - \bar{t}^k - r(x^k)). \tag{6.9}$$

Second possibility. If (6.8) is violated and $x^k \in K$, then take m_k being the smallest nonnegative integer such that there is $\hat{t}^k$ satisfying

$$\hat{t}^k \in T(x^k - \lambda^{m_k}(x^k - \bar{x}^k)), \tag{6.10}$$

$$\left\langle t^k - \hat{t}^k, r(x^k) \right\rangle \leq L\|r(x^k)\|^2. \tag{6.11}$$

Set $\theta^k = \frac{1}{\lambda^{m_k}}$ and

$$\gamma_k = \frac{\|r(x^k)\|^2 - \langle t^k - \hat{t}^k, r(x^k)\rangle}{\|t^k - \theta_k \hat{t}^k - r(x^k)\|^2},$$

$$x^{k+1} := x^k + \gamma_k(t^k - \theta_k \hat{t}^k - r(x^k)). \tag{6.12}$$

Third possibility. If (6.8) is violated and $x^k \notin K$, *then take* $x^{k+1} = P_K(x^k)$.

Remark 2.1 *(i) We can replace (6.8) and (6.9) by the Lipschitz condition, e.g. (6.8) by* $\|t^k - \bar{t}^k\| \leq L\|r(x^k)\|$, *but this is stricter than (6.8) and restricts the use of the first possibility, which is simpler than the last two.*

(ii) The linesearch (6.10) may lead to the evaluation of several values of multifunction T, but not to performing several projections as the linesearch by choosing α_k *inside* $P_K(x^k - \alpha_k Tx^k)$ *used in several existing algorithms.*

(iii) If T is single-valued, Algorithm 2.1 is still a new alternative to many known projection-type algorithms. The computational complexity here is less or not more than for almost all the existing algorithms. In particular, we need two projections only in one of three subcases. Denote $x^k - \lambda^{m_k}(x^k - \bar{x}^k) := y^k(m_k)$, *the direction from* x^k *to* x^{k+1} *in Algorithm 2.1 is* $Tx^k - T\bar{x}^k - r(x^k)$ *(by (9)) or* $Tx^k - \theta_k Ty^k(m_k) - r(x^k)$ *(by (12)). For the sake of comparison, we give the directions of a number of existing projection-type algorithms:*

- *the direction* $-Ty^k(m_k)$ *in Iusem et al (1997) , Solodov et al (1999), and Wang et al (2001a), where* m_k *is chosen by a linesearch different from (6.10);*
- *the direction* $z^k(\alpha_k) + \alpha_k Tx^k - \alpha_k T(z^k(\alpha_k)) - x^k$ *in Solodov et al (1996), where* $z^k(\alpha) = P_K(x^k - \alpha T(x^k))$ *and* α_k *is chosen by the linesearch:* α_k *is the largest* $\alpha \in \{\alpha_{k-1}, \alpha_{k-1}\beta, \alpha_{k-1}\beta^2, ...\}$ *satisfying*

$$\left\langle \alpha(x^k - z^k(\alpha)), Tx^k - Tz^k(\alpha) \right\rangle \leq L\|x^k - x^k(\alpha)\|^2 \tag{6.13}$$

with β *and L are given parameters in* (0,1);

- *the direction* $-r(x^k) - \theta_k Ty^k(m_k)$, *where* θ_k *and* m_k *are chosen as in (6.10), in Wang et al (2001b);*
- *the direction* $-\theta_k r(x^k) - Tx^k - Ty^k(m_k)$, *where* θ_k *and* m_k *are chosen as in (6.10), in Noor et al (2002);*
- *the direction* $-\theta_k r(x^k) + \theta_k Tx^k - Ty^k(m_k)$, *where* θ_k *and* m_k *are chosen as in (6.10), in Noor et al (2003);*
- *the direction* $-Tz^k(\alpha_k)$, *where* α_k *is chosen to satisfy a condition similar to (6.13) but in a different set, or* $-Ty^k(m_k)$ *where* m_k *is chosen similarly as in (6.10), or* $-r(x^k) + \alpha_k Tx^k - \alpha_k Tz^k(\alpha_k)$ *where* α_k *is chosen as in (6.13) but in a different set, in Noor (2003a).*
- *the direction* $-Tz^k(\alpha_k)$, *where* α_k *satisfies*

$$\|\alpha_k(Tx^k - Tz^k(\alpha_k))\| \leq L\|x^k - z^k(\alpha_k)\|,$$

similarly as (6.13) but is not restricted to a given set, in He and Liao (2002);

- *the classical direction* $-Tx^k$, *as in Goldstein (1964), Levitin et al (1966), Sibony (1970) and He et al (2002). However, here the authors proposed a self-adaptive technique to choose the stepsizes;*
- *the direction* $-T(x^k) + e^k$, *where* e^k *(called the error vector) can be chosen in various ways in Xiu et al (2002) (This general algorithmic model includes many other known algorithms and requires monotonicity and Lipschitz conditions for the convergence);*
- *the direction* $-Tu^k(\alpha_k)$, *where* $u^k(\alpha_k) = x^k - \alpha_k r(x_k, \beta_k)$ *with* $r(x^k, \beta_k) = x^k - P_K(x^k - \beta^k T x^k)$ *and* $\alpha_k = \lambda^{m_k}$, m_k *being the smallest integer satisfying (together with* β_k*)*

$$\beta_k \lambda^{m_k} \left\langle Tx^k - \lambda^{m_k} T u^k(\alpha_k), r(\alpha^k, \beta_k) \right\rangle \leq L \| r(x^k, \beta_k) \|^2,$$

in Noor (2003c);

- *the direction* $-r(x^k, \beta_k) - \beta_k T x^k + \beta_k T \bar{x}^k$ *in He (1997).*

Note that the above mentioned directions were combined with various rules of choosing the stepsize γ_k, and that most of the above mentioned algorithms used two or more projections at each iteration. In fact, many among these recent algorithms were known to us after we had completed this paper. Fortunately, we could employ them in revising.

3. Global convergence

To establish a global convergence of Algorithm 2.1 we need several propositions.

Proposition 3.1 *Assume that T is pseudomonotone at a solution* x^* *of (VI). If* x^{k+1} *is defined by (6.9), then*

$$\|x^{k+1} - x^*\|^2 \leq \|x^k - x^*\|^2 - \frac{(1-L)^2 \|r(x^k)\|^4}{\|t^k - \bar{t}^k - r(x^k)\|^2}.$$

Proof. Observe first that $\gamma_k > 0$ since

$$\|r(x^k)\|^2 - \left\langle t^k - \bar{t}^k, r(x^k) \right\rangle \geq \|r(x^k)\|^2 - L\|r(x^k)\|^2 > 0.$$

One has

$$
\begin{aligned}
&\|x^{k+1} - x^*\|^2 \\
&= \|x^k - x^* + \gamma_k(t^k - \bar{t}^k - r(x^k)\|^2 \\
&= \|x^k - x^*\|^2 + \gamma_k^2\|t^k - \bar{t}^k - r(x^k)\|^2 + 2\gamma_k \left\langle t^k - \bar{t}^k - r(x^k), x^k - x^* \right\rangle \\
&= \|x^k - x^*\|^2 + \gamma_k^2\|t^k - \bar{t}^k - r(x^k)\|^2 \\
&\quad +2\gamma_k \left\langle t^k - r(x^k), x^k - x^* \right\rangle - 2\gamma_k \left\langle \bar{t}^k, x^k - x^* \right\rangle \\
&= \|x^k - x^*\|^2 + \gamma_k^2\|t^k - \bar{t}^k - r(x^k)\|^2 + 2\gamma_k \left\langle t^k - r(x^k), r(x^k) \right\rangle \\
&\quad -2\gamma_k \left\langle \bar{t}^k, r(x^k) \right\rangle + 2\gamma_k \left\langle t^k - r(x^k), \bar{x}^k - x^* \right\rangle - 2\gamma_k \left\langle \bar{t}^k, \bar{x}^k - x^* \right\rangle \\
&\leq \|x^k - x^*\|^2 + \gamma_k^2\|t^k - \bar{t}^k - r(x^k)\|^2 \\
&\quad +2\gamma_k \left\langle t^k - r(x^k), r(x^k) \right\rangle - 2\gamma_k \left\langle \bar{t}^k, r(x^k) \right\rangle \\
&= \|x^k - x^*\|^2 + \gamma_k^2\|t^k - \bar{t}^k - r(x^k)\|^2 - 2\gamma_k(\|r(x^k)\|^2 - \left\langle t^k - \bar{t}^k, r(x^k) \right\rangle) \\
&= \|x^k - x^*\|^2 - \frac{(\|r(x^k)\|^2 - \left\langle t^k - \bar{t}^k, r(x^k) \right\rangle)^2}{\|t^k - \bar{t}^k - r(x^k)\|^2} \\
&\leq \|x^k - x^*\|^2 - \frac{(1-L)^2\|r(x^k)\|^4}{\|t^k - \bar{t}^k - r(x^k)\|^2}.
\end{aligned}
$$

The first inequality in the chain is due to the pseudomonotonicity of T at x^* and Lemma 1.2 (i) with $z = P_K(x^k - t^k)$, $x = \bar{x}^k$ and $y = x^*$. The second inequality holds by (6.8). □

Remark 3.1 *From the proof of Proposition 3.1 and also that of Proposition 3.3 below, we see that* γ_k *designed in Algorithm 2.1 are chosen optimally as minima of numerical quadratic functions.*

The following proposition asserts that the algorithm is well defined.

Proposition 3.2 *Assume that T is lhc. If* x^k *is not a solution of (VI) and (6.8) is not fulfilled, then there exist a nonnegative integer* m_k *and* $\hat{t}^k$ *satisfying (6.10) and (6.11).*

Proof. Suppose that for all m and all $t \in T(x^k - \lambda^m(x^k - \bar{x}^k))$ one has $\left\langle t^k - t, r(x^k) \right\rangle > L\|r(x^k)\|^2$. Since $x^k - \lambda^m(x^k - \bar{x}^k) \to x^k$ as $m \to \infty$, $t^k \in Tx^k$ and T is lhc, there exists a sequence $t_m \in T(x^k - \lambda^m(x^k - \bar{x}^k))$ such that $t_m \to t^k$. One has $\left\langle t^k - t_m, r(x^k) \right\rangle > L\|r(x^k)\|^2$, $\forall m$. Hence, $\|t^k - t_m\| > L\|r(x^k)\|$, $\forall n$ and in the limit $0 = \|t^k - t^k\| \geq L\|r(x^k)\| > 0$. This impossibility completes the proof. □

Proposition 3.3 *Assume that T is pseudomonotone at a solution x^* of (VI). If x^{k+1} is obtained from (6.12), then*

$$\|x^{k+1} - x^*\|^2 \le \|x^k - x^*\|^2 - \frac{(1-L)^2\|r(x^k)\|^4}{\|t^k - \theta_k \hat{t}^k - r(x^k)\|^2}.$$

Proof. Observe that in this case γ_k is also positive since

$$\|r(x^k)\|^2 - \left\langle t^k - \hat{t}^k, r(x^k)\right\rangle \ge (1-L)\|r(x^k)\|^2 > 0.$$

Similarly as for Proposition 3.1, one has

$$\begin{aligned}
&\|x^{k+1} - x^*\|^2 \\
&= \|x^k - x^*\|^2 + \gamma_k^2\|t^k - \theta_k\hat{t}^k - r(x^k)\|^2 \\
&\quad +2\gamma_k\left\langle t^k - r(x^k), x^k - x^*\right\rangle - 2\gamma_k\left\langle \theta_k\hat{t}^k, x^k - x^*\right\rangle \\
&= \|x^k - x^*\|^2 + \gamma_k^2\|t^k - \theta_k\hat{t}^k - r(x^k)\|^2 + 2\gamma_k\left\langle t^k - r(x^k), r(x^k)\right\rangle \\
&\quad +2\gamma_k\left\langle t^k - r(x^k), \bar{x}^k - x^*\right\rangle - 2\gamma_k\left\langle \theta_k\hat{t}^k, x^k - y^k\right\rangle - 2\gamma_k\left\langle \theta_k\hat{t}^k, y^k - x^*\right\rangle \\
&\le \|x^k - x^*\|^2 + \gamma_k^2\|t^k - \theta_k\hat{t}^k - r(x^k)\|^2 + 2\gamma_k\left\langle t^k, r(x^k)\right\rangle \\
&\quad -2\gamma_k\|r(x^k)\|^2 - 2\gamma_k\left\langle \theta_k\hat{t}^k, \lambda^{m_k} r(x^k)\right\rangle \\
&= \|x^k - x^*\|^2 - \frac{(\|r(x^k)\|^2 - \langle t^k - \hat{t}^k, r(x^k)\rangle)^2}{\|t^k - \theta_k\hat{t}^k - r(x^k)\|^2} \\
&\le \|x^k - x^*\|^2 - \frac{(1-L)^2\|r(x^k)\|^4}{\|t^k - \theta_k\hat{t}^k - r(x^k)\|^2}.
\end{aligned}$$

The first inequality holds by the pseudomonotonicity of F at x^*, $\hat{t}^k \in Ty^k$, $y^k = x^k - \lambda^{m_k}(x^k - \bar{x}^k)$ and Lemma 1.2 (i). The second inequality is due to (6.11). □

Proposition 3.4 *Assume that T is closed and bounded on each bounded subset of $\mathbf{R}^n$. If x^k is any sequence converging to $\hat{x}$ such that*

$$\sum_{k=0}^{\infty} \frac{(1-L)^2\|r(x^k)\|^4}{\|t^k - \bar{t}^k - r(x^k)\|^2} < +\infty, \tag{6.14}$$

then $\hat{x}$ is a solution of (VI).

Proof. Since $\{t^k\} \subseteq T\{x^k\}$ is bounded, there exists a convergent subsequence $t^{k_i} \to \hat{t}$. By the closedness of T, one has $\hat{t} \in T\hat{x}$. By the nonexpansivity of the projection $P_K(.)$ and the assumed boundedness of

T, the set $\|t^k - \bar{t}^k - r(x^k)\|$ is bounded. Therefore, the convergence of the series (6.14) implies that

$$0 = \lim_{i\to\infty} r(x^{k_i}) = \lim_{i\to\infty} (x^{k_i} - P_K(x^{k_i} - t^{k_i})) = \hat{x} - \bar{\hat{x}},$$

which means that $\hat{x}$ is a solution of (VI) by Lemma 1.1. □

Proposition 3.5 *Assume that T is closed, lhc and bounded on each bounded subset of* R^n. *Assume further that* x^k *is produced by (6.12) and tends to some* $\hat{x} \in R^n$. *If*

$$\sum_{k=0}^{\infty} \frac{(1-L)^2\|r(x^k)\|^4}{\|t^k - \theta_k \hat{t}^k - r(x^k)\|^2} < +\infty, \tag{6.15}$$

then $\hat{x}$ *is a solution of (VI).*

Proof. If $\hat{x}$ is a solution of *(VI)* then we are done. We show first that m_k associated with x^k in (6.12) cannot tend to $+\infty$ when $\hat{x}$ is not a solution. By the definition of m_k, one has $\|t - t^k\| > L\|x^k - \bar{x}^k\|$ for all $t \in T(x^k - \lambda^{m_k-1}(x^k - \bar{x}^k))$. By the assumed boundedness there exists a convergent subsequence $t^{k_i} \to \hat{t}$. Since T is closed, $\hat{t} \in T\hat{x}$. If $m_k \to +\infty$, then $x^{k_i} - \lambda^{m_{k_i}-1}(x^{k_i} - \bar{x}^{k_i}) \to \hat{x}$. The lhc of T, in turn, implies the existence of $\hat{t}^{k_i} \in T(x^{k_i} - \lambda^{m_{k_i}-1}(x^{k_i} - \bar{x}^{k_i}))$ such that $\hat{t}^{k_i} \to \hat{t}$. Now passing $\|\hat{t}^{k_i} - t^{k_i}\| > L\|x^{k_i} - \bar{x}^{k_i}\|$ to limit, one sees the contradiction

$$0 = \|\hat{t} - \hat{t}\| \geq L\|\hat{x} - \bar{\hat{x}}\| > 0.$$

Therefore, $\{m_k\}$ is bounded and so is $\{\theta_k\}$. It follows that there is $M > 0$ such that $\|t^k - \theta_k\hat{t}^k - (x^k - \bar{x}^k)\|^2 < M$. Then, (6.15) implies that $r(x^k) \to 0$ and, as for Proposition 3.4, $\hat{x}$ is a solution of (VI). □

Now we can establish a global convergence of Algorithm 2.1 as follows.

Theorem 3.1 *Assume that T is closed, lhc, bounded on each bounded subset of* R^n *and pseudomonotone at a solution* x^* *of (VI). Then, any sequence* x^k *generated by Algorithm 2.1 is either finitely terminated or converges to a solution of (VI).*

Proof. By the Proposition 3.1, 3.3 and Lemma 1.2 (ii) one has, for all k,

$$\|x^{k+1} - x^*\|^2 \leq \|x^k - x^*\| - r_k^{i_k}, \tag{6.16}$$

where $i_k = 1$, 2 or 3 with

$$r_k^1 = \frac{(1-L)^2\|r(x^k)\|^4}{\|t^k - \bar{t}^k - r(x^k)\|^2}, \tag{6.17}$$

$$r_k^2 = \frac{(1-L)^2\|r(x^k)\|^4}{\|t^k - \theta_k \hat{t}^k - r(x^k)\|^2},$$

$$r_k^3 = \|x^k - P_k(x^k)\|^2.$$

Adding (6.16) for $k = 0, 1, \ldots$ one obtains

$$\sum_{k=0}^{\infty} r_k^{i_k} \leq \|x^0 - x^*\|^2.$$

Observe that one of the two r_k^1 and r_k^2 must happen for an infinite number of times. Indeed, it is the case if r_k^3 happens finitely. Otherwise, there are infinitely many x^k corresponding to r_k^3 (the third possibility in Algorithm 2.1), hence $x^{k+1} = P_K(x^k)$ satisfies the first or second possibility.

Now assume further that an infinite subsequence $r_{k_j}^{i_{k_j}}$ is of the form (6.17) (for the form of r_k^2 the argument is similar). Since $\{x^k\}$ is bounded (by (6.16)), we can assume that the subsequence x^{k_j} (with these indices k_j) converges to some $\hat{x}$. Proposition 3.4, in turn, asserts that $\hat{x}$ is a solution of *(VI)*. It follows from Proposition 3.1 that $\|x^{k+1} - \hat{x}\| \leq \|x^k - \hat{x}\|$. Therefore, the whole sequence x^k converges to $\hat{x}$. □

Remark 3.2 *If T is single-valued, we can modify the third possibility of Algorithm 2.1 as follows to make the projection needed in this case easier. Choose another parameter $r > 0$. If x^k is close enough to K in the sense that*

$$\left\langle T\bar{x}^k, x^k - \bar{x}^k \right\rangle \leq r\left\|x^k - \bar{x}^k\right\|^2,$$

take $x^{k+1} = P_{K\cap Q}(x^k)$, where $Q = \{x \in R^n : \langle T\bar{x}^k, x - \bar{x}^k\rangle \leq 0\}$ is a halfspace. If x^k is not so close to K, then take x^{k+1} as the projection of x^k on the hyperplane $H = \{x \in R^n : \langle T\bar{x}^k, x - \bar{x}^k\rangle = 0\}$, i.e.

$$x^{k+1} = x^k - \frac{\langle T\bar{x}^k, x^k - \bar{x}^k\rangle}{\|T\bar{x}^k\|^2} T\bar{x}^k.$$

Clearly performing a projection on H is easier and on $K \cap Q$ may be easier (since $K \cap Q$ has a face being a part of H) than on K.

Remark 3.3 *Algorithm 2.1 uses two projections only in a number of iterations (one of the three subcases). The remaining iterations contain* *only one projection. Moreover, while choosing a point in an image, like* $t^k \in Tx^k$, *we take arbitrarily and do not have to solve additional minimization problems. If we want to combine the three subcases to make Algorithm 2.1 simpler in formulating, we have to use two projections as follows. (Then the algorithm becomes a so called double-projection algorithm.)*

Modified Algorithm 2.1 Two parameters λ and L are taken in (0,1).

1. Initialization. $x^0 \in K$.

2. Iteration. Given x^k. If $r(x^k) = 0$, then stop; x^k is a solution of (VI). Otherwise, take arbitrarily $t^k \in Tx^k$ and $\bar{t}^k \in T\bar{x}^k =: TP_K(x^k - t^k)$. Find m_k being the smallest nonnegative integer such that there exists $\hat{t}^k$ satisfying (6.10) and (6.11). Set θ_k and γ_k as in the second possibility of Algorithm 2.1 and

$$x^{k+1} = P_K(x^k + \gamma_k(t^k - \theta_k \hat{t}^k - r(x^k))).$$

It is easy to see that Theorem 3.1 remains true for this modified algorithm.

Remark 3.4 *A multifunction satisfying the assumptions of Theorem 3.1 does not need to be continuous. Indeed, let* $T : R^2 \to 2^R$ *be defined by*

$$T(x,y) = \left[\begin{array}{l} [0, \frac{1}{2}] \ if \ x = y = 0, \\ [0, \frac{|x \cdot y|}{x^2+y^2}] \ otherwise. \end{array} \right.$$

Then T is clearly closed, lhc and bounded on each bounded subset. But T is not lsc at (0,0) and then not continuous. In fact, take $x^k = \frac{1}{k}$, $y^k = \frac{4}{k}$, $u = \frac{1}{2} \in T(0,0)$. *Since* $T(x^k, y^k) = [0, \frac{4}{17}]$, *there does not exist* $u^k \in T(x^k, y^k)$ *which tends to* u. *Note that in this example we take the image space to be R, not* R^2, *for the sake of simplicity.*

4. A computational example

The computational results presented here have been obtained by using MATLAB to implement the algorithm (applying the quadratic - program solver quadprog.m from the MATLAB Optimization Toolbox to perform the projections).

Example 4.1. Consider problem

$$minimize \qquad \varphi(x) = \max_{i=1,\ldots,n} |x_i|$$

$$s.t. \qquad -2 \leq x_i \leq 2,\ i = 1, ..., n,$$

where n (10 or 20) is the dimension of the problem.

It is obvious that the minimizer is $x^* = (0, ..., 0)^T$ and the optimum value is $f(x^*) = 0$, and that the above problem is equivalent to the following multivalued variational inequality: finding $x^* \in K$ such that there is $t^* \in Tx^*$ satisfying

$$\langle t^*, x - x^* \rangle \geq 0, \forall x \in K,$$

where $T = \partial\varphi$, the subdifferential of φ, and $K = \{(x_1, ..., x_n)^T : -2 \leq x_i \leq 2,\ i = 1, ..., n\}$. The parameters of Algorithm 2.1 have the following values: $\lambda = 0.9$ and $L = 0.4$. The starting points are $x^0 = (1, ..., 1, -1, ..., -1)^T$ and $x^{00} = (1, -1, 1, -1, 0, 1, -1, 1, -1, 0, ...)^T$. The Algorithm stops when $\varphi(x)$ is less than 10^{-7}. The result is summarized in the following table.

Dimension	$n = 10$	$n = 20$
Startingpoint	x^0	x^0
Iterations	10	35
Startingpoint	x^{00}	x^{00}
Iterations	8	31

Example 4.2. Consider problem *(VI)* with

$$T(x) = \begin{bmatrix} 0.9(5x_2 + 3x_3)/x_1 \\ 0.1(5x_2 + 3x_3)/x_2 - 5 \\ -3 \end{bmatrix},$$

$$K = \{x \in R^3_+ : x_1 + x_2 + x_3 = 1, x_1 - x_2 - x_3 \leq 0, x_1 \geq a\}.$$

This test problem was discussed in e.g. Iusem et al (2000); Noor (2001a) with a = 0. We add the constraint $x_1 \geq a$ with several cases of a. The parameters of Algorithm 2.1 are taken as follows: $\lambda = \frac{1}{2}$ and L has various values from 0.4 to 0.01. We adopt to stop the computation when the tolerance $r(x^k) = \epsilon$ is achieved for various values of ϵ. The result of the test is encouraging. The following tables give the number of iterations to get an approximate solution with a given tolerance ϵ.

$a = 0.05$

L	0.4	0.3	0.2	0.1	0.05
$\epsilon = 10^{-3}$	30	20	12	6	2
$\epsilon = 10^{-5}$	52	34	22	10	2
$\epsilon = 10^{-7}$	72	48	30	16	2

$a = 0.001$

L	0.4	0.3	0.2	0.1	0.05
$\epsilon = 10^{-3}$	30	20	14	6	2
$\epsilon = 10^{-5}$	52	34	22	10	2
$\epsilon = 10^{-7}$	72	50	32	16	2

Observe that the smaller L is, the less iterations are needed.

References

Alber Y. I. (1983), Recurence relations and variational inequalities, it Soviet Mathematics, Doklady, Vol. 27, pp. 511 - 517.

Goldstein A. A. (1964), Convex programming in Hilbert space, *Bulletin of The American Mathematical Society,* Vol. 70, pp. 709 - 710.

He B. (1997), A class of projection and contraction methods for monotone variational inequalities, *Applied Mathematical Optimization,* Vol. 35, pp. 69 - 76.

He B. S., Liao L. Z. (2002), Improvements of some projection methods for monotone nonlinear variational inequalities, *Journal of Optimization Theory and Applications,* Vol. 112, pp. 111 - 128.

He B. S., Yang H., Meng Q. and Han D. R. (2002), Modified Goldstein - Levitin - Polyak projection method for asymmetric strongly monotone variational inequalities, *Journal of Optimization Theory and Applications,* Vol. 112, pp. 129 - 143.

Iusem A. N., Pérez L. R. L. (2000), An extragradient type algorithm for nonsmooth variational inequalities, *Optimization,* Vol. 48, pp. 309 - 332.

Iusem A. N., Svaiter B. F. (1997), A variant of Korpelevich's method for variational inequalities with a new search strategy, *Optimization,* Vol. 42, pp. 309 - 321.

Korpelevich G. M. (1976), The extragradient method for finding saddle points and other problems, *Ecomomika i Matematicheskie Metody,* Vol. 12, pp. 747 - 756.

Levitin E. S., Polyak B. T. (1966), Constrained minimization problem, *USSR Computational Mathematics and Mathematical Physics,* Vol. 6, pp. 1 -50.

Martinet B. (1970), Regularization d'inéquations variationelles par approximations successives, *Revue d' Automatique Informatique et Recherche Opérationelle,* Vol. 4, pp. 154 -158.

Martinet B. (1972), Determination approchée d'un point fixed d'une application pseudo-contractante, *C. R. Academic Science Paris Ser. A-B,* Vol. 274, pp. 163 - 165.

Noor M. A. (1999), A modified extragradient method for general monotone variational inequalities", *Computers and Mathematics with Applications,* Vol. 38, pp. 19 - 24.

Noor M. A. (2001a), Some predictor-corrector algorithms for multivalued variational inequalities, *Journal of Optimization Theory and Applications,* Vol. 108, pp. 659 - 671.

Noor M. A. (2001b), Iterative schemes for quasimonotone mixed variational inequalities, *Optimization,* Vol. 50, pp. 29 - 44.

Noor M. A. (2001c), Modified resolvent splitting algorithms for general mixed variational inequalities, *Journal of Computational and Applied Mathematics,* Vol. 135, pp. 111-124.

Noor M. A. (2003a), New extragradient-type methods for general variational inequalities, *Journal of Mathematical Analysis and Applications,* Vol. 277, pp. 379 - 394.

Noor M. A. (2003b), Pseudomonotone general mixed variational inequalities, *Applied Mathematics and Applications,* Vol. 141, pp. 529 - 540.

Noor M. A. (2003c), Extragradient methods for pseudomonotone variational inequalities, *Journal of Optimization Theory and Applications,* Vol. 117, pp. 475 - 488.

Noor M. A., Rassias T. M. (1999), Projection methods for monotone variational inequalities, *Journal of Optimization Theory and Applications,* Vol. 137, pp. 405 - 412.

Noor M. A., Wang Y. J., Xiu N. H. (2002), Projection iterative schemes for general variational inequalities, *Journal ofInequalities in Pure and Applied Mathematics,* Vol. 3, pp. 1-8.

Noor M. A., Wang Y. J., Xiu N. H. (2003), Some new projection methods for variational inequalities, *Applied Mathematics and Computation,* Vol. 137, pp. 423 - 435.

Rockafellar R. T. (1976a), Monotone operators and the proximal point algorithm, *SIAM Journal on Control and Optimization,* Vol. 14, pp. 877-898.

Rockafellar R. T. (1976b), Augmented Lagrangians and applications of the proximal point algorithm in convex programming, *Mathematics Operations Research,* Vol. 1, pp. 97 - 116

Sibony M. (1970), Méthodes itératives pour les équations et inéquations aux dérivées partielles non linéaires de type monotone, *Calcolo,* Vol. 7, pp. 65 - 183.

Solodov M. V., Svaiter B. F. (1999), A new projection method for variational inequality problems, *SIAM Journal on Control and Optimization,* Vol. 37, pp. 165 - 176.

Solodov M. V., Tseng P. (1996), Modified projection - type methods for monotone variational inequalities, *SIAM Journal on Control and Optimization,* Vol. 34, pp. 1814 - 1830.

Tseng P. (1997), Alternating projection - proximal methods for convex programming and variational inequalities, *SIAM Journal on Optimization,* Vol. 7, pp. 951 - 965.

Wang Y. J., Xiu N. H., Wang C. Y. (2001a), Unified framework of extragradient-type methods for pseudomonotone variational inequalities, *Journal of Optimization Theory and Applications,* Vol. 111, pp. 641 - 656.

Wang Y. J., Xiu N. H., Wang C. Y. (2001b), A new version of the extragradient method for variational inequality problems, *Computers and Mathematics with Applications,* Vol. 42, pp. 969 - 979.

Xiu N. H., Zhang J. Z. (2002), Local convergence analysis of projection-type algorithms: unified approach, *Journal of Optimization Theory and Applications,* Vol. 115, pp. 211 - 230.

Zhao Y. B. (1997), The iterative methods for monotone generalized variational inequalities, *Optimization,* Vol. 42, pp. 285 - 307.

Zhao Y. B. (1999), Extended projection methods for monotone variational inequalities, *Journal of Optimization Theory and Applications,* Vol. 100, pp. 219 - 231.

Chapter 7

DUALITY IN MULTIOBJECTIVE OPTIMIZATION PROBLEMS WITH SET CONSTRAINTS

Riccardo Cambini and Laura Carosi*

Dept. of Statistics and Applied Mathematics

University of Pisa, ITALY

Abstract We propose four different duality problems for a vector optimization program with a set constraint, equality and inequality constraints. For all dual problems we state weak and strong duality theorems based on different generalized concavity assumptions. The proposed dual problems provide a unified framework generalizing Wolfe and Mond-Weir results.

Keywords: Vector Optimization, Duality, Maximum Principle Conditions, Generalized Convexity, Set Constraints.

MSC2000: 90C29, 90C46, 90C26
Journal of Economic Literature Classification (1999): C61

1. Introduction

Vector optimization programs are extremely useful in order to model real life problems where several objectives conflict with one another, and so the interest of this topics crosses many different fields such as operation research, economic theory, location theory and management science. During the last decades the analysis of duality in multiobjective theory has been a focal issue. We can find papers dealing with duality

*This research has been partially supported by M.I.U.R. and C.N.R.
email: cambric@ec.unipi.it, lcarosi@ec.unipi.it

under smooth and non smooth assumptions for both the objective and constraint functions, some other papers consider particular objective functions such as vector fractional ones (see for example the recent contributions by Bathia and Pankaj (1998); Patel (2000); Zalmai (1997)). Moreover many different kinds of generalized convexity properties have been investigated in order to get the usual duality results. Despite of a very large number of papers on duality the most part of the recent literature deals with vector optimization problems where the feasible region is defined by equality and inequality constraint or by a compact set (for this latter case the reader can see for example the leading article by Tanino and Sawaragy (1979)).

In this paper we aim to deal with a vector optimization problem where the feasible region is defined by equality constraint, inequality and set constraint and we do not require any topological properties on the set constraint. Since our duality results are related to the concepts of C-maximal and weakly C-maximal point we first recall these definitions and then we propose some necessary optimality conditions which can be classified as a maximum principle condition. These suggest the introduction of the first dual $\boldsymbol{D_1}$ which is a generalization of the Wolfe-dual problem ([1]). Then we propose three further dual programs which are called $\boldsymbol{D_2}$, $\boldsymbol{D_3}$ and $\boldsymbol{D_4}$. While problem $\boldsymbol{D_4}$ can be classified as a generalization of the Mond-Weir dual problem (see Mond and Weir (1981); Weir et al (1986)), $\boldsymbol{D_2}$, $\boldsymbol{D_3}$ are a sort of mixed duals. In the recent literature (see for example Aghezzaf and Hachimi (2001); Mishra (1996)) similar mixed dual have been proposed, but they refer to a primal problem with feasible region defined only by equality and inequality constraints. For all our dual programs, duality theorems are stated and for each one, different generalized convexity properties are assumed. For a feasible region without set constraint, there are many duality results dealing with several kind of generalized convexity properties such as invexity, generalized invexity (see for all Bector et al (1993); Bector et al (1994); Bector (1996); Giorgi and Guerraggio (1998); Hanson and Mond (1987); Kaul et al (1994); Rueda et al (1995)), or **(F, ρ)-convexity** (see for example Aghezzaf and Hachimi (2001); Bhatia and Jain (1994); Bathia and Pankaj (1998); Gulati and Islam (1994); Mishra (1996); Preda (1992)). In our case the objective function $\boldsymbol{f}$ is C-concave or $(\text{Int}(C),\text{Int}(C))$-pseudoconcave while the inequality constraint function $\boldsymbol{g}$ is assumed to be V-concave or polarly V-quasiconcave and the equality constraint function $\boldsymbol{h}$ is affine or polarly quasiaffine.

[1] For a different duality approach when the feasible region is a subset of an arbitrary set, the reader can see for example Jahn (1994); Luc (1984); Zalmai (1997).

Finally, we compare the four dual programs in order to analyze them in a unified framework and to appreciate the differences among them.

2. Definitions and preliminary results

We consider the following multiobjective nonlinear programming P.

Definition 2.1 (Primal Problem)

$$P : \begin{cases} C\_\max \ f(x) \\ x \in S_P \end{cases} \equiv \begin{cases} C\_\max \ f(x) & \\ g(x) \in V & \textit{inequality constraints} \\ h(x) = 0 & \textit{equality constraints} \\ x \in X & \textit{set constraint} \end{cases}$$

where

$$S_P = \{x \in A : \ g(x) \in V, \ h(x) = 0, \ x \in X\},$$

$A \subseteq \mathbb{R}^n$ is an open convex set, $f : A \to \mathbb{R}^s$ and $g : A \to \mathbb{R}^m$ are Gâteaux differentiable functions, $h : A \to \mathbb{R}^p$ is a Fréchet differentiable function with a continuous Jacobian matrix $J_h(x)$. Moreover $C \subset \mathbb{R}^s$ and $V \subset \mathbb{R}^m$ are closed convex pointed cones with nonempty interior (that is to say convex pointed solid cones), and $X \subseteq A$ is a set verifying no particular topological properties. In other words, X is not required to be open or convex or with nonempty interior.

Throughout the paper we will denote with C^+ and V^+ the positive polar cones of C and V, respectively. For a better understanding of the paper, we recall some useful definitions and notations.

Definition 2.2 *Let $f : A \to \mathbb{R}^s$, $A \subseteq \mathbb{R}^n$, let $C \subset \mathbb{R}^s$ be a closed convex pointed cone with nonempty interior and let $S \subset \mathbb{R}^n$ be a set. Consider the following multiobjective problem:*

$$P : \begin{cases} C\_\max / C\_\min \ f(x) \\ x \in S \end{cases}$$

Using the notation $C^0 = C \setminus \{0\}$, a feasible point $x_0 \in S$ is said to be:

- a C-maximal [C-minimal] point for P *if:*

$$\nexists y \in S \textit{ such that } f(y) \in f(x_0) + C^0 \quad [f(y) \in f(x_0) - C^0]$$

in this case we will say that

$$x_0 \in C^0\_\arg\max(P) \quad [x_0 \in C^0\_\arg\min(P)],$$

- a weak C-maximal [weak C-minimal] point for P *if:*

$$\nexists y \in S \text{ such that } f(y) \in f(x_0) + \mathrm{Int}(C) \quad [f(y) \in f(x_0) - \mathrm{Int}(C)]$$

in this case we will say that

$$x_0 \in \mathrm{Int}(C)\_\arg\max(P) \quad [x_0 \in \mathrm{Int}(C)\_\arg\min(P)].$$

The following necessary optimality condition of the maximum principle type holds for problem P (see Cambini (2001)) ([2]).

Theorem 2.1 *Consider problem P and let $x_0 \in X$ be a local C-maximal point. Suppose also that X is convex with $\mathrm{Int}(X) \neq \emptyset$.*
Then $\exists \alpha_f \in C^+$, $\exists \alpha_g \in V^+$, $\exists \alpha_h \in \mathbb{R}^p$, $(\alpha_f, \alpha_g, \alpha_h) \neq 0$, such that $\alpha_g^T g(x_0) = 0$ and:

$$[\alpha_f^T J_f(x_0) + \alpha_g^T J_g(x_0) + \alpha_h^T J_h(x_0)](x - x_0) \leq 0 \ \ \forall x \in \mathrm{Cl}(X).$$

If in addition a constraint qualification holds then $\alpha_f \neq 0$.

As it is well known, a constraint qualification is any condition guaranteeing that $\alpha_f \neq 0$ ([3]). The following proposition presents a constraint qualification condition for problem P.

Proposition 2.1 *Consider problem P and let $x_0 \in X$ be a feasible local C-maximal point. Suppose also that X is convex with $\mathrm{Int}(X) \neq \emptyset$.*
The condition $\mathrm{Co}[\mathcal{L}_{x_0}] = \mathbb{R}^{m+p}$, where ([4]):

$$\mathcal{L}_{x_0} = \{(t_g, t_h) \in \mathbb{R}^{m+p} : \ (t_g, t_h) = [J_g(x_0), J_h(x_0)](x - x_0), \ x \in \mathrm{Cl}(X)\}$$

is a constraint qualification.

Proof. For the first part of Theorem 2.1 $\exists \alpha_f \in C^+$, $\exists \alpha_g \in V^+$, $\exists \alpha_h \in \mathbb{R}^p$, $(\alpha_f, \alpha_g, \alpha_h) \neq 0$, such that $\alpha_g^T g(x_0) = 0$ and:

$$[\alpha_f^T J_f(x_0) + \alpha_g^T J_g(x_0) + \alpha_h^T J_h(x_0)](x - x_0) \leq 0 \ \ \forall x \in \mathrm{Cl}(X).$$

Suppose now by contradiction that $\alpha_f = 0$, then $(\alpha_g, \alpha_h) \neq 0$ and:

$$(\alpha_g^T t_g + \alpha_h^T t_h) \leq 0 \ \ \forall (t_g, t_h) \in \mathcal{L}_{x_0}$$

[2] In the case f and g are Lipschitz and h are Fréchet differentiable, another necessary optimality conditions for Problem P can be found in Jiménez and Novo (2002).

[3] Among the wide literature on this subject many constraint qualification conditions have been stated with various approaches and for different kind of problems (see for example Clarke (1983); Giorgi and Guerraggio (1994); Jahn (1994); Jiménez and Novo (2002); Luc (1989)).

[4] We denote with $Co(X)$ the convex hull of a set X.

This implies also that:

$$(\alpha_g^T t_g + \alpha_h^T t_h) \leq 0 \ \forall (t_g, t_h) \in \mathrm{Co}[\mathcal{L}_{x0}] = \mathbb{R}^{m+p}$$

and hence $(\alpha_g, \alpha_h) = 0$, which is a contradiction. □

The maximum principle condition of Theorem 2.1 will suggest the definition of some dual problems for P.

3. Duality

In this section we aim to provide different kinds of dual problems for P and to study them in a unified framework. Starting from the necessary optimality condition of Theorem 2.1 we are able to define four dual problems D_1, D_2, D_3 and D_4. As the reader will see, D_1 is a Wolfe-type dual problem, D_4 is a Mond-Weir-type dual while D_2 and D_3 can be classified as a sort of mixed dual problems.

3.1 Dual problems

Definition 3.1 (1st Dual Problem) *Consider problem P and let $c \in \mathrm{Int}(C)$. The following Dual problem can be introduced:*

$$D_1 : \begin{cases} C\text{\_min} \ \ L_1(x, \alpha_f, \alpha_g, \alpha_h) = f(x) + \frac{c}{\alpha_f^T c}[\alpha_g^T g(x) + \alpha_h^T h(x)] \\ \qquad (x, \alpha_f, \alpha_g, \alpha_h) \in S_{D_1} \end{cases},$$

where

$$S_{D_1} = \left\{ \begin{array}{c} (x, \alpha_f, \alpha_g, \alpha_h) \in (A \times C^+ \times V^+ \times \mathbb{R}^p), \ \alpha_f \neq 0, \\ \left[\alpha_f^T J_f(x) + \alpha_g^T J_g(x) + \alpha_h^T J_h(x)\right](y - x) \leq 0 \ \forall y \in \mathrm{Cl}(X) \end{array} \right\}$$

Some other different duals can be proposed, with different objective functions, different feasible regions and different generalized concavity properties of the functions.

Definition 3.2 (2nd Dual Problem) *Consider problem P and let $c \in \mathrm{Int}(C)$. The following Dual problem can be introduced:*

$$D_2 : \begin{cases} C\text{\_min} \ \ L_2(x, \alpha_f, \alpha_g, \alpha_h) = f(x) + \frac{c}{\alpha_f^T c}[\alpha_h^T h(x)] \\ \qquad (x, \alpha_f, \alpha_g, \alpha_h) \in S_{D_2} \end{cases},$$

where $S_{D_2} = \{(x, \alpha_f, \alpha_g, \alpha_h) \in S_{D_1} : \ \alpha_g^T g(x) \leq 0\}$

Definition 3.3 (3rd Dual Problem) *Consider problem P and let $c \in \mathrm{Int}(C)$. The following Dual problem can be introduced:*

$$D_3 : \begin{cases} C\text{\_min} \ \ L_3(x, \alpha_f, \alpha_g, \alpha_h) = f(x) + \frac{c}{\alpha_f^T c}[\alpha_g^T g(x)] \\ \qquad (x, \alpha_f, \alpha_g, \alpha_h) \in S_{D_3} \end{cases},$$

where $S_{D_3} = \{(x, \alpha_f, \alpha_g, \alpha_h) \in S_{D_1} : \alpha_h^T h(x) = 0\}$

Definition 3.4 (4th Dual Problem) *Consider problem P. The following Dual problem can be introduced:*

$$D_4 : \begin{cases} C\_\min \ L_4(x, \alpha_f, \alpha_g, \alpha_h) = f(x) \\ (x, \alpha_f, \alpha_g, \alpha_h) \in S_{D_4} \end{cases},$$

where $S_{D_4} = \{(x, \alpha_f, \alpha_g, \alpha_h) \in S_{D_1} : \alpha_g^T g(x) \le 0, \ \alpha_h^T h(x) = 0\}$

In order to prove weak and strong duality results for the introduced pairs of primal-dual problems some generalized convexity properties are needed.

Definition 3.5 *Consider the primal problem P and the dual problems* D_j, $j \in \{1, 2, 3, 4\}$. *We say that functions* f, g *and* h *verify the generalized convexity properties* (GC_j) *if:*

- *in the case* $j = 1$, f *is C-concave in A,* g *is V-concave in A and* h *is affine in A,*
- *in the case* $j = 2$, f *is C-concave in A,* g *is polarly V-quasiconcave in A and* h *is affine in A,*
- *in the case* $j = 3$, f *is C-concave in A,* g *is V-concave in A and* h *is polarly quasiaffine in A,*
- *in the case* $j = 4$, f *is (Int(C),Int(C))-pseudoconcave in A,* g *is polarly V-quasiconcave in A and* h *is polarly quasiaffine in A.*

3.2 Weak Duality

Let us now prove weak duality results for the pairs of dual problems introduced so far. With this aim, it is worth noticing that we do not need to assume the convexity of the set X.

Theorem 3.1 *Let us consider the primal problem P and the dual problems* D_j, $j \in \{1, 2, 3, 4\}$. *If* (GC_j) *property holds for* $j \in \{1, 2, 3, 4\}$ *then:*

$$f(x_1) \notin L_j(x_2, \alpha_f, \alpha_g, \alpha_h) + \mathrm{Int}(C)$$

$\forall x_1 \in S_P$ *and* $\forall (x_2, \alpha_f, \alpha_g, \alpha_h) \in S_{D_j}$.

Proof. *Case* $j = 1$*)* Suppose by contradiction that

$$f(x_1) \in f(x_2) + \frac{1}{\alpha_f^T c} c[\alpha_g^T g(x_2) + \alpha_h^T h(x_2)] + \mathrm{Int}(C)$$

so that, being $\alpha_f \in C^+$, $\alpha_f \neq 0$, it is

$$\alpha_f^T f(x_1) > \alpha_f^T f(x_2) + \alpha_g^T g(x_2) + \alpha_h^T h(x_2). \tag{7.1}$$

Since f is C-concave it is:

$$f(x_1) \in f(x_2) + J_f(x_2)(x_1 - x_2) - C,$$

so that, since $\alpha_f \in C^+$,

$$\alpha_f^T f(x_1) \leq \alpha_f^T f(x_2) + \alpha_f^T J_f(x_2)(x_1 - x_2); \tag{7.2}$$

from the V-concavity of g it is:

$$g(x_1) \in g(x_2) + J_g(x_2)(x_1 - x_2) - V,$$

so that, since $\alpha_g \in V^+$ and $g(x_1) \in V$,

$$0 \leq \alpha_g^T g(x_1) \leq \alpha_g^T g(x_2) + \alpha_g^T J_g(x_2)(x_1 - x_2); \tag{7.3}$$

finally, being h affine it is:

$$h(x_1) = h(x_2) + J_h(x_2)(x_1 - x_2),$$

so that, $h(x_1) = 0$ implies

$$0 = \alpha_h^T h(x_1) = \alpha_h^T h(x_2) + \alpha_h^T J_h(x_2)(x_1 - x_2). \tag{7.4}$$

Adding the leftmost and rightmost components of inequalities (7.2), (7.3) and (7.4) we then have, for the definition of S_{D_1} and since $x_1 \in X$:

$$\begin{aligned} \alpha_f^T f(x_1) &\leq \alpha_f^T f(x_2) + \alpha_g^T g(x_2) + \alpha_h^T h(x_2) + \\ &\quad + [\alpha_f^T J_f(x_2) + \alpha_g^T J_g(x_2) + \alpha_h^T J_h(x_2)](x_1 - x_2) \\ &\leq \alpha_f^T f(x_2) + \alpha_g^T g(x_2) + \alpha_h^T h(x_2) \end{aligned}$$

which contradicts condition (7.1).

Case $j = 4$*)* Suppose by contradiction that $f(x_1) \in f(x_2) + \text{Int}(C)$; for the (Int($C$),Int($C$))-pseudoconcavity of f it follows that $J_f(x_2)(x_1 - x_2) \in \text{Int}(C)$; being $\alpha_f \in C^+$, $\alpha_f \neq 0$, it then results:

$$\alpha_f^T J_f(x_2)(x_1 - x_2) > 0. \tag{7.5}$$

For the hypotheses we have $g(x_1) \in V$, $\alpha_g \in V^+$, $\alpha_g^T g(x_2) \leq 0$, so that $\alpha_g^T g(x_2) \leq 0 \leq \alpha_g^T g(x_1)$; if $\alpha_g \neq 0$ then the polar V-quasiconcavity of g implies that

$$\alpha_g^T J_g(x_2)(x_1 - x_2) \geq 0, \tag{7.6}$$

while if $\alpha_g = 0$ then (7.6) holds trivially. For the hypotheses we have $h(x_1) = 0$ and $\alpha_h^T h(x_2) = 0$, so that $\alpha_h^T h(x_1) = 0 = \alpha_h^T h(x_2)$; if $\alpha_h \neq 0$ then the polar quasiaffinity of h implies that

$$\alpha_h^T J_h(x_2)(x_1 - x_2) = 0, \tag{7.7}$$

while if $\alpha_h = 0$ then (7.7) holds trivially. Adding the leftmost and rightmost components of inequalities (7.5), (7.6) and (7.7) we then have:

$$[\alpha_f^T J_f(x_2)^T + \alpha_g^T J_g(x_2) + \alpha_h^T J_h(x_2)](x_1 - x_2) > 0$$

so that, since $x_1 \in X$, it is $(x_2, \alpha_f, \alpha_g, \alpha_h) \notin S_{D_4}$ which is a contradiction.

Case $j = 2, 3$*)* The proofs are analogous to those of cases $j = 1, 4$. □

In the same way, the following stronger version of the weak duality theorem can be proved just changing the generalized convexity assumptions of function f.

Theorem 3.2 *Let us consider the primal problem P and the dual problems* D_j, $j \in \{1, 2, 3, 4\}$. *The following statements hold:*

i) in the case of $j \in \{1, 2, 3\}$, *if* (GC_j) *property holds and* f *is* Int(C)-*concave then:*

$$f(x_1) \notin L_j(x_2, \alpha_f, \alpha_g, \alpha_h) + C$$

$\forall x_1 \in S_P$ *and* $\forall (x_2, \alpha_f, \alpha_g, \alpha_h) \in S_{D_j}$ *such that* $x_1 \neq x_2$,

ii) in the case of $j = 4$, *if* (GC_4) *property holds and* f *is* (C,Int(C))-*pseudoconcave then:*

$$f(x_1) \notin L_4(x_2, \alpha_f, \alpha_g, \alpha_h) + C$$

$\forall x_1 \in S_P$ *and* $\forall (x_2, \alpha_f, \alpha_g, \alpha_h) \in S_{D_4}$ *such that* $x_1 \neq x_2$,

iii) in the case of $j = 4$, *if* (GC_4) *property holds and* f *is* $(C^0, \text{Int}(C))$-*pseudoconcave then:*

$$f(x_1) \notin L_4(x_2, \alpha_f, \alpha_g, \alpha_h) + C^0$$

$\forall x_1 \in S_P$ *and* $\forall (x_2, \alpha_f, \alpha_g, \alpha_h) \in S_{D_4}$ *such that* $x_1 \neq x_2$.

3.3 Strong Duality

We are now ready to prove the following results related to strong duality. With this aim, from now on we will assume the set X to be convex and with nonempty interior.

Theorem 3.3 *Let us consider the primal problem P and the dual problems D_j, $j \in \{1,2,3,4\}$. Suppose that X is convex with nonempty interior and a constraint qualification holds for problem P. If (GC_j) property holds for $j \in \{1,2,3,4\}$ then $\forall x \in C^0\text{-}\arg\max(P)$ $\exists \alpha_f \in C^+ \setminus \{0\}$, $\exists \alpha_g \in V^+$, $\exists \alpha_h \in \mathbb{R}^p$ such that:*

$$(x, \alpha_f, \alpha_g, \alpha_h) \in \text{Int}(C)\text{-}\arg\min(D_j) \quad \textit{and} \quad f(x) = L_j(x, \alpha_f, \alpha_g, \alpha_h)$$

Proof. Let $x \in C^0\text{-}\arg\max(P)$; by means of Theorem 2.1 $\exists \alpha_f \in C^+ \setminus \{0\}$, $\exists \alpha_g \in V^+$, $\exists \alpha_h \in \mathbb{R}^p$ such that $\alpha_g^T g(x) = 0$ and

$$[\alpha_f^T J_f(x) + \alpha_g^T J_g(x) + \alpha_h^T J_h(x)](y - x) \leq 0 \ \forall y \in \text{Cl}(X).$$

Since $h(x) = 0$ and $\alpha_g^T g(x) = 0$ it results $f(x) = L_j(x, \alpha_f, \alpha_g, \alpha_h)$ for all $j \in \{1,2,3,4\}$. It results also that $(x, \alpha_f, \alpha_g, \alpha_h) \in S_{D_4}$ and hence $(x, \alpha_f, \alpha_g, \alpha_h) \in S_{D_j}$ for all $j \in \{1,2,3,4\}$ since $S_{D_4} \subseteq S_{D_2} \subseteq S_{D_1}$ and $S_{D_4} \subseteq S_{D_3} \subseteq S_{D_1}$. Let $j \in \{1,2,3,4\}$, for the weak duality theorem $\nexists(\hat{x}, \hat{\alpha_f}, \hat{\alpha_g}, \hat{\alpha_h}) \in S_{D_j}$ such that

$$L_j(x, \alpha_f, \alpha_g, \alpha_h) = f(x) \in L_j(\hat{x}, \hat{\alpha_f}, \hat{\alpha_g}, \hat{\alpha_h}) + \text{Int}(C)$$

In other words, $\nexists(\hat{x}, \hat{\alpha_f}, \hat{\alpha_g}, \hat{\alpha_h}) \in S_{D_j}$ such that

$$L_j(\hat{x}, \hat{\alpha_f}, \hat{\alpha_g}, \hat{\alpha_h}) \in L_j(x, \alpha_f, \alpha_g, \alpha_h) - \text{Int}(C)$$

and hence $(x, \alpha_f, \alpha_g, \alpha_h) \in \text{Int}(C)\text{-}\arg\min(D_j)$. □

The following result follows directly from Theorem 3.3.

Corollary 3.1 *Let us consider the primal problem P and the dual problems D_j, $j \in \{1,2,3,4\}$. Suppose that X is convex with nonempty interior and a constraint qualification holds for problem P. If there exists an index $j \in \{1,2,3,4\}$ such that (GC_j) property holds and*

$$\text{Int}(C)\text{-}\arg\min(D_j) = \emptyset$$

then $C^0\text{-}\arg\max(P) = \emptyset$.

The following further duality result follows from the weak and the strong duality theorems.

Corollary 3.2 *Let us consider the primal problem P and the dual problems D_j, $j \in \{1,2,3,4\}$. Suppose that X is convex with nonempty interior and a constraint qualification holds for problem P. If (GC_j) property holds for $j \in \{1,2,3,4\}$ then*

$$f(x_1) - L_j(x_2, \alpha_f, \alpha_g, \alpha_h) \notin (\mathrm{Int}(C) \cup \mathrm{Int}(-C))$$

$\forall x_1 \in C^0\text{\_}\arg\max(P)$ *and* $\forall (x_2, \alpha_f, \alpha_g, \alpha_h) \in \mathrm{Int}(C)\text{\_}\arg\min(D_j)$.

Proof. Let $j \in \{1,2,3,4\}$, $x_1 \in C^0\text{\_}\arg\max(P)$ and $(x_2, \alpha_f, \alpha_g, \alpha_h) \in \mathrm{Int}(C)\text{\_}\arg\min(D_j)$; for the weak duality theorem it is

$$f(x_1) - L_j(x_2, \alpha_f, \alpha_g, \alpha_h) \notin \mathrm{Int}(C)$$

For the strong duality theorem $\exists \alpha_f \in C^+ \setminus \{0\}$, $\exists \alpha_g \in V^+$, $\exists \alpha_h \in \mathbb{R}^p$ such that

$$(x_1, \alpha_f, \alpha_g, \alpha_h) \in \mathrm{Int}(C)\text{\_}\arg\min(D_j) \text{ and } f(x_1) = L_j(x_1, \alpha_f, \alpha_g, \alpha_h).$$

Hence, condition $(x_2, \alpha_f, \alpha_g, \alpha_h) \in \mathrm{Int}(C)\text{\_}\arg\min(D_j)$ implies

$$L_j(x_1, \alpha_f, \alpha_g, \alpha_h) \notin L_j(x_2, \alpha_f, \alpha_g, \alpha_h) - \mathrm{Int}(C)$$

so that, for the equality $f(x_1) = L_j(x_1, \alpha_f, \alpha_g, \alpha_h)$, we have

$$f(x_1) - L_j(x_2, \alpha_f, \alpha_g, \alpha_h) \notin \mathrm{Int}(-C)$$

which proves the result. □

4. Final remarks

Comparing the introduced dual programs it can be easily seen that problem D_1 (the Wolfe-type dual problem) has the most "complex" objective function while problem D_4 (the Mond-Weir type) has the simplest one. Furthermore as you move from the dual program D_1 to D_4 you can require weaker generalized concavity assumptions in order to prove duality theorems. Finally, the feasible region of D_4 is the smallest, S_{D_1} is the biggest and $S_{D_4} \subseteq S_{D_2} \subseteq S_{D_1}$ and $S_{D_4} \subseteq S_{D_3} \subseteq S_{D_1}$. As the reader has already noted, whenever you get duality results by defining a simpler objective function and by requiring weaker generalized concavity properties (see Problem D_4), the feasible region of the dual problem is smaller and viceversa a bigger feasible region (see Problem D_1) is "paid" by a more complex objective function and stronger generalized concavity assumptions. The described behavior is represented in Figure 7.1.

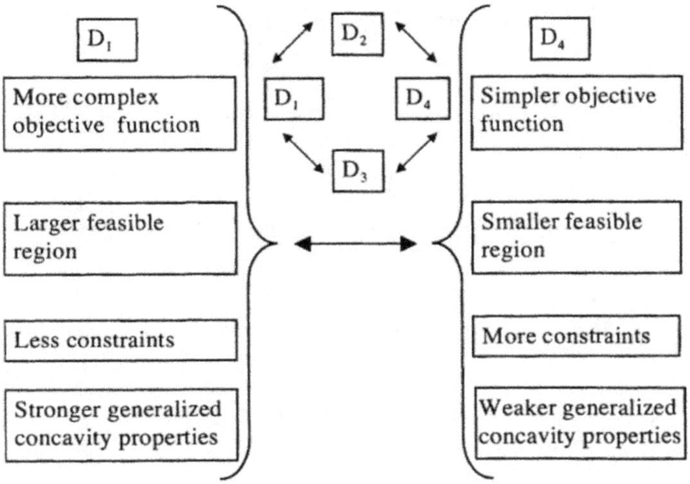

Figure 7.1.

Appendix - Generalized Concave Functions

The following classes of vector valued functions have been defined and studied in Cambini (1996); Cambini (1998); Cambini (1998).

Definition 4.1 *Let $f : A \to \mathbb{R}^m$, where $A \subset \mathbb{R}^n$ is an open convex set, be a differentiable vector valued function and let $C \subset \mathbb{R}^m$ be a closed convex cone with nonempty interior. Let also $C^0 = C \setminus \{0\}$ and C^+ the positive polar cone of C. Function f is said to be:*

- *C-concave if and only if* $\forall x, y \in A,\ x \neq y$, *it holds:*

$$f(y) - f(x) - J_f(x)(y - x) \in -C,$$

- C^0*-concave if and only if* $\forall x, y \in A,\ x \neq y$, *it holds:*

$$f(y) - f(x) - J_f(x)(y - x) \in -C^0,$$

- Int(C)*-concave if and only if* $\forall x, y \in A,\ x \neq y$, *it holds:*

$$f(y) - f(x) - J_f(x)(y - x) \in -\text{Int}(C),$$

- (Int(C),Int(C))*-pseudoconcave if and only if* $\forall x, y \in A,\ x \neq y$, *it holds:*

$$f(y) \in f(x) + \text{Int}(C) \quad \Rightarrow \quad J_f(x)(y - x) \in \text{Int}(C),$$

- $(C^0, \text{Int}(C))$**-*pseudoconcave*** *if and only if* $\forall x, y \in A$, $x \neq y$, *it holds:*

$$f(y) \in f(x) + C^0 \quad \Rightarrow \quad J_f(x)(y-x) \in \text{Int}(C),$$

- *(C,* Int*(C))-pseudoconcave if and only if* $\forall x, y \in A$, $x \neq y$, *it holds:*

$$f(y) \in f(x) + C \quad \Rightarrow \quad J_f(x)(y-x) \in \text{Int}(C).$$

See Cambini (1998); Cambini and Komlósi (1998); Cambini and Komlósi (2000) for the definition and the study of the following classes of functions.

Definition 4.2 *Let* $f : A \to \mathbb{R}^m$, *where* $A \subset \mathbb{R}^n$ *is an open convex set, be a differentiable vector valued function and let* $C \subset \mathbb{R}^m$ *be a closed convex cone with nonempty interior. Let also* $C^0 = C \setminus \{0\}$ *and* C^+ *the positive polar cone of C. Function* f *is said to be:*

- *polarly C-quasiconcave if and only if* $\phi(x) = \alpha^T f(x)$ *is quasiconcave* $\forall \alpha \in C^+$, $\alpha \neq 0$, *that is to say if and only if* $\forall \alpha \in C^+$, $\alpha \neq 0$, $\forall x, y \in A$, $x \neq y$, *it holds:*

$$\alpha^T f(y) \geq \alpha^T f(x) \quad \Rightarrow \quad \alpha^T J_f(x)(y-x) \geq 0,$$

- *polarly C-pseudoconcave if and only if* $\phi(x) = \alpha^T f(x)$ *is pseudoconcave* $\forall \alpha \in C^+$, $\alpha \neq 0$, *that is to say if and only if* $\forall \alpha \in C^+$, $\alpha \neq 0$, $\forall x, y \in A$, $x \neq y$, *it holds:*

$$\alpha^T f(y) > \alpha^T f(x) \quad \Rightarrow \quad \alpha^T J_f(x)(y-x) > 0,$$

- *polarly* C^0**-*pseudoconcave*** *if and only if* $\forall \alpha \in C^+$, $\alpha \neq 0$, $\forall x, y \in A$, $x \neq y$, *it holds:*

$$\alpha^T f(y) \geq \alpha^T f(x) \text{ with } f(y) \neq f(x) \quad \Rightarrow \quad \alpha^T J_f(x)(y-x) > 0,$$

- *polarly* Int*(C)-pseudoconcave if and only if* $\phi(x) = \alpha^T f(x)$ *is strictly pseudoconcave* $\forall \alpha \in C^+$, $\alpha \neq 0$, *that is to say if and only if* $\forall \alpha \in C^+$, $\alpha \neq 0$, $\forall x, y \in A$, $x \neq y$, *it holds:*

$$\alpha^T f(y) \geq \alpha^T f(x) \quad \Rightarrow \quad \alpha^T J_f(x)(y-x) > 0,$$

- *polarly quasiaffine if and only if* $\phi(x) = \alpha^T f(x)$ *is both quasiconvex and quasiconcave* $\forall \alpha \in \mathbb{R}^m$, $\alpha \neq 0$, *that is to say if and only if* $\forall \alpha \in \mathbb{R}^m$, $\alpha \neq 0$, $\forall x, y \in A$, $x \neq y$, *it holds:*

$$\alpha^T f(y) = \alpha^T f(x) \quad \Rightarrow \quad \alpha^T J_f(x)(y-x) = 0.$$

Note that the characterization of polarly quasiaffine functions follows from the properties of scalar generalized concave functions and scalar generalized affine functions studied in Cambini (1995). Let us finally recall that (see Cambini and Komlósi (1998); Cambini and Komlósi (2000)):

- If f is polarly C-pseudoconcave then it is also (Int(C),Int(C))-pseudoconcave
- If f is polarly C^0**-pseudoconcave** then it is also $(C^0$, Int(C))-pseudoconcave
- If f is polarly Int(C)-pseudoconcave then it is also (C, Int(C))-pseudoconcave

Acknowledgments

Careful reviews by the anonymous referees are gratefully acknowledged.

References

Aghezzaf, B. and Hachimi, M. (2001), Sufficiency and Duality in Multiobjective Programming Involving Generalized (F,ρ)**−convexity**, *Journal of Mathematical Analysis and Applications,* Vol. 258, pp. 617-628.

Bhatia, D. and Jain, P. (1994), Generalized (F,ρ)**-convexity** and duality for non smooth multi-objective programs, *Optimization,* Vol. 31, pp. 153-164.

Bathia, D. and Pankaj, K. G. (1998), Duality for non-smooth nonlinear fractional multiobjective programs via (F,ρ)**-convexity,** Optimization, Vol. 43, pp. 185-197.

Bector, C.R., Suneja, S.K. and Lalitha, C.S. (1993), Generalized B-Vex Functions and Generalized B-Vex Programming, *Journal of Optimization Theory and Application,* Vol. 76, pp. 561-576.

Bector, C.R., Bector, M.K., Gill, A. and Singh, C. (1994), Duality for Vector Valued B-invex Programming, in *Generalized Convexity,* edited by S. Komlósi, T. Rapcsák and S. Schaible, Lecture Notes in Economics and Mathematical Systems, Vol. 405, Springer-Verlag, Berlin, pp. 358-373.

Bector, C.R. (1996), Wolfe-Type Duality involving (B,η)**-invex** Functions for a Minmax Programming Problem, *Journal of Mathematical Analysis and Application,* Vol. 201, pp. 114-127.

Cambini, R. (1995), Funzioni scalari affini generalizzate , *Rivista di Matematica per le Scienze Economiche e Sociali,* year 18th, Vol.2, pp. 153-163.

Cambini, R. (1996), Some new classes of generalized concave vector-valued functions, *Optimization,* Vol. 36, pp. 11-24.

Cambini, R. (1998), Composition theorems for generalized concave vector valued functions, *Journal of Information and Optimization Sciences,* Vol. 19, pp. 133-150.

Cambini, R. (1998), Generalized Concavity for Bicriteria Functions, in *Generalized Convexity, Generalized Monotonicity: Recent Results,* edited by J.-P. Crouzeix, J.-E. Martinez-Legaz and M. Volle, Nonconvex Optimization and Its Applications, Vol. 27, Kluwer Academic Publishers, Dordrecht, pp. 439-451.

Cambini, R. and Komlósi, S. (1998), On the Scalarization of Pseudoconcavity and Pseudomonotonicity Concepts for Vector Valued Functions", in *Generalized Convexity, Generalized Monotonicity: Recent Results,* edited by J.-P. Crouzeix, J.-E. Martinez-Legaz and M. Volle, Nonconvex Optimization and Its Applications, Vol. 27, Kluwer Academic Publishers, Dordrecht, pp. 277-290.

Cambini, R. and Komlósi S. (2000), On Polar Generalized Monotonicity in Vector Optimization, *Optimization,* Vol. 47, pp. 111-121.

Cambini, R. (2001), Necessary Optimality Conditions in Vector Optimization, Report n.212, Department of Statistics and Applied Mathematics, University of Pisa.

Clarke, F.H. (1983), *Optimization and Nonsmooth Analysis,* John-Wiley & Sons, New York.

Giorgi, G. and Guerraggio, A. (1994), First order generalized optimality conditions for programming problems with a set constraint, in *Generalized Convexity,* edited by S. Komlósi, T. Rapcsák and S. Schaible, Lecture Notes in Economics and Mathematical Systems, Vol. 405, Springer-Verlag, Berlin, pp. 171-185.

Giorgi, G. and Guerraggio, A. (1998), The notion of invexity in vector optimization: smooth and nonsmooth case, in *Generalized Convexity, Generalized Monotonicity: Recent Results,* edited by J.-P. Crouzeix, J.-E. Martinez-Legaz and M. Volle, Nonconvex Optimization and Its Applications, Vol. 27, Kluwer Academic Publishers, Dordrecht, pp. 389-405.

Göpfert, A. and Tammer, C. (2002), Theory of Vector Optimization, in *Multiple Criteria Optimization,* edited by M. Ehrgott and X. Gandibleux, International Series in Operations Research and Management Science, Vol. 52, Kluwer Academic Publishers, Boston.

Gulati, T. R. and Islam, M.A. (1994), Sufficiency and Duality in Multiobjective Programming Involving Generalized **(F,ρ)-convexity,** *Journal of Mathematical Analysis and Applications,* Vol. 183, pp. 181-195.

Hanson, M.A. and Mond, B. (1987), Necessary and Sufficiency Conditions in Constrained Optimization, *Mathematical Programming,* Vol. 37, pp. 51-58.

Jahn, J. (1994), *Introduction to the Theory of Nonlinear Optimization,* Springer-Verlag, Berlin, 1994.

Jiménez, B. and Novo, V. (2002), A finite dimensional extension of Lyusternik theorem with applications to multiobjective optimization, *Journal of Mathematical Analysis and Applications,* Vol. 270, pp. 340-356.

Kaul R.N., Suneja S.K. and Srivastava M.K. (1994), Optimality Criteria and Duality in Multiobjective Optimization Involving Generalized Invexity, *Journal of Optimization Theory and Application,* Vol. 80, pp. 465-481.

Luc, D. T. (1984), On Duality Theory in Multiobjective Programming, *Journal of Optimization Theory and Application,* Vol. 43, pp. 557-582.

Luc, D.T. (1989), *Theory of vector optimization,* Lecture Notes in Economics and Mathematical Systems, Vol. 319, Springer-Verlag, Berlin, 1989.

Maeda, T. (1994), Constraint Qualifications in Multiobjective Optimization Problems: Differentiable Case, *Journal of Optimization Theory and Application,* Vol. 80, pp. 483-500.

Mangasarian, O.L. (1969), *Nonlinear Programming,* McGraw-Hill, New York.

Mishra, S.K. (1996), On Sufficiency and Duality for Generalized Quasiconvex Nonsmooth Programs, *Optimization,* Vol. 38, pp. 223-235.

Mond, B. and Weir, T. (1981), Generalized concavity and duality, in *Generalized Concavity and Duality in Optimization and Economics,* edited by S. Schaible and W.T. Ziemba, Academic Press, New York, pp. 263-279.

Patel, R.B. (2000), On efficiency and duality theory for a class of multiobjective fractional programming problems with invexity, *Journal of Statistics and Management Systems,* Vol. 3, pp. 29-41.

Preda, V. (1992), On Efficiency and Duality in Multiobjective Programs, *Journal of Mathematical Analysis and Applications,* Vol. 166, pp. 365-377.

Rueda, N.G., Hanson, M.A. and Singh C., (1995), Optimality and Duality with Generalized Convexity, *Journal of Optimization Theory and Application,* Vol. 86, pp. 491-500.

Tanino, T. and Sawaragy, Y. (1979), Duality Theory in Multiobjective Programming, *Journal of Optimization Theory and Application,* Vol. 27, pp. 509-529.

Weir, T., Mond, B. and Craven B.D. (1986), On duality for weakly minimized vector-valued optimization problems, *Optimization,* Vol. 17, pp. 711-721.

Zalmai, G.J. (1997), Efficiency criteria and duality models for multiobjective fractional programming problems containing locally subdifferentiable and ρ-**convex** functions, *Optimization,* Vol. 41, pp. 321-360.

Chapter 8

DUALITY IN FRACTIONAL PROGRAMMING PROBLEMS WITH SET CONSTRAINTS

Riccardo Cambini, Laura Carosi*
Dept. of Statistics and Applied Mathematics
University of Pisa, ITALY

Siegfried Schaible
A. Gary Anderson Graduate School of Management
University of California at Riverside, U.S.A.

Abstract Duality is studied for a minimization problem with finitely many inequality and equality constraints and a set constraint where the constraining convex set is not necessarily open or closed. Under suitable generalized convexity assumptions we derive a weak, strong and strict converse duality theorem. By means of a suitable transformation of variables these results are then applied to a class of fractional programs involving a ratio of a convex and an affine function with a set constraint in addition to inequality and equality constraints. The results extend classical fractional programming duality by allowing for a set constraint involving a convex set that is not necessarily open or closed.

Keywords: Duality, Set Constraints, Fractional Programming.

MSC2000: 90C26, 90C32, 90C46
Journal of Economic Literature Classification (1999): C61

---

*This research has been partially supported by M.I.U.R. and C.N.R.
email: cambric@ec.unipi.it, lcarosi@ec.unipi.it, siegfried.schaible@ucr.edu

1. Introduction

Duality in mathematical programming has been studied extensively. Often solution methods are based on duality properties. Most of the existing results deal with problems where the feasible region is defined by finitely many inequalities and/or equalities. Duality results for problems with a non-open set constraint in addition to inequality constraints can be found in Giorgi and Guerraggio (1994).

In this paper, we aim at studying duality for minimization problems with finitely many inequalities and/or equalities and a set constraint involving a convex set that is not necessarily open or closed. A necessary optimality condition of the minimum-principle type holds for this class of problems; see for all Mangasarian (1969). This allows us to introduce a Wolfe-type dual problem and to derive weak, strong and strict converse duality theorems.

Furthermore we consider fractional programs with a set constraint in addition to inequality and equality constraints where again the constraining set is not necessarily open or closed. The function to be minimized is the ratio of a convex and an affine function. For fractional programs without a set constraint a variety of approaches have been proposed; see for example Barros et al (1996); Barros (1998); Bector (1973); Bector et al (1977); Craven (1981); Dinkelbach (1967); Jagannathan (1973); Liang et al (2001); Liu (1996); Mahajan and Vartak (1977); Schaible (1973); Schaible (1976a); Schaible (1976b); Scott and Jefferson (1996).

The dual in fractional programming with a set constraint is obtained from the general duality results with help of a suitable variable transformation. The objective function of the dual program turns out to be linear. Our results can be viewed as an extension of classical duality results without a set constraint; see Mahajan and Vartak (1977); Schaible (1973); Schaible (1974); Schaible (1976a); Schaible (1976b).

2. General duality results

Let us consider the following primal problem:

$$P : \begin{cases} \min \ f(x) \\ x \in S_P \end{cases} \equiv \begin{cases} \min \ f(x) & \\ g(x) \in -V & \textit{inequality constraints} \\ h(x) = 0 & \textit{equality constraints} \\ x \in X & \textit{set constraint} \end{cases}$$

where

$$S_P = \{x \in A : \ g(x) \in -V, \ h(x) = 0, \ x \in X\}$$

and

- the set $A \subseteq \mathbb{R}^n$ is open and convex,
- the functions $f : A \to \mathbb{R}$, $g : A \to \mathbb{R}^m$ and $h : A \to \mathbb{R}^p$ are differentiable with gradient $\nabla f(x)$ and Jacobians $J_g(x)$ and $J_h(x)$, respectively,
- the cone $V \subset \mathbb{R}^m$ is closed, convex, pointed and has a nonempty interior,
- the set $X \subseteq A$ is convex with nonempty interior and it is not necessarily open or closed,
- the set O_P is the (possibly empty) set of optimal solutions of P.

The following necessary optimality condition, known as minimum principle condition (see Mangasarian (1969)), holds for problem P. Recall that V^+ denotes the positive polar cone of V ([1]) while $\mathbb{R}_+$ is the set of nonnegative numbers.

Theorem 2.1 *If the vector* $\bar{x}$ *belongs to* O_P, *then there exists some nonzero vector* $(\alpha_f, \alpha_g, \alpha_h)$ *belonging to* $(\mathbb{R}_+ \times V^+ \times \mathbb{R}^p)$ *such that* $\alpha_g^T g(\bar{x}) = 0$ *and*

$$[\alpha_f \nabla f(\bar{x})^T + \alpha_g^T J_g(\bar{x}) + \alpha_h^T J_h(\bar{x})](x - \bar{x}) \geq 0 \ \forall x \in \mathrm{Cl}(X) \ \ (^2). \quad (8.1)$$

Moreover, if in addition a constraint qualification holds, then we may take $\alpha_f = 1$ *in relation (8.1).*

Remark 2.1 *It can easily be proved that* $\mathrm{Co}[W] = \mathbb{R}^{m+p}$ ([2]) *for*

$$W = \{(t_g, t_h) \in \mathbb{R}^{m+p} : \ (t_g, t_h) = [J_g(x_0), J_h(x_0)](x - x_0), \ x \in \mathrm{Cl}(X)\}$$

is a constraint qualification for problem P (see Cambini and Carosi (2002)). A comprehensive study of constraint qualifications for scalar problems with set constraints is given in Giorgi and Guerraggio (1994).

Theorem 2.1 suggests the following Wolfe-type dual problem of P:

$$D : \begin{cases} \max \ L(x, \alpha_g, \alpha_h) \\ (x, \alpha_g, \alpha_h) \in S_D \end{cases}$$

---

[1] The positive polar cone of a set $V \subset \mathbb{R}^m$ is given by

$$V^+ = \{y \in \mathbb{R}^m : \ y^T v \geq 0 \ \text{for every} \ v \in V\}.$$

[2] We denote by $Cl(X)$ the closure and by $Co(X)$ the convex hull of a set $X \subseteq \mathbb{R}^n$.

where

$$S_D=\{(x,\alpha_g,\alpha_h)\in(A\times V^+\times \mathbb{R}^p):\ \nabla_x L(x,\alpha_g,\alpha_h)(y-x)\geq 0,\ y\in \mathrm{Cl}(X)\}$$

and

- $L(x,\alpha_g,\alpha_h)=f(x)+\alpha_g^T g(x)+\alpha_h^T h(x)$,
- $\nabla_x L(x,\alpha_g,\alpha_h)=\nabla f(x)^T+\alpha_g^T J_g(x)+\alpha_h^T J_h(x)$,
- the set O_D is the (possibly empty) set of optimal solutions of D.

Remark 2.2 *Note that if X is open, then the dual problem D coincides with the one proposed in Mahajan and Vartak (1977). Moreover if $X=\mathbb{R}^n$, then D can be rewritten as*

$$\begin{cases} \max\ \ L(x,\alpha_g,\alpha_h) \\ \nabla_x L(x,\alpha_g,\alpha_h)=0 \\ x\in A,\ \ \alpha_g\in V^+,\ \ \alpha_h\in \mathbb{R}^p \end{cases}$$

which is the well known Wolfe dual problem; see for example Mangasarian (1969).

Actually, weak and strong duality results can be proved under the pseudoconvexity ([3]) of function $x\to L(x,\alpha_g,\alpha_h)$.

Theorem 2.2 (Weak Duality) *Let $x_1\in S_P$ and $(x_2,\alpha_g,\alpha_h)\in S_D$. If for every $\alpha_g\in V^+$ and $\alpha_h\in\mathbb{R}^p$ the function $x\to L(x,\alpha_g,\alpha_h)$ is pseudoconvex at x_2, then*

$$f(x_1)\geq L(x_2,\alpha_g,\alpha_h).$$

Proof. Since $S_P\subset \mathrm{Cl}(X)$, we have $\nabla_x L(x_2,\alpha_g,\alpha_h)(x_1-x_2)\geq 0$. From $\alpha_g\in V^+$, $\alpha_g^T g(x_1)\leq 0$, $h(x_1)=0$ and the pseudoconvexity of function $x\to L(x,\alpha_g,\alpha_h)$ it follows

$$f(x_1)\geq f(x_1)+\alpha_g^T g(x_1)+\alpha_h^T h(x_1)\geq f(x_2)+\alpha_g^T g(x_2)+\alpha_h^T h(x_2).$$

□

[3] Let $A\subset\mathbb{R}^n$ be an open convex set and $f:A\to\mathbb{R}$ a differentiable function. Function f is called [strictly] pseudoconvex at $x\in A$ if for all $y\in A$ it holds

$$f(y)<f(x)\ \ [f(y)\leq f(x)]\quad\Longrightarrow\quad \nabla f(x)^T(y-x)<0.$$

Function f is said to be pseudoconvex in A if it is pseudoconvex at every $x\in A$.

Theorem 2.3 (Strong Duality) *Assume that a constraint qualification holds and for every $\alpha_g \in V^+$ and $\alpha_h \in \mathbb{R}^p$ the function $x \to L(x, \alpha_g, \alpha_h)$ is pseudoconvex on A. It follows that for every $\overline{x}$ belonging to O_P there exists some $(\overline{\alpha}_g, \overline{\alpha}_h) \in (V^+ \times \mathbb{R}^p)$ such that $(\overline{x}, \overline{\alpha}_g, \overline{\alpha}_h) \in O_D$ and*

$$f(\overline{x}) = L(\overline{x}, \overline{\alpha}_g, \overline{\alpha}_h).$$

Proof. Since $\overline{x} \in O_P$, from Theorem 2.1 we obtain that there exists some $(\overline{\alpha}_g, \overline{\alpha}_h) \in (V^+ \times \mathbb{R}^p)$ satisfying $\overline{\alpha}_g^T g(\overline{x}) = 0$ and

$$[\nabla f(\overline{x})^T + \overline{\alpha}_g^T J_g(\overline{x}) + \overline{\alpha}_h^T J_h(\overline{x})](x - \overline{x}) \geq 0 \ \forall x \in \mathrm{Cl}(X).$$

Hence $(\overline{x}, \overline{\alpha}_g, \overline{\alpha}_h)$ belongs to S_D. Since $\overline{x} \in S_P$, we have $h(\overline{x}) = 0$. Thus the weak duality theorem yields

$$L(\overline{x}, \overline{\alpha}_g, \overline{\alpha}_h) = f(\overline{x}) + \overline{\alpha}_g^T g(\overline{x}) + \overline{\alpha}_h^T h(\overline{x}) =$$
$$= f(\overline{x}) \geq L(x_2, \alpha_g, \alpha_h) \quad \forall (x_2, \alpha_g, \alpha_h) \in S_D,$$

and the result follows. □

Theorem 2.3 allows us to prove the following results.

Corollary 2.1 *Assume a constraint qualification holds and for every $\alpha_g \in V^+$ and $\alpha_h \in \mathbb{R}^p$ the function $x \to L(x, \alpha_g, \alpha_h)$ is pseudoconvex on A. If $S_D = \emptyset$, then $O_P = \emptyset$.*

Corollary 2.2 *Assume a constraint qualification holds and for every $\alpha_g \in V^+$ and $\alpha_h \in \mathbb{R}^p$ the function $x \to L(x, \alpha_g, \alpha_h)$ is pseudoconvex on A. It follows that for every $\overline{x} \in O_P$ and for every $(\widehat{x}, \widehat{\alpha}_g, \widehat{\alpha}_h) \in O_D$ we have*

$$f(\overline{x}) = L(\widehat{x}, \widehat{\alpha}_g, \widehat{\alpha}_h).$$

Proof. According to the strong duality theorem there exists some $(\overline{x}, \overline{\alpha}_g, \overline{\alpha}_h) \in O_D$ such that $f(\overline{x}) = L(\overline{x}, \overline{\alpha}_g, \overline{\alpha}_h) = L(\widehat{x}, \widehat{\alpha}_g, \widehat{\alpha}_h)$. □

Finally, under the strict pseudoconvexity assumption on the dual objective function L we can prove a strict converse duality theorem.

Theorem 2.4 (Strict Converse Duality) *Let $(\widehat{x}, \widehat{\alpha}_g, \widehat{\alpha}_h) \in O_D$ and $\overline{x} \in O_P$. Assume that a constraint qualification holds and for every $\alpha_g \in V^+$ and $\alpha_h \in \mathbb{R}^p$ the function $x \to L(x, \alpha_g, \alpha_h)$ is pseudoconvex on A and strictly pseudoconvex at $\widehat{x}$. Then $\overline{x} = \widehat{x}$, i.e., $\widehat{x} \in O_P$.*

Proof. With $\overline{x} \in S_P$ and $(\widehat{x}, \widehat{\alpha}_g, \widehat{\alpha}_h) \in S_D$ we have $h(\overline{x}) = 0$, $g(\overline{x}) \in -V$, $\widehat{\alpha}_g \in V^+$ implying $\widehat{\alpha}_g^T g(\overline{x}) \leq 0$. From the previous corollary we get

$$L(\widehat{x}, \widehat{\alpha}_g, \widehat{\alpha}_h) = f(\overline{x}) \geq f(\overline{x}) + \widehat{\alpha}_g^T g(\overline{x}) + \widehat{\alpha}_h^T h(\overline{x}) = L(\overline{x}, \widehat{\alpha}_g, \widehat{\alpha}_h).$$

Suppose to the contrary that $\overline{x} \neq \hat{x}$. By the strict pseudoconvexity of $x \to L(x, \alpha_g, \alpha_h)$ at $\hat{x}$, condition $L(\widehat{x}, \widehat{\alpha}_g, \widehat{\alpha}_h) \geq L(\overline{x}, \widehat{\alpha}_g, \widehat{\alpha}_h)$ yields

$$\nabla_x L(\widehat{x}, \widehat{\alpha}_g, \widehat{\alpha}_h)^T(\overline{x} - \widehat{x}) = [\nabla f(\widehat{x})^T + \widehat{\alpha}_g^T J_g(\widehat{x}) + \widehat{\alpha}_h^T J_h(\widehat{x})](\overline{x} - \widehat{x}) < 0. \tag{8.2}$$

Since $\overline{x} \in X$, (8.2) implies $(\widehat{x}, \widehat{\alpha}_g, \widehat{\alpha}_h) \notin S_D$, a contradiction. □

According to the above results, pseudoconvexity of the function $x \to L(x, \alpha_g, \alpha_h)$ plays an important role in duality theory. Therefore we may ask which kind of (generalized) convexity assumptions on the functions f, g and h guarantee this property of L. It can easily be seen that if f is [strictly] convex at x_2, g is V-convex at x_2 ([4]) and h is affine, then for every fixed $(\alpha_g \times \alpha_h) \in (V^+ \times \mathbb{R}^p)$ the function $x \to L(x, \alpha_g, \alpha_h)$ is [strictly] pseudoconvex at x_2. We mention that these convexity assumptions have also been used in Lagrangean duality theory in Frenk and Kassay (1999) for example.

We mention that using the same pseudoconvexity assumptions on the Lagrangean function $x \to L(x, \alpha_g, \alpha_h)$, Mahajan and Vartak (1977) proved the duality results for a problem P whose feasible region is defined by equality and inequality constraints only. On the other hand, under pseudoinvexity properties, Giorgi and Guerraggio (1994) derive duality theorems for problems with a set constraint and inequality constraints only.

3. The fractional case

In this section we consider a fractional program where the objective function f is the ratio of a convex function and an affine function and the feasible region is defined as in Problem P, that is

$$P_F: \begin{cases} \min \ f(x) = \frac{n(x)}{d(x)} & \\ g(x) \in -V & \textit{inequality constraints} \\ h(x) = 0 & \textit{equality constraints} \\ x \in X & \textit{set constraint} \end{cases}$$

where

- functions $n : A \to \mathbb{R}$ and $g : A \to \mathbb{R}^m$ are differentiable,

[4] Let $A \subset \mathbb{R}^n$ be an open convex set and $V \subset \mathbb{R}^m$ be a convex cone. A differentiable function $g : A \to \mathbb{R}^m$ is said to be V-convex at $x \in A$ if

$$g(y) - g(x) - Jg(x)(y - x) \in V \quad \forall y \in A.$$

Function g is said to be V-convex in A if it is V-convex at every $x \in A$. For a complete study of this class of functions, even in the nondifferentiable case, see for example Cambini (1996); Cambini and Komlósi (1998); Frenk and Kassay (1999).

- the set $A \subseteq \mathbb{R}^n$ is open and convex,
- the cone $V \subset \mathbb{R}^m$ is closed, convex, pointed and has a nonempty interior,
- $d(x) = d^T x + d_0$ with $d \in \mathbb{R}^n$, $d_0 \in \mathbb{R}$ and $d^T x + d_0 > 0 \; \forall x \in A$,
- $h(x) = Bx + b_0$ with $B \in \mathbb{R}^{p \times n}$, $b_0 \in \mathbb{R}^p$,
- n is convex in A and g is V-convex in A,
- the set $X \subseteq A$ is convex with nonempty interior and is not necessarily open or closed.

Our goal is to show that the following problem D_F can be viewed as a dual of P_F, i.e., the various duality results of the previous section hold. We set

$$D_F : \begin{cases} \max \; -\lambda \; \equiv \; -\min \; \lambda \\ L_F(x) \geq 0 \\ \nabla_x L_F(x) \, (x_1 - x) \geq 0 \; \forall x_1 \in X \\ x \in A, \; \alpha_g \in V^+, \; \alpha_h \in \mathbb{R}^p, \; \lambda \in \mathbb{R} \end{cases}$$

where

$$L_F(x) = L_F(x, \alpha_g, \alpha_h, \lambda) = n(x) + \alpha_g^T g(x) + \alpha_h^T h(x) + \lambda d(x)$$
$$\nabla_x L_F(x)^T = \nabla_x L_F(x, \alpha_g, \alpha_h, \lambda)^T = \nabla n(x)^T + \alpha_g^T J_g(x) + \alpha_h^T B + \lambda d^T$$

Since we consider an arbitrary convex set X, we cannot apply the Wolfe-type duality results that can be found in the literature on duality in fractional programming. On the other hand, even though the objective function f is pseudoconvex (see Mangasarian (1969)), the function $x \to L(x, \alpha_g, \alpha_h)$ is not pseudoconvex in general. Hence we are not able to directly apply the duality results stated in the previous section. But along the lines proposed in Schaible (1976a) and Schaible (1976b) we can transform problem P_F into the following equivalent problem

$$\overline{P} : \begin{cases} \min \; \overline{n}(y, t) = t \cdot n\left(\frac{y}{t}\right) \\ \overline{g}(y, t) = t \cdot g\left(\frac{y}{t}\right) \in -V \\ \overline{h}(y, t) = t \cdot h\left(\frac{y}{t}\right) = 0 \\ \overline{d}(y, t) = t \cdot d\left(\frac{y}{t}\right) - 1 = 0 \\ -t \leq 0 \\ (y, t) \in \Pi_X \end{cases}$$

where $\Pi_X = \{(y, t) \in \mathbb{R}^n \times \mathbb{R} : \frac{y}{t} \in X, \; t > 0\}$ and the functions $\overline{n}, \overline{d}, \overline{g}$ and $\overline{h}$ are defined on the set $\Pi_A = \{(y, t) \in \mathbb{R}^n \times \mathbb{R} : \frac{y}{t} \in A, \; t > 0\}$.

Due to the performed transformation, the new problem $\overline{P}$ has the following convexity properties.

Lemma 3.1 *In problem* $\overline{P}$

i) Π_X *and* Π_A *are convex sets with nonempty interior,*

ii) $\overline{g}$ *is V-convex in* Π_A*,*

iii) $\overline{n}$ *is convex in* Π_A*,*

iv) $\overline{h}$ *and* $\overline{d}$ *are affine in* Π_A*.*

Proof. *i)* Consider $(y_1,t_1) \in \Pi_X$ and $(y_2,t_2) \in \Pi_X$. We want to prove that

$$(y_1,t_1) + \mu\left((y_2,t_2) - (y_1,t_1)\right) \in \Pi_X \ \forall \mu \in [0,1],$$

i.e.,

$$\frac{y_1 + \mu(y_2 - y_1)}{t_1 + \mu(t_2 - t_1)} \in X \text{ and } \ t_1 + \mu\,(t_2 - t_1) > 0,\ \forall \mu \in [0,1].$$

Simple calculations show that $t_1 + \mu\,(t_2 - t_1) > 0$ and

$$\frac{y_1 + \mu(y_2 - y_1)}{t_1 + \mu(t_2 - t_1)} = \frac{y_1}{t_1} + \beta\left(\frac{y_2}{t_2} - \frac{y_1}{t_1}\right)$$

where

$$\beta = \frac{1}{1 + \frac{t_1(1-\mu)}{t_2\mu}}.$$

Since X is convex and β belongs to [0, 1], $\frac{y_1+\mu(y_2-y_1)}{t_1+\mu(t_2-t_1)} \in X$. The convexity of Π_A follows along the same lines.

ii) Consider $(y_1,t_1) \in \Pi_A$ and $(y_2,t_2) \in \Pi_A$. We want to show that $\overline{g}(y_2,t_2) - \overline{g}(y_1,t_1) - J_{\overline{g}}(y_1,t_1)\begin{bmatrix} y_2 - y_1 \\ t_2 - t_1 \end{bmatrix} \in V.$

$$\begin{aligned} J_{\overline{g}}(y_1,t_1)\begin{bmatrix} y_2 - y_1 \\ t_2 - t_1 \end{bmatrix} &= J_g\left(\frac{y_1}{t_1}\right)(y_2 - y_1) + g\left(\frac{y_1}{t_1}\right)(t_2 - t_1) - \frac{1}{t_1}J_g\left(\frac{y_1}{t_1}\right) y_1\,(t_2 - t_1) \\ &= J_g\left(\frac{y_1}{t_1}\right)\left(y_2 - y_1 - \frac{y_1}{t_1}t_2 + y_1\right) + g\left(\frac{y_1}{t_1}\right)(t_2 - t_1) \\ &= t_2 J_g\left(\frac{y_1}{t_1}\right)\left(\frac{y_2}{t_2} - \frac{y_1}{t_1}\right) + g\left(\frac{y_1}{t_1}\right)(t_2 - t_1). \end{aligned} \tag{8.3}$$

Since $t_2 > 0$, $\frac{y_2}{t_2}, \frac{y_1}{t_1} \in X$ and g is V-convex in A, we have

$$t_2 g\left(\frac{y_2}{t_2}\right) - t_2 g\left(\frac{y_1}{t_1}\right) - t_2 J_g\left(\frac{y_1}{t_1}\right)\left(\frac{y_2}{t_2} - \frac{y_1}{t_1}\right) \in V. \tag{8.4}$$

From (8.3) we obtain

$$t_2 J_g\left(\frac{y_1}{t_1}\right)\left(\frac{y_2}{t_2}-\frac{y_1}{t_1}\right) = J_{\overline{g}}(y_1,t_1)\begin{bmatrix} y_2-y_1 \\ t_2-t_1 \end{bmatrix} - t_2 g\left(\frac{y_1}{t_1}\right) + t_1 g\left(\frac{y_1}{t_1}\right). \quad (8.5)$$

Substituting (8.5) in (8.4), we obtain

$$t_2 g\left(\frac{y_2}{t_2}\right) - t_2 g\left(\frac{y_1}{t_1}\right) - J_{\overline{g}}(y_1,t_1)\begin{bmatrix} y_2-y_1 \\ t_2-t_1 \end{bmatrix} + t_2 g\left(\frac{y_1}{t_1}\right) - t_1 g\left(\frac{y_1}{t_1}\right) \in V.$$

Hence

$$\overline{g}(y_2,t_2) - \overline{g}(y_1,t_1) - J_{\overline{g}}(y_1,t_1)\begin{bmatrix} y_2-y_1 \\ t_2-t_1 \end{bmatrix} \in V.$$

iii) It is a particular case of *ii)*.

iv) We have

$$\overline{d}(y,t) = t\cdot\left(d^T\left(\frac{y}{t}\right)+d_0\right) - 1 = d^T y + d_0 t - 1.$$

The affinity of $\overline{h}$ is obtained by the same argument. □

In view of the concluding remarks of the previous section and Lemma 3.1, the duality results proved in Section 2 can now be applied to $\overline{P}$, thus yielding the following dual problem:

$$\left\{\begin{array}{c} \max\ tZ(y,t) - t\delta - \lambda \\ t\nabla_y Z(y,t)^T(y_1-y) + [Z(y,t)-\delta+t\nabla_t Z(y,t)](t_1-t) \geq 0\ \forall (y_1,t_1)\in \Pi_X \\ (y,t)\in\Pi_A,\ \alpha_g\in V^+,\ \alpha_h\in \mathbb{R}^p,\ \lambda\in\mathbb{R},\ \delta\geq 0 \end{array}\right. \quad (8.6)$$

where $Z(y,t) = L_F(\frac{y}{t},\alpha_g,\alpha_h,\lambda)$. We are now left to show that problems D_F and (8.6) are equivalent. With this aim in mind, we first derive the following lemma.

Lemma 3.2 *The following conditions are equivalent:*

i) $\nabla_x L_F(x)(x_1-x) + [L_F(x)-\delta]\left(1-\frac{t}{t_1}\right) \geq 0\ \forall x_1\in X,\ \forall t_1>0,$

ii) $0\leq [\delta - L_F(x)] \leq \nabla_x L_F(x)(x_1-x)\ \forall x_1 \in X.$

Proof. *i)⇒ii)* Since *i)* holds $\forall t_1>0$, we have

$$\begin{aligned} 0 &\leq \lim_{t_1\to+\infty} \nabla_x L_F(x)(x_1-x) + [L_F(x)-\delta]\left(1-\frac{t}{t_1}\right) \\ &= \nabla_x L_F(x)(x_1-x) + [L_F(x)-\delta]. \end{aligned}$$

Suppose to the contrary that $[L_F(x)-\delta]>0$. Then $t>0$ implies

$$\lim_{t_1\to 0+} \nabla_x L_F(x)(x_1-x) + [L_F(x)-\delta]\left(1-\frac{t}{t_1}\right) = -\infty$$

which contradicts *i)*.

ii)⇒i) Since $t > 0$ and $t_1 > 0$, we have $\left(1 - \frac{t}{t_1}\right) < 1$, so that *ii)* implies

$$[\delta - L_F(x)]\left(1 - \frac{t}{t_1}\right) \le [\delta - L_F(x)] \le \nabla_x L_F(x)\,(x_1 - x)\ \ \forall x_1 \in X$$

and the result follows. □

Theorem 3.1 *Problems D_F and (8.6) are equivalent.*

Proof. Since

$$\nabla_y Z(y,t)^T = \frac{1}{t}\nabla_x L_F\left(\frac{y}{t}, \alpha_g, \alpha_h, \lambda\right)$$
$$\nabla_t Z(y,t) = -\frac{1}{t^2}\nabla_x L_F\left(\frac{y}{t}, \alpha_g, \alpha_h, \lambda\right) y$$

and using the notation $x = \frac{y}{t}$ and $x_1 = \frac{y_1}{t_1}$, the dual problem (8.6) can be rewritten as follows:

$$\begin{cases} \max\ tL_F(x) - t\delta - \lambda \\ \nabla_x L_F(x)\,(x_1 - x) + [L_F(x) - \delta]\left(1 - \frac{t}{t_1}\right) \ge 0\ \forall x_1 \in X,\ \forall t_1 > 0 \\ x \in A,\ t > 0,\ \alpha_g \in V^+,\ \alpha_h \in \mathbb{R}^p,\ \lambda \in \mathbb{R},\ \delta \ge 0. \end{cases} \tag{8.7}$$

From Lemma 3.2 problem (8.7) is equivalent to the following one:

$$\begin{cases} \max\ -t\,[\delta - L_F(x)] - \lambda \\ \nabla_x L_F(x)\,(x_1 - x) \ge [\delta - L_F(x)]\ \forall x_1 \in X \\ [\delta - L_F(x)] \ge 0 \\ x \in A,\ \alpha_g \in V^+,\ \alpha_h \in \mathbb{R}^p,\ t > 0,\ \lambda \in \mathbb{R},\ \delta \ge 0. \end{cases} \tag{8.8}$$

We can show that for any optimal solution of problem (8.8) we have $[\delta - L_F(x)] = 0$. Suppose to the contrary that there exists an optimal solution $(\tilde{x}, \tilde{\alpha}_g, \tilde{\alpha}_h, \tilde{\lambda}, \tilde{t}, \tilde{\delta})$ such that $\left[\tilde{\delta} - L_F(\tilde{x})\right] > 0$. Since $\tilde{t} > 0$, for any $0 < \epsilon < \tilde{t}$ the vector $(\tilde{x}, \tilde{\alpha}_g, \tilde{\alpha}_h, \tilde{\lambda}, \tilde{t} - \epsilon, \tilde{\delta})$ is feasible for problem (8.8) and it is better than $(\tilde{x}, \tilde{\alpha}_g, \tilde{\alpha}_h, \tilde{\lambda}, \tilde{t}, \tilde{\delta})$ which is a contradiction. Hence the result follows from (8.8) where $[\delta - L_F(x)] = 0$ and $\delta \ge 0$. □

In conclusion, it is worth mentioning that in the absence of a set constraint in P_F, i.e., $X = A = \mathbb{R}^n$, problem D_F coincides with the one already studied in the literature (see Jagannathan (1973); Schaible

(1973); Schaible (1976a); Schaible (1976b)), namely

$$\begin{cases} \max\ -\lambda \equiv -\min\ \lambda \\ L_F(x) \geq 0 \\ \nabla_x L_F(x) = 0 \\ x \in \mathbb{R}^n,\ \alpha_g \in V^+,\ \alpha_h \in \mathbb{R}^p,\ \lambda \in \mathbb{R}. \end{cases}$$

Acknowledgments

The authors wish to thank an anonimous referee for his valuable comments and suggestions which improved the presentation of the results.

References

Barros, A.I., Frenk, J.B.G., Schaible, S. and Zhang, S. (1996), Using duality to solve generalized fractional programming problems, *Journal of Global Optimization,* Vol. 8, pp. 139-170.

Barros A. I. (1998), *Discrete and Fractional Programming Techniques for Location Models,* Kluwer Academic Publishers, Dordrecht.

Bector, C.R. (1973), Duality in nonlinear fractional programming, *Zeitschrift fur Operations Research,* Vol. 17, pp. 183-193.

Bector, C.R., Bector, M.H. and Klassen, J.E. (1977), Duality for a nonlinear programming problem, *Utilitas Mathematicae,* Vol. 11, pp. 87-99.

Cambini R. (1996), Some new classes of generalized concave vector-valued functions, *Optimization,* Vol. 36, n. 1, pp. 11-24.

Cambini R. and S. Komlósi (1998), On the Scalarization of Pseudoconcavity and Pseudomonotonicity Concepts for Vector Valued Functions, in *Generalized Convexity, Generalized Monotonicity: Recent Results,* edited by J.-P. Crouzeix, J.-E. Martinez-Legaz and M. Volle, Nonconvex Optimization and Its Applications, Vol. 27, Kluwer Academic Publishers, Dordrecht, pp. 277-290.

Cambini, R. and Carosi, L. (2002), Duality in multiobjective optimization problems with set constraints, Report n. 233, Department of Statistics and Applied Mathematics, University of Pisa.

Chandra, S., Abha Goyal and Husain, I. (1998), On symmetric duality in mathematical programming with F-convexity, *Optimization,* Vol. 43, pp. 1-18.

Charnes, A. and Cooper, W.W. (1962), Programming with linear fractional functionals, *Naval Research Logistic Quarterly,* Vol. 9, pp. 181-196.

Craven, B.D. (1981), Duality for generalized convex fractional programs, in *Generalized Concavity in Optimization and Economics,* edited by S. Schaible and W.T. Ziemba, Academic Press, New York, pp. 473-489.

Crouzeix, J.P., Ferland, J.A. and Schaible, S. (1983), Duality in generalized linear fractional programming, *Mathematical Programming,* vol. 27, pp. 342-354.

Dinkelbach, W. (1967), On nonlinear fractional programming, *Management Science,* Vol. 13, pp. 492-498.

Frenk, J.B.G. and Kasssay, G. (1999), On classes of generalized convex functions, Gordan-Farkas type theorems and Lagrangean duality, *Journal of Optimization Theory and Applications,* Vol. 102, n. 2, pp. 315-343.

Geoffrion, A.M. (1971), Duality in nonlinear programming: a simplified applications-oriented development, *SIAM Review,* Vol. 12, pp. 1-37.

Giorgi, G. and Guerraggio, A. (1994), First order generalized optimality conditions for programming problems with a set constraint, in *Generalized Convexity,* edited by S. Komlósi, T. Rapcsák and S. Schaible, Lecture Notes in Economics and Mathematical Systems, Vol. 405, Springer-Verlag, Berlin, pp. 171-185.

Jagannathan, R. (1973), Duality for nonlinear fractional programs, *Zeitschrift fur Operations Research,* Vol. 17, pp. 1-3

Jahn, J. (1994), *Introduction to the Theory of Nonlinear Optimization,* Springer-Verlag, Berlin.

Liang, Z.A., Huang, H.X. and Pardalos, P.M. (2001), Optimality conditions and duality for a class of nonlinear fractional programming problems, *Journal of Optimization Theory and Applications,* Vol. 110, pp. 611-619.

Liu, J.C. (1996), Optimality and duality for generalized fractional programming involving nonsmooth pseudoinvex functions, *Journal of Mathematical Analysis and Applications,* Vol. 202, pp. 667-685.

Mahajan, D.G. and Vartak, M.N. (1977), Generalization of some duality theorems in nonlinear programming, *Mathematical Programming,* Vol. 12, pp. 293-317.

Mangasarian, O.L. (1969), *Nonlinear Programming,* McGraw-Hill, N.Y.

Mond, B. and Weir, T. (1981), Generalized concavity and duality, in *Generalized Concavity in Optimization and Economics,* edited by S. Schaible and W.T. Ziemba, Academic Press, New York, pp. 263-279.

Schaible, S. (1973), Fractional programming: transformations, duality and algorithmic aspects, Technical Report 73-9, Department of Operation Research, Stanford University, November 1973.

Schaible, S. (1974), Parameter-free convex equivalent and dual programs of fractional programming problems, *Zeitschrift fur Operations Research,* Vol. 18, pp. 187-196.

Schaible, S. (1976), Duality in fractional programming: a unified approach, *Operations Research,* Vol. 24, pp. 452-461.

Schaible, S. (1976), Fractional programming. I, duality, *Management Science,* Vol. 22, pp. 858-867.

Scott, C.H. and Jefferson, T.R. (1996), Convex dual for quadratic concave fractional programs, *Journal of Optimization Theory and Applications,* Vol. 91, pp. 115-122.

Chapter 9

ON THE PSEUDOCONVEXITY OF THE SUM OF TWO LINEAR FRACTIONAL FUNCTIONS

Alberto Cambini*
Department of Statistics and Applied Mathematics
University of Pisa - Italy

Laura Martein†
Department of Statistics and Applied Mathematics
University of Pisa - Italy

Siegfried Schaible‡
A. G. Anderson Graduate School of Management
University of California at Riverside - U. S. A.

Abstract Charnes and Cooper (1962) reduced a linear fractional program to a linear program with help of a suitable transformation of variables. We show that this transformation preserves pseudoconvexity of a function. The result is then used to characterize sums of two linear fractional functions which are still pseudoconvex. This in turn leads to a characterization of pseudolinear sums of two linear fractional functions.

Keywords: Fractional programming, sum of ratios, pseudoconvexity, pseudolinearity.

MSC2000: 26B25

---

* email:acambini@ec.unipi.it
† email:lmartein@ec.unipi.it
‡ email:siegfried.schaible@ucr.edu

1. Introduction

Fractional programming has often been studied in the context of generalized convex functions; see for example Martos (1975), Avriel et al (1988), Craven (1988). In a single-ratio linear fractional program the objective function is pseudoconvex. Hence a local is a global minimum. Furthermore a minimum is attained at an extreme point of a polyhedral convex feasible region since a linear fractional function is also pseudoconcave. These properties are valuable for solving such nonconvex minimization problems.

Linear fractional programs not only share the above two properties with linear programs. Each linear fractional program can also be directly related to a linear program with help of a suitable nonlinear transformation of variables proposed by Charnes and Cooper (1962).

Linear fractional functions are not only pseudoconvex, but also pseudoconcave; i.e., they are pseudolinear. Such functions have been analyzed extensively. For recent studies see for example Rapcsak (1991), Komlosi (1993), Jeyakumar and Yang (1995).

Many applications give rise to multi-ratio fractional programs; see for example Schaible (1995). The sum-of-ratios fractional program is a particular class of such problems. Compared with other multi-ratio problems, it is much more difficult to analyze and to solve. The current study focuses on generalized convexity properties of the sum of two linear fractional functions.

Coming from single-ratio linear fractional programming, a number of questions naturally arise. In the case of the sum of two linear fractional functions, is such a function still pseudoconvex or even pseudolinear? The answer to both questions is negative in general. In fact a local minimum is often not a global minimum and a minimum is often not attained at an extreme point of a polyhedral convex feasible region; see Schaible (1977).

Furthermore one could ask which role, if any, the Charnes-Cooper transformation plays in the analysis and solution of such problems. Cambini et al (1989) show that one of the two linear ratios can be reduced to a linear function. Pseudoconvexity of the resulting sum of a linear and linear fractional function is characterized in Cambini et al (2002). A more general question is whether the Charnes- Cooper transformation of variables preserves pseudoconvexity.

In Section 2 we show that indeed pseudoconvexity of a general function is preserved under the Charnes- Cooper transformation. This result is then applied in Section 3 to characterize a sum of two arbitrary linear fractional functions which is still pseudoconvex. Based on this characterization, a procedure for testing for pseudoconvexity is given in Section 4 and is illustrated by numerical examples. While Sections 3 and 4 deal with pseudoconvexity, Sections 5 and 6 present corresponding results for the pseudolinearity of the sum of two linear fractional functions.

2. Pseudoconvexity under the Charnes-Cooper transformation

The aim of this section is to show that pseudoconvexity is preserved by the Charnes-Cooper transformation of variables. Consider the following transformation $y = \frac{x}{b^T x + b_0}$ defined on the set $\Gamma = \{x \in \mathbb{R}^n : b^T x + b_0 > 0\}$ where $b \in \mathbb{R}^n$ and $b_0 \in \mathbb{R}$. This map is a diffeomorphism and its inverse is $x = \frac{b_0 y}{1 - b^T y}$ defined on the set $\Gamma^* = \{y \in \mathbb{R}^n : \frac{b_0}{1-b^T y} > 0\}$.

Let $f(x)$ be a twice differentiable real-valued function defined on an open subset of $\mathbb{R}^n$. Consider the function $\psi(y)$ obtained by applying the previous transformation to $f(x)$. Obviously we have $f(x(y)) = \psi(y)$ and $f(x) = \psi(y(x))$.

We introduce the following notations:
- $f'(x)$ ($\psi'(y)$), $f''(x)$ ($\psi''(y)$) is the gradient and the Hessian matrix of f (ψ), respectively;
- J is the Jacobian matrix of the transformation $y(x)$;
- $y_i''(x)$ is the Hessian matrix of the i-th component of the map $y(x)$, that is $y_i = \frac{x_i}{b^T x + b_0}$.

The relationships between the gradients and between the Hessian matrices of the functions f and ψ are expressed in the following theorem whose proof follows directly from differential calculus rules.

Theorem 2.1 *We have*
i) $f'(x) = J^T \psi'(y(x))$;
ii) $f''(x) = J^T \psi''(y(x)) J + \sum_{i=1}^n \psi_i'(y(x)) y_i''(x)$.

The following lemma shows the relationship between the gradient of f and $\sum_{i=1}^n \psi_i'(y(x)) y_i''(x)$.

Lemma 2.1 *We have* $\sum_{i=1}^n \psi_i'(y(x)) y''_i(x) = Z + Z^T$ *where* Z *is the* $n \times n$ *matrix whose* j*-th column is* $-\frac{b_j}{b^T x + b_0} f'(x)$.

Proof. It can be shown that for each i=1,..n, we have

$$\frac{\partial^2 y_i}{\partial x_k \partial x_j} = -\frac{b_j}{b^T x + b_0} \frac{\partial y_i}{\partial x_k} - \frac{b_k}{b^T x + b_0} \frac{\partial y_i}{\partial x_j}, \ \forall k, j \in \{1, .., n\}$$

so that the j-th column of the Hessian matrix $y_i''(x)$ is

$$-\frac{b_j}{b^T x + b_0} y_i'(x) - \frac{\partial y_i}{\partial x_j} \frac{b}{b^T x + b_0}.$$

As a consequence, the j-th column of the matrix $\sum_{i=1}^{n} \psi_i'(y(x))y_i''(x)$ is given by

$$\sum_{i=1}^{n} \psi_i'(y(x))[-\frac{b_j}{b^T x + b_0} y_i'(x) - \frac{\partial y_i}{\partial x_j} \frac{b}{b^T x + b_0}] =$$

$$= -\frac{b_j}{b^T x + b_0} \sum_{i=1}^{n} \psi_i'(y(x))y_i'(x) - \frac{b}{b^T x + b_0} \sum_{i=1}^{n} \psi_i'(y(x)) \frac{\partial y_i}{\partial x_j}$$

$$= -\frac{b_j}{b^T x + b_0} f'(x) - \frac{b}{b^T x + b_0} \sum_{i=1}^{n} \psi_i'(y(x)) \frac{\partial y_i}{\partial x_j}.$$

Since $-\frac{b_j}{b^T x + b_0} f'(x)$ is the j-th column of Z and

$$-\frac{b}{b^T x + b_0} \sum_{i=1}^{n} \psi_i'(y(x)) \frac{\partial y_i}{\partial x_j}$$

is the transpose of the j-th row of Z, the result follows. □

Now we are able to prove the main result of this section related to pseudoconvexity. We assume that the function f is defined on a convex set $S \subseteq \mathbb{R}^n$ and, consequently, ψ is defined on a convex set $S^* \subseteq \mathbb{R}^n$.

Theorem 2.2 *The function $f(x)$ is pseudoconvex if and only if the function $\psi(y)$ is pseudoconvex.*

Proof. We will prove that the pseudoconvexity of ψ implies the pseudoconvexity of the function f. The converse follows by noting that $\psi(y) = f(x(y))$ where the tranformation $x(y)$ is of the same kind as the tranformation $y(x)$.

It is known that a twice differentiable function g is pseudoconvex on an open convex set $X \subseteq \mathbb{R}^n$ if and only if the following two conditions hold (see Crouzeix (1998)):

$$z \in X \text{ and } d^T g'(z) = 0 \Rightarrow d^T g''(z)d \geq 0 \tag{9.1}$$

$$\bar{z} \in X \text{ and } g'(\bar{z}) = 0 \Rightarrow g(\bar{z}) \leq g(z) \ \forall z \in X. \tag{9.2}$$

Assume that $\psi(y)$ is pseudoconvex. Since $f'(x) = J^T \psi'(y(x))$ and J is a nonsingular matrix, we have $f'(x_0) = 0$ if and only if $\psi'(y(x_0)) = \psi'(y_0) = 0$. Since $\psi(y_0) \leq \psi(y) \ \forall y \in S^*$, we have $f(x_0) = \psi(y_0) \leq \psi(y(x)) = f(x) \ \forall x \in S$, so that f satisfies (9.2).

Let d be an orthogonal direction to $f'(x)$. We have $d^T f'(x) =$ $= d^T J^T \psi'(y) = (Jd)^T \psi'(y)$, so that $h = Jd$ is an orthogonal direction

to $\psi'(y)$. The pseudoconvexity of ψ implies $h^T\psi''(y)h \geq 0$. From ii) of Theorem 2.1 and from Lemma 2.1 we have $d^T f''(x)d = h^T\psi''(y)h + d^T Zd + d^T Z^T d$.
Taking into account that $d^T Z = d^T f'(x)(-\frac{b^T}{b^T x+b_0}) = 0 = Z^T d$, we have $d^T f''(x)d = h^T\psi''(y)h \geq 0$ and thus f satisfies (9.1). □

Taking into account that a function f is pseudoconcave if and only if $-f$ is pseudoconvex, we obtain the following corollary:

Corollary 2.1 *The function $f(x)$ is pseudoconcave if and only if the function $\psi(y)$ is pseudoconcave.*

3. Pseudoconvexity of the sum of two linear fractional functions

The results obtained in the previous section allow us to study the pseudoconvexity of the sum of two linear fractional functions. Applying the Charnes-Cooper transformation, this sum can be transformed into a sum of a linear and a linear fractional function (see Cambini et al (1989)). The pseudoconvexity of such a function has been characterized in the following theorem (see Cambini et al (2002); for related earlier results see Schaible (1977)).

Theorem 3.1 *Consider the function $z(x) = q^T x + \frac{d^T x+\alpha}{b^T x+\beta}$ on the set $X = \{x : b^T x + \beta > 0\}$ with $b, d, q \in \mathbb{R}^n$ and $\alpha, \beta \in \mathbb{R}$. $z(x)$ is pseudoconvex if and only if one of the following conditions holds:*
i) $q = kb,\ k \geq 0$;
ii) there is $t \in \mathbb{R}$ such that $d = tb$ and $\alpha \geq t\beta$.

Consider now the function

$$h(x) = \frac{m^T x + m_0}{p^T x + p_0} + \frac{c^T x + c_0}{a^T x + a_0}$$

defined on $X = \{x \in \mathbb{R}^n : p^T x + p_0 > 0,\ a^T x + a_0 > 0,\ p_0 \neq 0\}$ where $m, p, c, a \in \mathbb{R}^n$ and $m_0, p_0, c_0, a_0 \in \mathbb{R}$.
The following theorem characterizes the pseudoconvexity of the function h.

Theorem 3.2 *The function h is pseudoconvex if and only if one of the following conditions holds:*
case $p_0 > 0$
i) there exists $k \geq 0$ such that

$$m - \frac{m_0}{p_0} p = k(p_0 a - a_0 p); \tag{9.3}$$

ii) there exists $t \in \mathbb{R}^n$ *such that* $c_0 \geq ta_0$ *and*

$$p_0 c - c_0 p = t(p_0 a - a_0 p). \tag{9.4}$$

case $p_0 < 0$
i) there exists $k \leq 0$ *such that*

$$m - \frac{m_0}{p_0} p = k(p_0 a - a_0 p); \tag{9.5}$$

ii) there exists $t \in \mathbb{R}^n$ *such that* $c_0 \leq ta_0$ *and*

$$p_0 c - c_0 p = t(p_0 a - a_0 p). \tag{9.6}$$

Proof. Consider the Charnes-Cooper transformation $y = \frac{x}{p^T x + p_0}$ and its inverse $x = \frac{p_0 y}{1 - p^T y}$. The function $h(x)$ is transformed into the function

$$\psi(y) = (m - \frac{m_0}{p_0} p)^T y + \frac{m_0}{p_0} + \frac{(p_0 c - c_0 p)^T y + c_0}{(p_0 a - a_0 p)^T y + a_0},$$

and we have $a^T x + a_0 = \frac{(p_0 a - a_0 p)^T y + a_0}{1 - p^T y}$.

The assumption $p^T x + p_0 > 0$, $p_0 \neq 0$, implies $\frac{p_0}{1 - p^T y} > 0$, while $a^T x + a_0 > 0$ implies $\frac{(p_0 a - a_0 p)^T y + a_0}{1 - p^T y} > 0$. As a consequence, if $p_0 > 0$, then $1 - p^T y > 0$ and $(p_0 a - a_0 p)^T y + a_0 > 0$. Applying Theorem 3.1, we obtain the result.

On the other hand, if $p_0 < 0$, then $1 - p^T y < 0$ and $(p_0 a - a_0 p)^T y + a_0 < 0$. In order to apply Theorem 3.1, the denominator in $\psi(y)$ must be positive. This can be achieved by changing the sign of the numerator and the denominator of the linear fractional term in $\psi(y)$. Applying Theorem 3.1, condition i) becomes $m - \frac{m_0}{p_0} p = -\alpha(p_0 a - a_0 p)$, $\alpha \geq 0$, that is (9.5), while in condition ii) we have $-c_0 \geq t(-a_0)$, that is (9.6). □

Taking into account that a function is pseudoconcave if and only if its negative is pseudoconvex, we can characterize the pseudoconcavity of the function $h(x)$ with help of the previous theorem.

Corollary 3.1 *The function* h *is pseudoconcave if and only if one of the following conditions holds:*
case $p_0 > 0$
i) there exists $k \leq 0$ *such that*

$$m - \frac{m_0}{p_0} p = k(p_0 a - a_0 p); \tag{9.7}$$

ii) there exists $t \in \mathbb{R}^n$ *such that* $c_0 \leq ta_0$ *and*

$$p_0 c - c_0 p = t(p_0 a - a_0 p). \tag{9.8}$$

case $p_0 < 0$
i) there exists $k \geq 0$ *such that*

$$m - \frac{m_0}{p_0} p = k(p_0 a - a_0 p); \tag{9.9}$$

ii) there exists $t \in \mathbb{R}^n$ *such that* $c_0 \geq ta_0$ *and*

$$p_0 c - c_0 p = t(p_0 a - a_0 p). \tag{9.10}$$

Remark 3.1 *If i) and ii) of Theorem 3.2 and Corollary 3.1 hold with* $k = 0$ *and* $c_0 = ta_0$, *respectively, the function* $h(x)$ *reduces to a linear fractional function which is both pseudoconvex and pseudoconcave (see Martos (1975)).*

A particular case

Consider now the function $h(x)$ where $a = p$ and $a_0 \neq p_0$, that is

$$g(x) = \frac{m^T x + m_0}{p^T x + p_0} + \frac{c^T x + c_0}{p^T x + a_0}$$

with $p^T x + p_0 > 0$, $p_0 \neq 0$, $p^T x + a_0 > 0$. The results given in Theorem 3.2 can be specialized as follows.

Theorem 3.3 *The function* g *is pseudoconvex if and only if one of the following conditions holds:*
case $p_0 > 0$
i) there exists $k^* \in \mathbb{R}$ *such that* $m = k^* p$ *with* $k^* \geq \frac{m_0}{p_0}$ *if* $p_0 > a_0$ *and* $k^* \leq \frac{m_0}{p_0}$ *if* $p_0 < a_0$.
ii) there exists $t^* \in \mathbb{R}$ *such that* $c = t^* p$ *with* $c_0 \geq ta_0$ *if* $p_0 > a_0$ *and* $c_0 \leq ta_0$ *if* $p_0 < a_0$.
case $p_0 < 0$
i) there exists $k^* \in \mathbb{R}$ *such that* $m = k^* p$ *with* $k^* \leq \frac{m_0}{p_0}$ *if* $p_0 > a_0$ *and* $k^* \geq \frac{m_0}{p_0}$ *if* $p_0 < a_0$.
ii) there exists $t^* \in \mathbb{R}$ *such that* $c = t^* p$ *with* $c_0 \leq ta_0$ *if* $p_0 > a_0$ *and* $c_0 \geq ta_0$ *if* $p_0 < a_0$.

As a direct consequence of Theorem 3.3 we obtain the following canonical form for the pseudoconvexity of g.

Corollary 3.2 *The function $g(x)$ is pseudoconvex if and only if it can be rewritten in the following way:*

$$g(x) = \frac{\alpha}{p^T x + p_0} + \frac{c^T x + c_0}{p^T x + a_0} + \beta$$

where $\alpha > 0$ if $p_0 < a_0$ and $\alpha < 0$ if $p_0 > a_0$, $\beta \in \mathbb{R}$.

4. An algorithm to test for pseudoconvexity

The results obtained in the previous section allow us to introduce the following algorithm to test for pseudoconvexity of the function h.

STEP 0: Calculate $m^* = m - \frac{m_0}{p_0}p$, $a^* = p_0 a - a_0 p$. If $p_0 > 0$ go to STEP 1; otherwise go to STEP 3.

STEP 1: If $m^* = ka^*$, $k \geq 0$, STOP: h is pseudoconvex; otherwise go to STEP 2.

STEP 2: Calculate $c^* = p_0 c - c_0 p$. If c^* and a^* are linearly independent, STOP: h is not pseudoconvex; otherwise let t be such that $c^* = ta^*$. If $c_0 \geq ta_0$, STOP: h is pseudoconvex; otherwise h is not pseudoconvex.

STEP 3: If $m^* = ka^*$, $k < 0$, STOP: h is pseudoconvex; otherwise go to STEP 4.

STEP 4: Calculate $c^* = p_0 c - c_0 p$. If c^* and a^* are linearly independent, STOP: h is not pseudoconvex; otherwise let t be such that $c^* = ta^*$. If $c_0 \leq ta_0$, STOP: h is pseudoconvex; otherwise h is not pseudoconvex.

Example 4.1 *(case $p_0 > 0$)*
Consider the function

$$h(x_1, x_2) = \frac{x_1 + 2x_2 + 5}{3x_1 + 4x_2 + 8} + \frac{c_1 x_1 + c_2 x_2 + c_0}{7x_1 + 12x_2 + 28}.$$

Step 0. We have $m^* = m - \frac{m_0}{p_0}p = (-\frac{7}{8}, -\frac{1}{2})$, $a^* = p_0 a - a_0 p =$ (–28,–16). Since $p_0 = 8 > 0$, go to Step 1.
Step 1. We have $k = \frac{1}{32} > 0$. Hence the function is pseudoconvex for all $c_1, c_2, c_0 \in \mathbb{R}$.

Example 4.2 *(case $p_0 < 0$)*
Consider the function

$$h(x_1, x_2) = \frac{x_1 + 5x_2 + 13}{2x_1 + x_2 - 1} + \frac{x_1 - x_2 + 1}{x_1 + 2x_2 + 4}.$$

Step 0. *We have* $m^* = m - \frac{m_0}{p_0}p = (27, 18)$, $a^* = p_0 a - a_0 p = (-9, -6)$. *Since* $p_0 = -1 < 0$, *go to Step 3.*
Step 3. *We have* $k = -3 < 0$. *Hence the function is pseudoconvex.*

Note that we can also apply the Charnes-Cooper transformation $y = \frac{x}{x_1+2x_2+4}$; *in such a case we have* $m = (1, -1)$, $m_0 = 1$, $p = (1, 2)$, $p_0 = 4$, $a = (2, 1)$, $a_0 = -1$, $c = (1, 5)$, $c_0 = 13$. *Hence* $m^* = (\frac{3}{4}, -\frac{3}{2})$, $a^* = (9, 6)$. *Since* m^* *and* a^* *are linearly independent, we calculate* $c^* = p_0 c - c_0 p = (-9, -6)$. *We have* $c^* = -a^*$ *so that (9.4) holds with* $t = -1$ *and furthermore* $c_0 = 13 > ta_0 = 1$. *Consequently* h *is pseudoconvex.*

Example 4.3 *Consider the function*

$$h(x_1, x_2) = \frac{6x_1 + 9x_2 + 14}{2x_1 + 3x_2 + 5} + \frac{5x_1 + 7x_2 + 1}{2x_1 + 3x_2 + 2}.$$

The function can be rewritten in the following way

$$h(x_1, x_2) = 3 - \frac{1}{2x_1 + 3x_2 + 5} + \frac{5x_1 + 7x_2 + 1}{2x_1 + 3x_2 + 2}.$$

Referring to Corollary 3.2, we have $\alpha = -1 < 0$, $p_0 = 5 > a_0 = 2$. *Hence* $h(x_1, x_2)$ *is pseudoconvex.*

5. Pseudolinearity of the sum of two linear fractional functions

The results obtained in the previous section allow us to characterize the pseudolinearity of the function $h(x)$ of Section 3.

Theorem 5.1 *The function* $h(x)$ *is pseudolinear if and only if one of the following conditions holds:*
i) $h(x)$ *is a linear fractional function;*
ii) there exists $k > 0$ *such that* $m - \frac{m_0}{p_0}p = k(p_0 a - a_0 p)$ *and there exists* $t \in \mathbb{R}$ *such that* $c_0 < ta_0$ *and* $p_0 c - c_0 p = t(p_0 a - a_0 p)$;
iii) there exists $k < 0$ *such that* $m - \frac{m_0}{p_0}p = k(p_0 a - a_0 p)$ *and there exists* $t \in \mathbb{R}$ *such that* $c_0 > ta_0$ *and* $p_0 c - c_0 p = t(p_0 a - a_0 p)$.

Proof. The function $h(x)$ is pseudolinear if and only if it satisfies the conditions given in Theorem 3.2 and in Corollary 3.1. If $k = 0$ in (9.3) or $c_0 = ta_0$ in (9.4), then $h(x)$ reduces to a linear fractional function.
Assertion (ii) follows by noting that this condition is equivalent to condition i) of Theorem 3.2 which ensures the pseudoconvexity of $h(x)$

and to condition ii) of Corollary 3.1 which ensures the pseudoconcavity of $h(x)$.
Analogously, assertion iii) is equivalent to condition ii) of Theorem 3.2 and to condition i) of Corollary 3.1. □

In the particular case $a = p$, $p_0 \neq a_0$, we have the following theorem.

Theorem 5.2 *Consider the function*

$$g(x) = \frac{m^T x + m_0}{p^T x + p_0} + \frac{c^T x + c_0}{p^T x + a_0}$$

with $p^T x + p_0 > 0$, $p_0 \neq 0$, $p^T x + a_0 > 0$, $p_0 \neq a_0$.
The function $g(x)$ *is pseudolinear if and only if* $m = k^* p$, $c = t^* p$ *and* $m_0 - k^* p_0$, $c_0 - t^* a_0$ *have the same sign.*

The previous theorem allows us to obtain a canonical form for the pseudolinearity of the function g.

Corollary 5.1 *Consider the function*

$$g(x) = \frac{m^T x + m_0}{p^T x + p_0} + \frac{c^T x + c_0}{p^T x + a_0}$$

with $p^T x + p_0 > 0$, $p_0 \neq 0$, $p^T x + a_0 > 0$, $p_0 \neq a_0$.
The function $g(x)$ *is pseudolinear if and only if it can be reduced to the form*

$$g(x) = \frac{A}{p^T x + p_0} + \frac{B}{p^T x + a_0} + \beta$$

where $A, B \in \mathbb{R}$ *have the same sign.*
In particular $g(x)$ *is convex if* $A, B > 0$ *and it is concave if* $A, B < 0$.

6. An algorithm to test for pseudolinearity

The results obtained in the previous section allow us to introduce the following algorithm to check for pseudolinearity of the function h.

Step 0: If (m, m_0) and (p, p_0) are linearly dependent or (c, c_0) and (a, a_0) are linearly dependent, STOP: h is pseudolinear; otherwise calculate $m^* = m - \frac{m_0}{p_0} p$, $a^* = p_0 a - a_0 p$, $c^* = p_0 c - c_0 p$ and go to STEP 1.

Step 1: If m^* and a^* are linearly independent or c^* and a^* are linearly independent, STOP: h is not pseudolinear; otherwise $m^* = k a^*$, and $c^* = t a^*$. If $k > 0$, go to STEP 2; if $k < 0$, go to STEP 3.

Step 2: If $c_0 < ta_0$, STOP: h is pseudolinear; otherwise h is not pseudolinear.

Step 3: If $c_0 > ta_0$, STOP: h is pseudolinear; otherwise h is not pseudolinear.

Example 6.1 *Consider the function*

$$h(x_1, x_2) = \frac{22x_1 + 5x_2 - 4}{4x_1 + x_2 - 2} + \frac{-9x_1 - 2x_2 + 1}{x_1 + 3}.$$

Step 0. We have $m = (22, 5)$, $m_0 = -4$, $p = (4, 1)$, $p_0 = -2$, $c = (-9, -2)$, $c_0 = 1$, $a = (1, 0)$, $a_0 = 3$. *Since* (m, m_0), (p, p_0) *are linearly independent like* (c, c_0), (a, a_0), *we calculate* $m^* = m - \frac{m_0}{p_0}p = (14, 3)$, $a^* = p_0 a - a_0 p = (-14, -3)$, $c^* = p_0 c - c_0 p = (14, 3)$. *Go to Step 1.*
Step 1. We have $m^* = ka^*$ *with* $k = -1 < 0$. *Go to Step 3.*
Step 3. We have $t = -1 < 0$, $c_0 = 1 > ta_0 = -1$ *and thus the function is pseudolinear.*

Example 6.2 *Consider the function*

$$h(x_1, x_2) = \frac{10x_1 + 14x_2 + 23}{5x_1 + 7x_2 + 8} + \frac{20x_1 + 28x_2 + -45}{5x_1 + 7x_2 - 12}.$$

The function can be rewritten in the following form

$$h(x_1, x_2) = \frac{7}{5x_1 + 7x_2 + 8} + \frac{3}{5x_1 + 7x_2 - 12} + 6.$$

Referring to Corollary 5.1, we have A = 7, B = 3. Hence $h(x_1, x_2)$ *is pseudolinear and, in particular, it is convex.*

References

Avriel M., Diewert W. E., Schaible S. and Zang I., *Generalized concavity,* Plenum Press, New York, 1988.

Cambini A. and Martein L., *A modified version of Martos's algorithm for the linear fractional problem,* Methods of Operations Research, 53, 1986, 33-44.

Cambini A., Crouzeix J.P. and Martein L., *On the pseudoconvexity of a quadratic fractional function,* Optimization, 2002, vol. 51 (4), 677-687.

Cambini A., Martein L. and Schaible S., *On maximizing a sum of ratios,* J. of Information and Optimization Sciences, 10, 1989, 65-79.

Cambini A. and Martein L., *Generalized concavity and optimality conditions in vector and scalar optimization,* "Generalized convexity" (Komlosi et al. eds.), Lect. Notes Econom. Math. Syst., 405, Springer-Verlag, Berlin, 1994, 337-357.

Cambini R. and Carosi L., *On generalized convexity of quadratic fractional functions,* Technical Report n.213, Dept. of Statistics and Applied Mathematics, University of Pisa, 2001.

Charnes A. and Cooper W. W., *Programming with linear fractional functionals,* Nav. Res. Logist. Quart., 9, 1962, 181-196.

Craven B. D., *Fractional programming,* Sigma Ser. Appl. Math. 4, Heldermann Verlag, Berlin, 1988.

Crouzeix J.P., *Characterizations of generalized convexity and monotonicity, a survey,* Generalized convexity, generalized monotonicity (Crouzeix et al. eds.), Kluwer Academic Publisher, Dordrecht, 1998, 237-256.

Jeyakumar V. and Yang X. Q., *On characterizing the solution sets of pseudolinear programs,* J. Optimization Theory Appl., 87, 1995, 747-755.

Komlosi S., *First and second-order characterization of pseudolinear functions,* Eur, J. Oper. Res., 67, 1993, 278-286.

Martos B., *Nonlinear programming theory and methods,* North-Holland, Amsterdam, 1975.

Rapcsak T., *On pseudolinear functions,* Eur. J. Oper. Res., 50, 1991, 353-360.

Schaible S., *A note on the sum of a linear and linear-fractional function,* Nav. Res. Logist. Quart., 24, 1977, 691-693.

Schaible S., *Fractional programming,* Handbook of global optimization (Horst and Pardalos eds.), Kluwer Academic Publishers, Dordrecht, 1995, 495-608.

Chapter 10

BONNESEN-TYPE INEQUALITIES AND APPLICATIONS

A. Raouf Chouikha*
Université Paris 13
France

Abstract In this paper we discuss about conjectures ennounced in A.R. Chouikha (1999) and produce significant examples to underline the interest of the problem.

Keywords: Plane isoperimetric inequalities, Bonnesen inequality, Polygons, Pseudo-perimeter.

MSC2000: 51M10, 51M25, 52A402

1. Introduction

For a simple closed curve C (in the euclidian plane) of length L enclosing a domain of area A, inequalities of the form

$$L^2 - 4\pi A \geq K, \tag{10.1}$$

are called Bonnesen-type isoperimetric inequalities if equality is only attained for the euclidean circle. In the other words, K is positive and satisfies the condition

$$K = 0 \quad \text{implies} \quad L^2 - 4\pi A = 0.$$

Let an n-gon Π_n (a polygon with n sides of length $a_1, a_2, ..., a_n$) of perimeter L_n and area A_n. Consider the so called pseudo-perimeter

*chouikha@math.univ-paris13.fr

of second kind $\hat{L}_n$ defined by

$$\hat{L}_n = L_n(\frac{n}{n-2})[(1-\frac{2a_1}{L_n})(1-\frac{2a_2}{L_n})(1-\frac{2a_3}{L_n}).....(1-\frac{2a_n}{L_n})]^{1/n}, \quad (10.2)$$

and the ratios

$$\nu_n = (\frac{L_n}{\hat{L}_n})^{\frac{n}{2}-2}, \qquad \tau_n = \frac{4(n\tan\frac{\pi}{n})A_n}{L_n^2}\,(\nu_n)^{\frac{n}{n-4}}. \quad (10.3)$$

In A.R. Chouikha (1999) we proposed the following

Conjecture : *For any* **n-gon** Π_n*, we have the inequalities*

$$1 \le \tau_n \le \nu_n,$$

with $1 = \tau_n = \nu_n$ *if and only if* Π_n*, is regular.*

More generally, we may ask

Problem 1.1 *Let us consider a piecewise smooth closed curve C in the euclidean plane, of length L and area A. Let* $(\Pi_n)_n$ *be a sequence of* **n-gons** *approaching C.* L_n*,* $\hat{L}_n$ *and* A_n *are respectively the perimeter, the pseudo-perimeter and the area of* Π_n*. Supposing that* $\hat{L} = \lim_{n\to\infty}\hat{L}_n$ *exists, do we have the Bonnesen-type inequality*

$$\hat{L}^2 - 4\pi A \ge 0\ ?$$

These questions seem difficult to resolve with classical methods. Nevertheless, in using *Mathematica* we are able to give significant examples illustrating the interest of these problems.

Let C be a closed convex curve in the plane. Let R is the circumradius and r is the inradius of the curve. We get an isoperimetric inequality known as the (classical) Bonnesen inequality:

$$L^2 - 4\pi A \ge \pi^2(R-r)^2. \quad (10.4)$$

Note that if the right side of (10.4) equals zero, then $R = r$. This means that C is a circle and $L^2 - 4\pi A = 0$.

(See R. Osserman (1979) for a general discussion and different generalisations).

For an **n-gon** Π_n (a polygon with n sides) of perimeter L_n and area A_n, the following inequality is known

$$L_n^2 - 4(n \tan \frac{\pi}{n}) A_n \geq 0. \tag{10.5}$$

Equality is attained if and only if the n-gon is regular. Thus, if we consider a smooth curve as a polygon with infinitely many sides, it appears that inequality

$$L^2 - 4\pi A \geq 0$$

is a limiting case of (10.5).

2. Isoperimetric constants

We can ask if it is possible to get an analogous formula for other plane polygons (not necessarilly inscribed in a circle). More precisely, is the area A_n of the n-gon is close to the following expression

$$P_n = \frac{L_n^2}{4}\sqrt{(1 - \frac{2a_1}{L_n})(1 - \frac{2a_2}{L_n})(1 - \frac{2a_3}{L_n}).....(1 - \frac{2a_n}{L_n})}\ ? \tag{10.6}$$

This question has been considered by many geometers who tried to compare A_n with P_n. One of them, P. Levy (1966) was interested in this problem and more precisely he expected the following

Conjecture 2.1 *Define the ratio* $\varphi_n = \dfrac{A_n}{P_n}$. *For any n-gon* Π_n, *with sides* $a_1, a_2, ..., a_n$ *enclosing an area* A_n, *and* P_n *defined as above, this ratio verifies*

$$\text{a)}\ \ \frac{e}{\pi} \leq \varphi_n \qquad \textit{and} \qquad \textit{b)}\ \ \varphi_n \leq 1.$$

For regular n-gons we get $a_i = L_n/n$. The associated value of φ_n is given by

$$\varphi_n{}^0 = \frac{1}{n \tan \frac{\pi}{n}\ (1 - \frac{2}{n})^{n/2}} \tag{10.7}$$

and satisfies the inequalities of Conjecture 2.1. Moreover, it allows one to estimate the defect between any n-gon Π_n and the regular one. This defect may be measured by the quotient

$$\tau_n = \frac{\varphi_n}{\varphi_n{}^0} \tag{10.8}$$

which tends to 1 whenever Π_n is close to being regular. Moreover, τ_n is related to a new Bonnesen-type inequality for plane polygons.

Consider now the so called pseudo-perimeter of second kind $\hat{L}_n$ introduced by (H.T. Ku, M.C. Ku, X.M. Zhang (1995)), and defined by

$$\hat{L}_n = L_n(\frac{n}{n-2})[(1-\frac{2a_1}{L_n})(1-\frac{2a_2}{L_n})(1-\frac{2a_3}{L_n}).....(1-\frac{2a_n}{L_n})]^{1/n}. \quad (10.9)$$

They proposed the following

Conjecture 2.2 *For any cyclic n-gon Π_n, we have*

$$\hat{L}_n^2 - 4(n\tan\frac{\pi}{n})A_n \geq 0.$$

Equality holds if and only if Π_n is regular.

For any n-gon Π_n, we have the natural inequality $\hat{L}_n \leq L_n$. The equality $\hat{L}_n = L_n$ holds if and only if Π_n is regular (see Lemma (4-6) of A.R. Chouikha (1988)).

More generally, we need to introduce the following ratio

$$\nu_n = (\frac{L_n}{\hat{L}_n})^{\frac{n}{2}-2}. \quad (10.10)$$

We proved the following results (A.R. Chouikha (1988)), which give another more general Bonnesen-type inequality

Theorem 2.1 *Let $\tau_n = \frac{\varphi_n}{\varphi_n{}^0}$ and $\nu_n = (\frac{L_n}{\hat{L}_n})^{\frac{n}{2}-2}$ be the constants associated to any cyclic n-gon Π_n, with sides $a_1, a_2, ..., a_n$. L_n and $\hat{L}_n$ are respectively the perimeter and the pseudo-perimeter. We then have*
(i) The inequality $\tau_n \leq 1$ implies conjecture 2.1 b) and conjecture 2.2. Moreover, this implication is strict.
(ii) The inequalities $1 \leq \tau_n \leq \nu_n$ imply conjecture 2.1 a) and conjecture 2.2.
(iii) The inequality $\nu_n < \tau_n$ contradicts conjecture 2.2.
In these three cases, equality $1 = \tau_n = \nu_n$ holds if and only if Π_n is regular.

Corollary 2.1 *Suppose $\tau_n \leq 1$ is verified by a cyclic n-gon; we then have the following Bonnesen-type isoperimetric inequality:*

$$\hat{L}_n^2 - 4(n\tan\frac{\pi}{n})A_n \;\geq\; \hat{L}_n^2(1-\frac{\varphi_n}{\varphi_n{}^0}).$$

Equality holds if and only if Π_n is regular (i.e., $\varphi_n = \varphi_n{}^0$). Moreover, this inequality implies Conjecture 2.2.

Thus, the preceding results lead to the more general conjecture (A.R. Chouikha (1988)).

Thus, it is natural to expect that the hypothesis (ii) of Theorem 2.1 is verified for any cyclic n-gon. We then may propose the following

Conjecture 2.3 *For any n-gon Π_n, we have the inequalities*

$$1 \leq \tau_n \leq \nu_n,$$

with $1 = \tau_n = \nu_n$ if and only if Π_n is regular.

Obviously, this implies Conjecture 2.2 and Conjecture 2.1 a). Thus, Conjecture 2.3 appears to be more significant than the previous conjectures. Notice that by Theorem 2.1, $\nu_n = 1 \Rightarrow \tau_n = 1$.

3. Description of examples

In this part, we shall see that Hypothesis (ii) of Theorem 2.1, which implies Conjecture 2.2, is in fact verified by many instructive examples.

3.1 Example 1

Let us consider the Macnab polygon, which is a cyclic equiangular alternate-sided $2n$-gon with n sides of length a and n sides of length b. We showed that this polygon verified Conjecture 2.3. Indeed, we get

Proposition 3.1 *Let $\Pi_{n,n}$ be a cyclic $2n$-gon with n sides of length a alternatively with n sides of length b and $\varphi_{n,n}$ its associated function. Then, we have*

$$1 \leq \frac{\varphi_{n,n}}{\varphi^0} \leq \left(\frac{L_{n,n}}{\hat{L}_{n,n}}\right)^{\frac{n}{2}-2}.$$

3.2 Example 2

Let Π_n^0 denote the regular n-gon whose sides a_i^0 are subtended by angles $\frac{\pi}{n}; i = 1, ..., n$. Consider a polygon Π_n^ε obtained from Π_n^0 by variations of a_1, a_2 which are subtended respectively by $\frac{\pi}{n} - \varepsilon$ and $\frac{\pi}{n} + \varepsilon$. The other sides of length $a_i^0 (3 \leq i \leq n)$ are unchanged. We prove that hypothesis (ii) is verified by Π_n^ε.

Proposition 3.2 *Let Π_n^ε be the n-gon defined above for $n \geq 4$, φ_n^ε being its associated function. Then, for $\varepsilon > 0$ small, we have $1 \leq \frac{\varphi_n^\varepsilon}{\varphi_n^0} \leq (\frac{L_{n,n}}{\hat{L}_{n,n}})^{\frac{n}{2}-2}$.*

Thus, it seems that the function $\varphi(a_1, a_2, \ldots, a_n)$ for an n-gon possesses a local minimum for the regular polygons.

Proof. Let $L_n^\varepsilon, \hat{L}_n^\varepsilon, A_n^\varepsilon$ be respectively perimeter, pseudo-perimeter and enclosing area of the polygon Π_n^ε defined above. We get $a_1 = 2R\sin(\frac{\pi}{n} - \varepsilon)$ and $a_2 = 2R\sin(\frac{\pi}{n} + \varepsilon)$. After calculation, we obtain the following expression

$$A_n^\varepsilon = A_n^0(1 - \frac{4\varepsilon^2}{n}) \quad \text{and} \quad L_n^\varepsilon = L_n^0(1 - \frac{\varepsilon^2}{n}).$$

On the other hand,

$$\begin{aligned}(1 - \frac{2a_1}{L_n^\varepsilon})(1 - \frac{2a_2}{L_n^\varepsilon}) &= 1 - \frac{4}{n}[1 - \varepsilon^2(\frac{1}{2} - \frac{1}{n})] + \frac{4R^2[\sin^2\frac{\pi}{n} - \varepsilon^2)}{(L_n^0)^2(1 - \frac{\varepsilon^2}{n})} \\ &= (1 - \frac{2}{n})^2[1 + \frac{4\varepsilon^2}{(n-2)^2}(\frac{n-2}{2} + \frac{2}{n} - \frac{1}{\sin^2\frac{\pi}{n}})].\end{aligned}$$

Also, we get

$$(1 - \frac{2a_i^0}{L_n^\varepsilon}) = (1 - \frac{2}{n})(1 - \frac{2\varepsilon^2}{n(n-2)}).$$

After simplification, we find the expression

$$\frac{\varphi_n^\varepsilon}{\varphi_n^0} = 1 - \frac{2\varepsilon^2}{(n-2)^2}[\frac{(n-2)^2}{2n} + \frac{n-2}{2} + \frac{2}{n} - \frac{1}{\sin^2\frac{\pi}{n}}],$$

which verifies $\frac{\varphi_n^\varepsilon}{\varphi_n^0} \geq 1$.
Notice that the factor ε^2 vanishes for $n = 4$.
From the expression

$$(\frac{L_{n,n}}{\hat{L}_{n,n}})^{\frac{n}{2}-2} = 1 - \frac{2(n-4)}{n(n-2)^2}\varepsilon^2[-\frac{(n-2)^2}{2n} + \frac{n-2}{2} + \frac{2}{n} - \frac{1}{\sin^2\frac{\pi}{n}}],$$

we also prove that

$$\frac{\varphi_n^\varepsilon}{\varphi_n^0} \leq (\frac{L_{n,n}}{\hat{L}_{n,n}})^{\frac{n}{2}-2}.$$

□

4. Other interesting examples

4.1 Example 3

Also, P. Levy tried to find these bounds and tested Conjecture 2.1 on a special curve polygon denoted by $\Pi(\alpha)$, inscribed in the Euclidean circle of radius 1. It is bounded by a circular arc with length $2(\pi-\alpha)$, and a chord of length $l = 2\sin\alpha$, where $0 \le \alpha \le \pi$. $\Pi(\alpha)$ can be considered as limit of an $(n+1)$-gon with n sides of length $2\sin\frac{2\pi}{n}$ while only one has a fixed length $l = 2\sin\alpha$. Let $\varphi_n(\alpha)$ be the corresponding ratio and $\varphi(\alpha)$ its limit value when n tends to infinity. In this case, ${\varphi^0}_\infty = \frac{e}{\pi}$ is the limit value of ${\varphi^0}_n = \frac{1}{n\tan\frac{\pi}{n}\,(1-\frac{2}{n})^{n/2}}$.

We get the following

Proposition 4.1 *Let $L(\alpha)$, $\hat{L}\alpha)$, $A(\alpha)$ be respectively the perimeter, the pseudo-perimeter and the enclosing area of the "polygon" $\Pi(\alpha)$, with $0 \le \alpha \le \pi$. We then obtain the inequalities*
a) $1 \le \frac{\varphi(\alpha)}{\varphi^0} \le \frac{\pi}{e}\sqrt{\frac{e}{3}}$ *with* $\varphi(\pi) = \frac{e}{\pi}$ *and* $\varphi(0) = \sqrt{\frac{e}{3}}$.
b) $\hat{L}^2(\alpha) - 4\pi A(\alpha) \ge 0$.
Equality holds if and only if $\alpha = 0$.

Thanks to *Mathematica* we shall show that $\Pi(\alpha)$ verifies Conjecture 2.1 and Problem 1.1 stated in Introduction.
Proof. We may calculate the exact value of the function $\varphi(\alpha)$. We refer for that to P. Levy (1966) and A.R. Chouikha (1988) for details. Here $L = 2(\alpha + \sin\alpha)$ and $A = \alpha - \sin\alpha\cos\alpha$ so that

$$\varphi(\alpha) = \frac{(\pi-\alpha-\sin\alpha\cos\alpha)}{(\pi-\alpha+\sin\alpha)^{\frac{3}{2}}\sqrt{\pi-\alpha-\sin\alpha}}e^{\frac{\pi-\alpha}{\pi-\alpha+\sin\alpha}}$$

$$\tau = \frac{\pi(\pi-\alpha-\sin\alpha\cos\alpha)e^{\frac{\sin\alpha}{\pi-\alpha+\sin\alpha}}}{(\pi-\alpha+\sin\alpha)^{\frac{3}{2}}\sqrt{\pi-\alpha-\sin\alpha}},$$

and

$$\nu = \sqrt{\frac{(\pi-\alpha+\frac{1}{2}\sin\alpha)}{(\pi-\alpha+\sin\alpha)}}e^{\frac{\pi-2\alpha-\sin\alpha}{\alpha+\sin\alpha}}.$$

Thus, for $0 \le \alpha \le \pi$ we obtain the double inequality

$$\frac{e}{\pi} \le \varphi(\alpha) \le \sqrt{\frac{e}{3}}.$$

These inequalities may be verified by *Mathematica.* On the other hand, we may also deduce the expression $\frac{4\pi A}{\hat{L}^2}$ in terms of α :

$$\frac{4A}{\hat{L}^2} = \frac{(\pi - \alpha - \frac{1}{2}\sin 2\alpha)}{(\pi - \alpha + \sin\alpha)^2}.$$

We can prove easily that the right side of the above expression is a decreasing function of α, and for $\alpha = 0$, its value is 1. We then obtain part b) of the Proposition. □

4.2 Example 4

P. Levy considered also another curvilinear polygon. Let us denote by $\Pi(\alpha, \theta)$ the polygon obtained from $\Pi(\alpha)$ by replacing the side with length $l = 2\sin\alpha$ by two sides. One of them has a length $2\sin\theta$. Then we get the expression of the perimeter and the area of the new polygon $\Pi(\alpha, \theta)$:

$$\begin{aligned} L(\alpha, \theta) &= 2[\pi - \alpha + \sin\theta + \sin(\alpha + \theta)], \\ A(\pi - \alpha, \theta) &= \pi - \alpha + \sin\alpha\cos(\alpha + 2\theta), \\ & 0 \le \theta \le \alpha. \end{aligned}$$

For $\theta = 0$ we get of course, $\Pi(\alpha, 0) \equiv \Pi(\alpha)$.

Proposition 4.2 *Let $L(\alpha,\theta)$, $\hat{L}(\alpha,\theta)$, $A(\alpha,\theta)$ be respectively the perimeter, the pseudo-perimeter and the enclosing area of the "polygon" $\Pi(\alpha,\theta)$, with $0 \le \alpha \le \pi$, and $0 \le \theta \le \pi - \alpha$. We then obtain the inequalities*

a) $\varphi(\alpha, \theta_0) \le \varphi(\alpha, \theta) \le \varphi(\alpha, \frac{\pi-\alpha}{2}) \le 1$ *for certain* $\theta_0 > 0$.

b) $1 \le \frac{\varphi(\alpha, \frac{\pi-\alpha}{2})}{\varphi^0} \le \frac{\pi}{e}$ *with* $\varphi(\pi, 0) = 1$ *and* $\varphi(0, \frac{\pi}{2}) = \frac{\pi}{e}$.

c) $\hat{L}^2(\alpha, \frac{\pi-\alpha}{2}) - 4\pi A(\alpha, \frac{\pi-\alpha}{2}) \ge 0$. *Equality holds if and only if* $\alpha = \pi$.

Proof. We calculate the following expression for the function $\varphi(\alpha, \theta)$ defined above

$$\varphi(\alpha,\theta) = \frac{(\alpha - \sin\alpha\cos(\alpha + 2\theta))\, e^{\frac{\alpha}{[\alpha + \sin\theta + \sin(\alpha+\theta)]}}}{[\alpha + \sin\theta + \sin(\alpha + \theta)]\sqrt{\alpha^2 - 4\sin^2\frac{\alpha}{2}\cos^2(\frac{\alpha}{2} + \theta)}}.$$

We may verify that for $0 \le \alpha \le \pi$ we have $\varphi(\alpha, \theta)$ admits a maximum $\theta_0 = \frac{\pi-\alpha}{2}$ and two minima θ_1, θ_2 symmetric with respect to θ_0, such that $\varphi(\alpha, \theta_1) = \varphi(\alpha, \theta_2)$.

Moreover, we may prove that $\frac{\varphi(\alpha,\frac{\pi-\alpha}{2})}{\varphi^0}$ is a decreasing function, $\varphi(\pi,0)=1$, and $\varphi(0,\frac{\pi}{2})=\frac{\pi}{e}$.

Furthermore, after simplifying the expression $\frac{4\pi A(\alpha,\frac{\pi-\alpha}{2})}{\hat{L}^2(\alpha,\frac{\pi-\alpha}{2})}$ we find the following :

$$4\frac{A(\alpha,\frac{\pi-\alpha}{2})}{\hat{L}^2(\alpha,\frac{\pi-\alpha}{2})}=\frac{\pi-\alpha+\sin\alpha}{(\pi-\alpha+2\cos\frac{\alpha}{2})^2}.$$

We may verify that a such function is decreasing and is less than 1. We have thus proved part c) of the Proposition. □

References

Chouikha, A.R., (1999), Problems on polygons and Bonnesen-type inequalities, *Indag. Mathem.*, vol 10 (4), pp. 495-506.

Chouikha, A.R., (1988), Problème de P. Levy sur les polygones articulés, *C. R. Math. Report, Acad of Sc. of Canada,* vol 10, pp. 175-180.

Ku, H.T., Ku, M.C., and Zhang, X.M. (1995), Analytic and geometric isoperimetric inequalities, *J. of Geometry,* vol 53, pp. 100-121.

Levy, P. (1966), Le problème des isoperimetries et des polygones articulés, *Bull. Sc. Math., 2eme serie,* 90, pp. 103-112.

Osserman, R. (1979), Bonnesen-style isoperimetric inequalities, *Amer. Math. Monthly,* vol 1, pp. 1-29.

Chapter 11

CHARACTERIZING INVEX AND RELATED PROPERTIES

B. D. Craven*
Dept of Mathematics
University of Melbourne, Australia

Abstract A characterization of *invex* , given by Glover and Craven, is extended to functions in abstract spaces. *Pseudoinvex* for a vector function coincides with *invex* in a restricted set of directions. The *V-invex* property of Jeyakumar and Mond is also characterized. Some differentiability properties of the invex *scale function* are also obtained.

Keywords: Invexity, Pseudoinvexity, V-invex and necessary Lagrangian conditions.

MSC2000: 26B25, 49J52, 90C26

1. Introduction

A differentiable vector function F is *invex* at a point p if

$$F(\cdot) - F(p) \geq F'(p)\eta(\cdot, p)$$

for some *scale function* η. As is well known, with this property necessary Lagragian optimization conditions are also sufficient, and various duality results hold. It is important to find when the *invex* property holds. This paper extends the characterizations of *invex* given by Craven and Glover (1985) and Craven (2002) to also characterize *V-invex* (Jeyakumar and Mond (1992)), and to show that a related *pseudoinvex* property of a vector function coincides with *invex* in a restricted set of directions.

*email: B.Craven@ms.unimelb.edu.au

Differentiability properties of the *scale function* can also be characterized.

2. Characterizing Invex

The differentiable vector function $F : \mathbb{R}^n \to \mathbb{R}^m$ is (globally) *invex* at $p \in \mathbb{R}^n$ if, for some differentiable *scale function* $\eta : \mathbb{R}^n \times \mathbb{R}^n \to \mathbb{R}^n$,

$$(\forall x \in \mathbb{R}^n) \quad F(x) - F(p) \geq F'(p)\eta(x,p) \ . \tag{11.1}$$

If F and ω are twice-differentiable, then they have Taylor expansions

$$\begin{aligned} F(x) &= F(p) + F'(p)(x-p) + F^{\#}(x-p;p); \\ \text{where} \quad \omega(x-p) &\equiv \eta(x,p) = x - p + \omega^{\#}(x-p). \end{aligned}$$

Substituting in the definition of invexity, invexity at p is equivalent to

$$(\forall x \in \mathbb{R}^n) \quad F^{\#}(x-p;p) \geq F'(p)\omega^{\#}(x-p;p). \tag{11.2}$$

This is applied to an optimization problem:

$$\text{MIN} \quad F_1(x) \text{ subject to } F_j(x) \leq 0 \quad (j = 2,3,\ldots,m). \tag{11.3}$$

Here, F is called *active-invex (a-invex)* at p if $F_1(p) = 0$ (by replacing $F_1(\cdot)$ by $F_1(\cdot) - F_1(p)$), and $F^A(\cdot)$, obtained from $F(\cdot)$ by omitting those components F_j for which $F_j(p) \neq 0$, is invex at p. If α is a vector of Lagrange multipliers, with $\alpha_1 = 1$, then the Lagrangian:

$$\begin{aligned} L(x) := \alpha^T F(x) &= \alpha^T(F(x) - F(p) - F'(p)(x-p)) && (11.4) \\ &\equiv \alpha^T F^{\#}(x-p;p) && (11.5) \end{aligned}$$

if $L'(p) = \alpha^T F'(p) = 0$.

Consider, more generally, an optimization problem:

$$\text{MIN} \quad F_1(x) \text{ subject to } G(x) := (F_2(x), \ldots, F_m(x)) \in -S, \tag{11.6}$$

where S is a closed convex cone. Let $E \subset \mathbb{R}^n$. By definition, $F(\cdot)$ is invex at p on E with respect to the convex cone $U := \mathbb{R}_+ \times S$ if:

$$(\exists \eta : \mathbb{R}^n \times \mathbb{R}^n \to \mathbb{R}^n) \, (\forall x \in E) \quad F(x) - F(p) - F'(p)\eta(x,p) \in U, \tag{11.7}$$

or equivalently if:

$$(\forall x \in E) F^{\#}(x-p;p) - F'(p)\omega^{\#}(x-p;p) \in U. \tag{11.8}$$

(Note that this definition restricts the set of points x for which the invex property is considered. If E is not stated, then $E = \mathbb{R}^n$ is assumed.)

For problem (11.6), F is called *a-invex* at p if $F_1(p) = 0$ and F is invex at p with respect to the convex cone $U^A := \{\gamma(u + F(p)) \mid u \in U, \gamma \geq 0\}$. This reduces to the previous case when $S = \mathbb{R}_+^{m-1}$. The dual cone of U is denoted by U^*; if U is pointed, the dual cone of U^A is:

$$U^{A*} = \{u^* \in U^* \mid u^{*T} F(p) = 0\}.$$

The characterization of *invex* depends on the following consequence (see Craven and Glover (1985); Craven (2002)) of Motzkin's alternative theorem. It is stated here in abstract spaces, so that it may also be applied to optimal control problems.

Theorem 2.1 *(Characterization) Let X and Y be normed spaces (or* lctvs*); let $M : X \to Y$ be a continuous linear mapping; let $V \subset Y$ be a closed convex cone; let $\theta \in Y$; let the convex cone $K^*(V^*)$ be weak * closed, where K^* is the adjoint of $K := [M, -\theta]$, and V^* is the dual cone of V. Then:*

$$\begin{aligned} (\exists \zeta : X \to X) \qquad & -M\zeta + \theta \in V \Leftrightarrow \\ \{[0 \neq \alpha \in V^*, \alpha M = 0] \quad & \Rightarrow \quad [0 \neq \alpha \in V^*, \langle \alpha, \theta \rangle \geq 0]\} \end{aligned} \qquad (11.9)$$

Proof. For a fixed $x \in \mathbb{R}^n$, set $\theta := F^{\#}(x - p; p)$ and $\zeta := \omega^{\#}(x - p; p)$, and set $M := F'(p)$. Then,

$$(\exists \zeta) - M\zeta + \theta \in V \quad \Leftrightarrow \quad (\exists z \in X) - Mz + t\theta \in V, \ t \in \operatorname{int} \mathbb{R}_+$$

and on substituting $\zeta = z/t, u = (z, t)$ this is equivalent to

$$\begin{aligned} & (\exists u \in X \times \mathbb{R}) - Ku \in V, \ -Nu \in \operatorname{int} \mathbb{R}_+ \quad \text{where} \quad N := [0, -1] \\ \Leftrightarrow \ & \textbf{NOT}(\ \exists \alpha \in V^*, \exists \beta \in \mathbb{R}\ ) \ \ \alpha K + \beta N = 0, \ \alpha \geq 0, \ 0 \neq \beta \geq 0 \end{aligned}$$

(by Motzkin's alternative theorem, since $K^*(V^*)$ is weak * closed)

$$\begin{aligned} \Leftrightarrow \quad & \textbf{NOT}\ (\ \exists \alpha \in V^*\ ) \ \ \alpha M = 0, \ 0 \neq \langle \alpha, \theta \rangle \leq 0 \\ \Leftrightarrow \quad & [0 \neq \alpha \in V*, \alpha M = 0] \quad \Rightarrow \quad \langle \alpha, \theta \rangle \geq 0. \end{aligned}$$

□

Theorem 2.2 *In the differentiable optimization problem (11.6), assume that the cone*

$$[F'(p), -F^{\#}(x-p; p)]^*(U^*) \ [\ \textit{alternatively} \ \ [F'(p), -F^{\#}(x-p; p)]^*(U^{A*})]$$

is closed. (This condition holds automatically for problem (11.3). Then F is invex [alternatively active-invex] at a point p satisfying $\alpha \in U$,

$\alpha^T F'(p) = 0$ *[alternatively* $\alpha \in U$, $\alpha^T F'(p) = 0$, $\alpha^T F(p) = 0\}$ *if and only if, for each* x, $\alpha^T F^{\#}(x-p;p) \geq 0$.

Proof. Apply Theorem 2.1 with $\theta = F^{\#}(x-p;p)$, for each fixed x, $M = F'(p)$, and $V = U$ [alternatively $V = U^A$]. For problem (11.3) the cone is polyhedral, hence closed. □

Remark 2.1 *Often* α *does not depend on* x, *in particular if* α *is a unique vector of Lagrange multipliers.*

Theorem 2.2 also applies to infinite-dimensional problems, such as optimal control in continuous time, provided that the mentioned cone is assumed to be weak * closed.

Consider the following small examples:

Example 2.1 *The point* (0,0) *is a Karush-Kuhn-Tucker (KKT) point for:*

$$\begin{aligned} &\textit{MIN} \quad F_1(x,y) = x^2 + y \\ &\textit{subject to} \\ &\quad F_2(x,y) = -y + \alpha y^2 \leq 0, \\ &\quad F_3(x,y) = -1 - x + \beta x^2 \leq 0, \end{aligned}$$

with Lagrange multipliers 1 *and 0. The function* $F = (F_1, F_2, F_3)$ *is invex at* (0,0) *if functions* $\rho(x,y)$ *and* $\sigma(x,y)$ *exist (they include the linear terms) for which:*

$$x^2 \geq 0\rho + 1\sigma, \; \alpha y^2 \geq 0 - 1\sigma, \; \beta x^2 \geq -1\rho + 0\sigma \; ;$$

which hold for $\alpha > 0$ *and* β *of either sign. The Lagrangian at* (0,0) *is:*

$$L(x,y) = x^2 + 1(\alpha y^2) + 0(\beta x^2),$$

so that $L(x,y) \geq L(0,0)$ *provided that* $\alpha \geq 0$. *If* $\alpha < 0$ *then* $L(x,y)$ *is not minimized at* (0,0), *and invexity fails, since* $(-\alpha)y^2 \leq \sigma \leq x^2$ *does not generally hold.*

Example 2.2 *The point* (0,0) *is a KKT point for:*

$$\begin{aligned} &\textit{MIN} \quad F_1(x,y) = x + x^2 + y \\ &\textit{subject to} \\ &\quad F_2(x,y) = -y + \alpha y^2 \leq 0, \\ &\quad F_3(x,y) = -1 - x + \beta x^2 \leq 0, \\ &\quad F_4(x,y) = -x \leq 0, \end{aligned}$$

with Lagrange multipliers 1, 0, 1. *However* $\alpha^T F'(0,0) = 0$, *with* $\alpha = (1, \lambda, \mu, v) \geq (0,0,0,0)$, *only requires that* $1 + 0\lambda - \mu - \nu = 0$ *and* $1 - \lambda + 0\mu + 0\nu = 0$, *hence* $\lambda = 1$, $\mu + \nu = 1$, *so the multiplier* μ *for the inactive constraint can be any value in* [0,1]. *The Lagrangian at* (0,0) *is:*

$$L(x,y) = x^2 + 1(\alpha y^2) + \mu(\beta x^2) - \mu + \nu(0) \geq L(0,0)$$

for each $\mu \in [0,1]$ *provided that* $\alpha > 0$ *and* $1+\mu\beta \geq 0$, *thus when* $\beta \geq -1$. *And* F *is invex at* (0,0) *when* $x^2 \geq \rho + \sigma$, $\alpha y^2 \geq -\sigma$, $\beta x^2 \geq -\rho$, $0 \geq -\rho$, *which hold when* $\alpha \geq 0$ *and* $(1+\beta)x^2 + \alpha y^2 \geq 0$, *thus when* $\beta \geq -1$.

However, $F^A = (F_1, F_2, F_4)$ *is invex when* F *is invex with respect to* U^A, *which requires* $\mu = 0$, *so here* β *is unrestricted.*

3. Vector Pseudoinvex

A differentiable vector function $F : \mathbb{R}^n \to \mathbb{R}^m$ is *vector pseudoinvex* (*vpi*) at p with respect to the convex cone U (Craven (2001)) if:

$$F(x) - F(p) \in -\operatorname{int} U \;\Rightarrow\; (\exists \eta : \mathbb{R}^n \times \mathbb{R}^n \to \mathbb{R}^n)\, F'(p)\eta(x,p) \in -\operatorname{int} U. \tag{11.10}$$

Theorem 3.1 *Let* F *be differentiable; let* $U = \mathbb{R}_+ \times S$ *be the convex cone in problem (11.6); for fixed* p, *denote* $E := \{x \mid F(x) - F(p) \in -\operatorname{int} U\}$; *assume that the cone* $[F'(p), -F(x) + F(p)](U^\#)$ *is closed, for each* $x \in E$. *Then* F *is vector pseudoinvex at* p *with respect to* U *if and only if* F *is invex at* p *on* E *with respect to* U.

Proof. For a given $x \in E$, denote $c := F(x) - F(p)$ and $d := F'(p)\eta(x,p)$. If (11.10) holds, and $c \in -\operatorname{int} U$, then $d + N \in -\operatorname{int} U$ for some open ball N. For some $\epsilon > 0$, $-\epsilon c \in N$. Hence $d - \epsilon c \in -\operatorname{int} U$. Conversely, if $c \in -\operatorname{int} U$ and $d - \epsilon c \in -\operatorname{int} U$, then $d \in -\operatorname{int} U - \operatorname{int} U = -\operatorname{int} U$. Hence (11.10), for a given x, is equivalent to $d - \epsilon c \in -\operatorname{int} U$, for some $\epsilon > 0$. Hence, for $k = 1/\epsilon > 0$, (11.10) is equivalent to:

$$c - kd \in \operatorname{int} U \subset U. \tag{11.11}$$

Applying Theorem 2.1 with $M = F'(p)$, $\theta = k^{-1}(F(x) - F(p)) \in -\operatorname{int} U$, and $V = U$ shows that η exists, satisfying (11.11), if and only if:

$$[\alpha \in U^*, \alpha^T F'(p) = 0] \;\Rightarrow\; \alpha^T k^{-1}(F(x) - F(p)) \geq 0\,;$$

or equivalently if and only if:

$$[\alpha \in U^*,\ \alpha^T F'(p) = 0] \;\Rightarrow\; \alpha^T F^\#(x - p; p) \geq 0. \tag{11.12}$$

From Theorem 2.2, (11.12) for each $x \in E$ holds if and only if F is invex at p on E with respect to U. □

Remark 3.1 *Thus* pseudoinvex *at a point reduces to* invex *at the point in a restricted set of directions, This result explains the scarcity in the literature of examples of functions that are pseudoinvex but not invex. However, such a function could be constructed by changing an invex function F at some points x for which $F(x) - F(p) \notin -\mathrm{int}\, U$.*

4. Description of V-invex

In (Jeyakumar and Mond, 1992), a vector function F is called *V-invex* if:

$$F_i(x) - F_i(p) \geq \gamma_i(x,p)F'(p)\eta(x,p) \tag{11.13}$$

holds for each x and each component F_i, with some positive scalar coefficients, here denoted $\gamma_i(x,p)$. They showed that property (11.13) can replace invex, in proving sufficient KKT conditions. Here, p is fixed, and a characterization is obtained for the property (11.13), using Theorem 2.1. For a given x, denote $r_i := \gamma_i(x,p)^{-1}$; let $v = x - p$. Then (11.13) may be written:

$$\theta_i := (r_i - 1)F_i'(p)v + r_iF_i^{\#}(v;p) \geq F_i'(p)\omega^{\#}(v;p). \tag{11.14}$$

Applying Theorem 2.1 with $\theta = (\theta_1, \ldots, \theta_m)$, $M = F'(p)$, $V = U$ gives the equivalent statement:

$$[0 \neq \alpha \in U^*,\ \alpha^T F'(p) = 0]$$
$$\Rightarrow \sum \alpha_i(r_i - 1)F_i''(p)v + \sum \alpha_i r_i F_i^{\#}(v,p) \geq 0. \tag{11.15}$$

Theorem 4.1 *Let F be differentiable; let $U = \mathbb{R}_+ \times S$ be the convex cone in problem (11.6); for each x, assume that the cone $[F'(p), -\theta](U^{\#})$, with θ from (11.14), is closed, Then F is V-invex at $\mathbf{p}$, with respect to the cone U, if and only if:*

$$[0 \neq \alpha \in U^*,\ \alpha^T F'(p) = 0] \Rightarrow \sum_i \beta_i(F_i'(p)(x-p) + F_i^{\#}(x-p;p) \geq 0, \tag{11.16}$$

for some coefficients $\beta_i \equiv \beta_i(x) \geq 0$, with

$$\beta_i > 0 \Leftrightarrow \alpha_i > 0. \tag{11.17}$$

Proof. From (11.15), with $\beta_i := \alpha_i r_i$. □

5. Properties of the Scale Function

Apply now the definition given in (11.8) of invex at p, but with cone U^A, in the form:

$$F^{\#}(\cdot;p) - \mathbf{M}\omega^{\#}(\cdot;p) \in \mathbf{U}^A. \tag{11.18}$$

considering E as a compact subset of $\mathbb{R}^n$, with $F^{\#}(\cdot : p) : C(E) \to \mathbb{R}^m$, $\omega^{\#}(\cdot;p) : C(E) \times C(E) \to C(E)$, and $\mathbf{U}^A := \{\mathbf{u} \in C(E) : (\forall x \in E),\ \mathbf{u}(x) \in U^A\}$, and define $\mathbf{M} : C(E, \mathbb{R}^m) \to C(E, \mathbb{R}^m)$ by $(\mathbf{M}\mathbf{u})(t) := M\mathbf{u}(t)$, with $M := F'(p)$. Define $\mathbf{K}$ by $(\mathbf{K}\mathbf{u})(x) := K(x)\mathbf{u}(x)$, with $K(x) := [M, -\theta(x)]$. Since Theorem 2.1 is formulated in abstract spaces, it can be applied to (11.18). Assume now that F is C^2, and express (11.18) as:

$$(\exists\sigma(\cdot) \in C^1(E)) \quad \theta(\cdot) - \mathbf{M}\sigma(\cdot) \in \mathbf{U}^A. \tag{11.19}$$

with $\theta(\cdot) := F^{\#}(\cdot;p)/\|x-p\|^2$, and $\sigma(\cdot) := \omega^{\#}(\cdot;p)/\|x-p\|^2$.

The elements of the dual cone $\mathbf{U}^{A*}$ are represented by signed vector measures μ. Then:

$$\mu \in \mathbf{U}^{A^*} \Leftrightarrow (\forall \mathbf{z} \in \mathbf{U}^A), 0 \leq \langle \mu, \mathbf{z} \rangle = \int_E \mu(x)\mathbf{z}(x)\,dx, \tag{11.20}$$

For each interval I, let φ be a smooth approximation to the indicator function of I. Substituting $\mathbf{z} = \mathbf{c}\varphi$, $0 \leq \int_E \mu(x)\mathbf{c}\varphi(x)\,dx$, for each $\mathbf{c} \in U^A$, hence $\int_E \mu(x)\varphi(x)\,dx \in U^{A*}$. Taking a limit of suitable φ,

$$\mu(E \cap I) \in U^{A^*}. \tag{11.21}$$

If $\mathbf{w} \in C(E, \mathbb{R}^m)$, then

$$\langle \mathbf{K}^*\mu, \mathbf{w} \rangle = \int_E \langle \mu(x), K(x)\mathbf{w}(x) \rangle\,dx = \int_E \langle \mathbf{q}(x), \mathbf{w}(x) \rangle\,dx, \tag{11.22}$$

where $\mathbf{q}(x) \in K(x)^T(U^{A*})$ from (11.21), may be approximated by the integral of a step-function, taking constant values on intervals I. If the cone $K(x)^T(U^{A*})$ is closed, then (as a limiting case) $\{\langle \mathbf{K}^*\mu, \mathbf{w} \rangle : \mu \in \mathbf{U}^{A*}\}$ is closed, for each $\mathbf{w}$. Thus $\mathbf{K}^*(\mathbf{U}^{A*})$ is weak $*$ closed.

Theorem 5.1 *(Property of scale function) Assume that:*

1 the function F is C^2 and a-invex at each point $x \in E$,

2 the convex cone $K(x)^T(U^{A^*})$ is closed, for each $x \in E$,

3 $[\alpha \in U, \alpha^T F'(p) = 0,\ \alpha^T F(p) = 0]$ defines a unique $\alpha = \hat{\alpha}$.

Then there exists a scale function ω such that:

$$\omega^{\#}(x-p;p) = \psi(x)\|x-p\|^2,$$

with $\psi(\cdot)$ continuous.

Proof. Since F is a-invex at each point x, and the cone $K(x)^T(U^{A*})$is closed, Theorem 2.2, applied to (11.19), shows that:

$$(\forall x \in E)\ [0 \neq \alpha(x) \in U^A,\ \ \alpha(x)^T F'(p) = 0] \ \Rightarrow\ \alpha(x)^T F^{\#}(x-p;p) \geq 0 \tag{11.23}$$

Define a signed vector measure μ by $\mu(I) = \int_I \hat{\alpha}(x)\,dx$, for intervals I. Then μ is unique, and (11.23) shows that

$$[0 \neq \mu \in \mathbf{U}^{A^*},\ \ \langle \mu, \mathbf{M}\rangle = 0] \ \Rightarrow\ \langle \alpha(\cdot), \theta(\cdot)\rangle \geq 0. \tag{11.24}$$

Since the cone $K(x)^T(U^{A*})$ is closed, the cone $\mathbf{K}^*(\mathbf{U}^{A*})$ is weak $*$ closed, by the earlier discussion. Hence Theorem 2.1 shows that (11.19) holds if and only if (11.24) holds; and (11.19) implies the existence of the stated ω. □

Remark 5.1 *Suppose now that $F^{\#}(\cdot;p)$ and $\omega^{\#}(\cdot;p)$ are defined on spaces $C^1(E)$ instead of C(E). If $\mathbf{U}^A$ is redefined with elements $\mathbf{u} \in C^1(E)$, then the dual cone $\mathbf{U}^{A*}$ is represented by Schwartz distributions μ, which are the weak derivatives of signed vector measures ρ. If ψ is a smooth vector function, then $\langle \mu, \psi\rangle = -\langle \rho, \psi'\rangle$, leading to $\rho(E \cap I) \in -U^{A*}$ instead of (11.21). A similar construction from (11.22), using ρ, shows again that the cone is weak $*$ closed. Hence the invex property (11.19) also holds with $\sigma(\cdot) \in C^1(E)$.*

The dependence of the scale function on the point p can also be analysed. The property:

$$(\forall v \in X)\ (\forall p \in P)\ F^{\#}(v;p) \geq F'(p)\omega^{\#}(v;p), \tag{11.25}$$

where X and P are suitable domains, may be expressed as:

$$\Phi(p)(v) \geq \mathbf{L}\Omega(p)(v), \tag{11.26}$$

in which $\Omega(p)(v) := \omega^{\#}(v;p)/\|v\|^2, \Phi(p)(v) := F^{\#}(v;p)/\|v\|^2$, and the linear mapping $\mathbf{L}$ is the Cartesian product of the mappings $F'(p)$, for $p \in P$. This construction may be illustrated by the following case, where p takes only two values p_1 and p_2:

$$\begin{bmatrix} F^{\#}(v;p_1) \\ F^{\#}(v;p_2) \end{bmatrix} \geq \begin{bmatrix} F'(p_1) & 0 \\ 0 & F'(p_2) \end{bmatrix} \begin{bmatrix} \omega^{\#}(v;p_1) \\ \omega^{\#}(v;p_2) \end{bmatrix} \tag{11.27}$$

If F is C^2, then $\mathbf{L}$ is a continuous mapping of $C^1(E) \times C^1(P)$ into itself. Theorem 2.1 may be used to characterize (11.26), provided that a certain convex cone is weak $*$ closed. (This can be described, similarly to Theorem 5.1). If F is assumed invex at each point p, then it follows that a scale function exists, with $\omega^{\#}(v,p) = \|v\|^2 \rho(v,p)$, where $\rho(\cdot,\cdot)$ is a C^1 function.

However, it does *not* follow that F is *convexifiable;* there need not exist any invertible transformation γ, such that $F \circ \gamma$ is convex at all points p.

References

Craven, B. D. and Glover, B. M. (1985), Invex functions and duality, *Journal of the Australian Mathematical Society,* Series A, Vol. 39, pp 1-20.

V. Jeyakumar and B. Mond (1992), On generalized convex mathematical programming, *Journal of the Australian Mathematical Society,* Series B, Vol. 34, pp 43-53.

Craven, B. D. (2001), Vector generalized invex, *Opsearch,* Vol. 38, no. 4, pp 345-361.

Craven, B. D. (2002), Global invexity and duality in mathematical programming, *Asia-Pacific Journal of Operational Research,* Vol. 19, pp 169-175.

Chapter 12

MINTY VARIATIONAL INEQUALITY AND OPTIMIZATION: SCALAR AND VECTOR CASE

Giovanni P. Crespi*
Faculty of Economics
Université de la Vallée d'Aoste, Italy

Angelo Guerraggio†
Department of Economics
University of Insubria, Italy

Matteo Rocca‡
Department of Economics
University of Insubria, Italy

Abstract Minty variational inequalities are considered as related to the scalar minimization problem in which the objective function is a primitive of the operator involved in the inequality itself. Well-posedness (in the sense of Tykhonov) of this primitive problem is proved as a consequence of the existence of a strict solution of a Minty variational inequality.
Further, the vector extension of Minty variational inequality proposed by F. Giannessi is considered. We observe that, in this case, the relationships with the primitive vector optimization problem extend those known for the scalar case only under convexity hypotheses. A notion of solution of a Minty vector inequality, stronger than that introduced by Giannessi, is presented to fulfill this gap.

*email:g.crespi@univda.it
†email:angelo.guerraggio@uni-bocconi.it
‡email:mrocca@eco.uninsubria.it

Keywords: Minty variational inequalities, vector variational inequalities, vector optimization, well-posedness.

MSC2000: 49J40, 90C29, 90C30

1. Introduction

Variational inequalities are known either in the form presented by Stampacchia (1960), or in the form introduced by Minty (1967). The well known Minty's Lemma states the equivalence of these two alternative formulations under (hemi)continuity and (pseudo)monotonicity of the operator involved.
Vector extensions of Stampacchia and Minty variational inequalities have been introduced in Giannessi (1980) and Giannessi (1998), respectively. Moreover it has been proved that these vector variational inequalities characterize (weakly) efficient solutions of a suitable (convex) vector minimization problem.
In this paper we focus on Minty variational inequalities. Starting from classical results for scalar variational inequalities, in Section 2 we point out that Minty variational inequality is a sufficient optimality condition for a primitive minimization problem (that is the problem of minimizing a function f such that $f' = F$, where F is the function involved in the inequality).
Moreover we observe that if a Minty variational inequality admits a solution, then for the primitive minimization problem some kind of regularity is implicit (star-shapedness of the level sets of the objective function and furthermore Tykhonov well-posedness, when the solution is strict).
In Section 3, we consider the vector extension of variational inequalities and point out that some of the most classical results recalled in Section 2 cannot be proved under the same set of hypotheses. Indeed C-convexity is needed, while convexity is not due in the scalar case. Hence we suggest an alternative (and stronger) formulation of the Minty vector variational inequality, which allows us to state vector results analogous to the scalar ones.
Section 4 is devoted to final remarks and comments.

2. Scalar case

If this section, unless otherwise specified, F will denote a function from $\mathbb{R}^n$ to $\mathbb{R}^n$ and K a nonempty convex subset of $\mathbb{R}^n$.

Definition 2.1 *A vector $x^* \in K$ is a solution of a Stampacchia variational inequality (for short, VI), when:*

$$VI(F, K) \qquad \langle F(x^*), y - x^* \rangle \geq 0, \qquad \forall y \in K,$$

where $\langle \cdot, \cdot \rangle$ denotes the inner product on $\mathbb{R}^n$.

Using the same setting, we can give the definition proposed in Minty (1967):

Definition 2.2 *A vector $x^* \in K$ is a solution of a Minty variational inequality (for short, MVI), when:*

$$MVI(F, K) \qquad \langle F(y), y - x^* \rangle \geq 0, \qquad \forall y \in K.$$

The relationships between VI and MVI are stated by Minty's Lemma.

Definition 2.3 *A function $F : K \subseteq \mathbb{R}^n \to \mathbb{R}^n$ is said to be hemicontinuous at $x^* \in K$ when its restriction along every ray with origin at x^* is continuous. When this property holds at any point $x^* \in K$, then we say that F is hemicontinuous.*

Definition 2.4 *A function $F : K \subseteq \mathbb{R}^n \to \mathbb{R}^n$ is said to be monotone when:*

$$\langle F(x) - F(y), x - y \rangle \geq 0, \qquad \forall x, y \in K.$$

Lemma 2.1 (Minty Lemma) *i) Let F be hemicontinuous at $x^* \in K$. If x^* is a solution of MVI(F, K), then it is also a solution of VI(F,K).*

ii) Let F be monotone. If $x^* \in K$ is a solution of VI(F,K), then it is also a solution of MVI(F,K).

Remark 2.1 The hypothesis of monotonicity in point ii) of Minty Lemma can be weakened to pseudo-monotonicity.

The easiest way to relate Definitions 2.1 and 2.2 to minimization problems, is to consider *integrable variational inequalities* (see Rockafellar (1967)), i.e. to assume there exists a function $f : \mathbb{R}^n \to \mathbb{R}$, differentiable on an open set containing *K*, which is a primitive of *F*, that is such that $f' = F$, $\forall x \in K$ (here f' denotes the gradient of f). Under this assumption we focus on the primitive (constrained) minimization problem:

$$P(f, K) \qquad \min_{x \in K} f(x)$$

The following results are known (Kinderlehrer et al (1980); Komlósi (1998); Crespi et al (2002)):

Proposition 2.1 *i) Let* x^* *be a solution of* $P(f, K)$*. Then* x^* *solves* $VI(f', K)$*.*

ii) If f *is convex and* x^* *is a solution of* $VI(f', K)$*, then* x^* *solves* $P(f, K)$*.*

Proposition 2.2 *i) Let* $x^* \in K$ *be a solution of* $MVI(f', K)$*. Then* x^* *solves* $P(f, K)$*.*

ii) If f *is convex and* x^* *is a solution of* $P(f, K)$*, then* x^* *solves* $MVI(f', K)$*.*

Remark 2.2 The hypothesis of convexity in the previous proposition can be weakened to pseudo-convexity.

Remark 2.3 If x^* is a "strict solution" of $MVI(f', K)$, i.e.:

$$\langle f'(y), y - x^* \rangle > 0, \qquad \forall y \in K,\ y \neq x^*,$$

then it is possible to prove that x^* is the unique solution of $P(f, K)$. For a deeper analysis of strict solutions of *MVI(F, K)*, one can see John (1998).

The result in Proposition 2.2 leads to some deeper relationships between the solutions of a MVI and the corresponding primitive minimization problem. It seems that an "equilibrium" modelled through a MVI is more regular than one modelled through a VI (see for instance John (1998) and John (2001)). Here we recall the following result from Crespi et al (2002).

Proposition 2.3 *Let* $f : K \subseteq \mathbb{R} \to \mathbb{R}$ *(K convex) and assume there exists a solution* x^* *of* $MVI(f', K)$*. Then* f' *is quasi-monotone and hence* f *is quasi-convex.*

However, an example given in the same paper denies the possibility of stating the same conclusion for $f : K \subseteq \mathbb{R}^n \to \mathbb{R}$, $n > 1$.
In this case, the following result is obtained:

Proposition 2.4 *If* $f : K \subseteq \mathbb{R}^n \to \mathbb{R}$, $n > 1$, *is such that there exists a solution* x^* *of* $MVI(f', K)$ *and K is star-shaped at* x^**, then all the nonempty level sets of* f*:*

$$lev_c f := \left\{ x \in K \mid f(x) \leq c \right\}$$

are star-shaped at x^*.

Now we show that the existence of a solution of $MVI(f', K)$ is somehow related to the well-posedness of the primitive minimization problem.

Definition 2.5 *Problem $P(f, K)$ is said to be Tykhonov well-posed when:*

i) there exists a unique $\bar{x} \in K$ s.t. $f(\bar{x}) \leq f(x)$ for all $x \in K$;

ii) for any sequence $\{x_k\} \subseteq K$, $f(x_k) \to \inf_K f$ implies $x_k \to \bar{x}$.

A sequence $\{x_k\}$ which satisfies property ii) of the previous definition will be called a minimizing sequence.

Definition 2.6 *A set $A \subseteq \mathbb{R}^n$ is said locally compact at $x^* \in A$, when there exists a closed ball centered at x^* with radius δ, say $B(x^*, \delta)$, such that $A \cap B(x^*, \delta)$ is a compact set.*

Theorem 2.1 *Let x^* be a solution of $MVI(f', K)$ and K be star-shaped at x^*. Then, one and only one of the following alternatives holds:*

i) problem $P(f, K)$ admits infinitely many solutions;

ii) problem $P(f, K)$ admits the unique solution x^*. Moreover if K is locally compact at x^*, then problem $P(f, K)$ is Tykhonov well-posed.

Proof. From Proposition 2.2 we know that x^* is a solution of problem $P(f, K)$.

i) Let us assume there exists $x_1 \in K$, $x_1 \neq x^*$, such that $f(x_1) = f(x^*)$. Hence $x_1 \in lev_{f(x^*)} f$ and, by Proposition 2.4, it holds $tx^* + (1-t)\, x_1 \in lev_{f(x^*)} f$, $\forall t \in [0, 1]$, i.e.:

$$f\big(tx^* + (1-t)x_1\big) \leq f(x^*)$$

Hence we have, for all $t \in [0, 1]$, $f\big(tx^* + (1-t)x_1\big) = f(x^*)$ and the thesis follows.

ii) Assume now, by contradiction, that x^* is the unique solution of $P(f, K)$ but the problem is not Tykhonov well-posed. Hence there exists a sequence $\{x_k\} \subseteq K$ which does not converge to x^*, but with $f(x_k) \to f(x^*)$. We assume that the minimizing sequence $\{x_k\}$ is bounded. The proof is similar if $\{x_k\}$ is unbounded. For every $\varepsilon > 0$ and for k large enough we have:

$$f(x^*) < f(tx^* + (1-t)x_k) \leq f(x_k) \leq f(x^*) + \varepsilon,$$

for all $t \in [0,1)$, and without loss of generality we can assume that $\{x_k\}$ converges to a point $\overline{x} \neq x^*$.
Hence there exists a closed ball $B(x^*, \delta_1)$ such that $x_k \notin B(x^*, \delta_1)$, at least for k sufficiently large. If we set $\overline{\delta} := min\{\delta, \delta_1\}$, we obtain the existence of a sequence $\{t_k\} \subseteq [0,1)$ such that, for k large enough, it holds:

$$y_k = t_k x^* + (1-t_k)x_k \in K \cap S(x^*, \overline{\delta}),$$

where $S(x^*, \delta) := \{x \in \mathbb{R}^n \mid \|x - x^*\| = \delta\}$. Since this set is compact by assumption, without loss of generality we can think that $y_k \to y \in K \cap S(x^*, \overline{\delta})$, and hence $y \neq x^*$. This is absurd, since the continuity of f would imply:

$$f(x^*) < \lim_{k \to +\infty} f(y_k) = f(y) \leq f(x^*) + \varepsilon$$

and hence, since ε is arbitrary, $f(y) = f(x^*)$, contradicting the uniqueness of the minimum point. □

Corollary 2.1 *If x^* is a "strict solution" of $MVI(f', K)$, then problem $P(f, K)$ is Tykhonov well-posed.*

Proof. It is straightforward from Theorem 2.1 and Remark 2.3. □

We end this section with some results which point out that the existence of a solution of a MVI has strong implications for the convergence of minimization algorithms. Consider the dynamical system:

$$DS \qquad x'(t) = -F(x(t)), \quad x \in K,$$

where $K \subseteq \mathbb{R}^n$ is open, and assume that F is continuous.

Definition 2.7 *i) A point $x^* \in K$ is said to be an equilibrium point of DS when $F(x^*) = 0$.*

ii) An equilibrium point x^* is said to be stable when for every $\varepsilon > 0$ there exists a $\delta > 0$ such that for every $x^0 \in K$ with $\|x^0 - x^*\| < \delta$, the solution $x(t)$ of DS with $x(0) = x^0$ is defined and $\|x(t) - x^*\| < \delta,\ \forall t > 0$.

iii) An equilibrium point x^* is asymptotically stable when there is a $\delta > 0$ such that, for every solution $x(t)$ with $\|x(0) - x^*\| < \delta$, one has $\lim_{t \to +\infty} x(t) = x^*$

The following theorem is known:

Theorem 2.2 (John (1998)) *Consider the dynamical system DS*

i) If x^* is a solution of MVI(F,K), then it is a stable equilibrium point of DS.

ii) If x^* is a strict solution of MVI(F,K), then it is an asymptotically stable equlibrium point of DS.

Definition 2.8 *Let $x(t)$ be a trajectory of DS. The sets:*

$$L(F) = \{a \in K | \exists t_k \to +\infty, x(t_k) \to a\}$$

and

$$L_1(F) = \{b \in K | \exists t_k \to -\infty, x(t_k) \to b\},$$

are called, respectively, the set of **ω-limit** *points and the set of* **α-limit** *points of $x(t)$.*

Now consider the gradient dynamical system:

$$GDS \qquad x'(t) = -f'(x(t)), \quad x \in K,$$

where K is an open convex subset of $\mathbb{R}^n$. Clearly *GDS* represents the continuous version of the gradient method.

As a corollary to the previous theorem we have:

Corollary 2.2 *Assume that f' is continuous.*

i) Let x^* be a solution of $MVI(f', K)$ and let $x(t)$ be a trajectory of GDS starting at a point (t_0, x_0), $x_0 \in K$, such that $\|x_0 - x^*\|$ is small enough. Then $L(f')$ and $L_1(f')$ are nonempty and every point $x \in L(f') \cup L_1(f')$ is a stationary point of f.

ii) Let x^* be a strict solution of $MVI(f', K)$. Then $L(f') = L_1(f') = \{x^*\}$ (hence the continuous gradient method converges to the unique minimum point of f over K).

Proof. *i)* From theorem 2.2, we know that x^* is a stable equilibrium point. The nonemptyness of $L(f')$ and $L_1(f')$ is straightforward. The stationarity of every $x \in L(f') \cup L_1(f')$ follows from Theorem 4, p. 203 in Hirsch et al (1974).

ii) It is easy to prove that x^* is the unique equilibrium point of *GDS* and hence the conclusion follows from point i). □

3. Vector case

In this section, C will denote a cone contained in $\mathbb{R}^l$, which is assumed to be closed, convex, pointed and with nonempty interior. The cone C clearly induces a partial order on $\mathbb{R}^l$, by means of which vector variational inequality (of Stampacchia type) has been first introduced in Giannessi (1980). Later a vector formualtion of Minty variational inequality has been proposed as well (see e.g. Giannessi (1998)). Both the inequalities involve a matrix valued function $F : \mathbb{R}^n \to \mathbb{R}^{l \times n}$ and a feasible region $K \subseteq \mathbb{R}^n$, assumed to be convex and nonempty. In the sequel, $\langle \cdot, \cdot \rangle_l$ denotes a vector of l inner products of $\mathbb{R}^n$. Moreover we will consider the following sets:

$$\Omega(x) := \left\{ u \in \mathbb{R}^l \mid u = \langle F(x), y - x \rangle_l, \; y \in K \right\},$$

$$\Theta(x) := \left\{ w \in \mathbb{R}^l \mid w = \langle F(y), y - x \rangle_l, \; y \in K \right\}.$$

Definition 3.1 *ii) A vector $x^* \in K$ is a solution of a strong vector variational inequality of Stampacchia type when:*

$$VVI^s(F, K) \qquad \Omega(x^*) \cap \left(-C\right) = \{0\}.$$

i) A vector $x^* \in K$ is a solution of a weak vector variational inequality of Stampacchia type when:

$$VVI(F, K) \qquad \Omega(x^*) \cap \left(-\text{int } C\right) = \emptyset,$$

where int A *denotes the interior of the set A.*

Definition 3.2 *i) A vector $x^* \in K$ is a solution of a strong vector variational inequality of Minty type when:*

$$MVVI^s(F, K) \qquad \Theta(x^*) \cap \left(-C\right) = \{0\}.$$

ii) A vector $x^* \in K$ is a solution of a weak vector variational inequality of Minty type when:

$$MVVI(F, K) \qquad \Theta(x^*) \cap \left(-\text{int } C\right) = \emptyset.$$

In the sequel we will deal with weak vector variational inequalities of Stampacchia and Minty type (for short VVI and MVVI, respectively). First we recall the definition of monotonicity for matrix-valued functions:

Definition 3.3 *Let $F : \mathbb{R}^n \to \mathbb{R}^{l \times n}$ be given. We say that F is C–monotone over K, when:*

$$\langle F(y) - F(x), y - x \rangle_l \in C, \qquad \forall x, y \in K.$$

The following result (see Giannessi (1998)) extends Minty Lemma to the vector case.

Lemma 3.1 *Let F be continuous and C-monotone. Then x^* is a solution of MVVI(F, K) if and only if it solves VVI(F, K).*

Similarly to the scalar case, now we consider a function $f : \mathbb{R}^n \to \mathbb{R}^l$, differentiable on an open set containing *K*, such that $f' = F$, for all $x \in K$ (here f' denotes the Jacobian of f). Then we introduce the following primitive vector minimization problem, depending on the ordering cone *C:*

$$VP(f, K) \qquad\qquad \text{int } C - \min_{x \in K} f(x).$$

A solution of $VP(f, K)$ (see e.g. Luc (1989)) is any vector $x^* \in K$ such that:

$$[f(K) - f(x^*)] \cap (-\text{ int } C) = \emptyset.$$

The vector x^* is called a weak efficient point for f over *K.*
We remember that $x^* \in K$ is said an efficient point for f over K when:

$$[f(K) - f(x^*)] \cap (-C) = 0.$$

Now we recall some basic definitions and results about vector–valued convex functions:

Definition 3.4 *The function $f : K \subseteq \mathbb{R}^n \to \mathbb{R}^l$ is said to be C–convex when:*

$$f(\lambda x + (1-\lambda)y) - [\lambda f(x) + (1-\lambda) f(y)] \in -C \quad \forall x, y \in K, \quad \forall \lambda \in [0, 1].$$

The following result is classical (see e.g. Karamardian et al (1990); Luc et al (1993)).

Proposition 3.1 *If f is differentiable, the following statements are equivalent:*

i) f *is C–convex;*

ii) $f(y) - f(x) - \langle f'(x), y - x \rangle_l \in C, \quad \forall x, y \in K;$

iii) f' *is C–monotone.*

The following results (see Giannessi (1980); Giannessi (1998); Komlósi (1998)) extend to the vector case Propositions 2.1 and 2.2.

Proposition 3.2 *Let $f : \mathbb{R}^n \to \mathbb{R}^l$ be differentiable on an open set containing K.*

i) If x^* is a a solution of $VP(f, K)$, then it solves also $VVI(f', K)$.

ii) If f is C-convex and x^* is a solution of $VVI(f', K)$, then it solves also $VP(f, K)$.

Proposition 3.3 *Let C be a polyhedral cone. If f is C–convex and differentiable on an open set containing K, then $x^* \in K$ is a solution of $VP(f, K)$ if and only if it is a solution of $MVI(f', K)$.*

In particular, Proposition 3.3 gives an extension to the vector case of Proposition 2.2. Anyway, in Proposition 3.3, convexity is needed also for proving that $MVVI(f', K)$ is a sufficient condition for optimality, while in the scalar case, convexity is needed only in the proof of the necessary part.
Some refinements of the relations between VVI and efficiency have been given in Crespi (2002). In this context, we focus on MVVI and believe that a suitable definition of it should extend Proposition 2.2 without any additional assumption.
First we show that Proposition 3.3 cannot be improved, at least until we keep Definition 3.2:

Example 3.1 *Let $C = \mathbb{R}^2_+$, $K := \left[-\frac{2}{\pi}, 0\right]$ and consider a function $f : \mathbb{R} \to \mathbb{R}^2$, $f(x) := \begin{bmatrix} f_1(x) \\ f_2(x) \end{bmatrix}$, defined as follows. We set:*

$$f_1(x) = \begin{cases} x^2 \sin \frac{1}{x} - x^2, & x \neq 0 \\ 0, & x = 0 \end{cases}$$

and observe that $-2x^2 \leq f_1(x) \leq 0$, $\forall x \in K$ and f_1 is differentiable on K; its graph is plotted in figure 12.1. Function f_1 has a countable number of local minimizers and of local maximizers over K. The local maximizers of f_1 are the points $y_k = -\frac{1}{\frac{\pi}{2}+2k\pi}$, $k = 0, 1, \ldots$ and $f_1(y_k) = 0$. If we denote by x_k, $k = 0, 1, \ldots$ the local minimizers of f over K, we have $y_k < x_k < y_{k+1}$, $\forall k = 0, 1, \ldots$.

The function f_2 *is defined on K as:*

$$f_2(x) = \begin{cases} -\frac{f_1(x_k)}{2}\left[\cos\left(\frac{\pi x}{x_k - y_k} + \frac{\pi(x_k - 2y_k)}{x_k - y_k}\right) - 1\right], & x \in [y_k, x_k) \\ -\frac{f_1(x_{k+1})}{2}\left[\cos\left(\frac{\pi x}{y_{k+1} - x_k} + \frac{\pi(2y_{k+1} - 3x_k)}{y_{k+1} - x_k}\right) - 1\right], & x \in [x_k, y_{k+1}) \\ 0, & x = 0 \end{cases}$$

for $k = 0, 1, \ldots$ *. It is easily seen that also* f_2 *is differentiable on K. The points* $x \in [-\frac{2}{\pi}, x_0]$ *are (weakly) efficient, while the other points in K are not efficient. In particular,* $x^* = 0$ *is an ideal maximal point (i. e.* $f(x) - f(x^*) \in \mathbb{R}^2_-$, $\forall x \in K$*). Anyway, it is easy to see that any point of K is a solution of* $MVVI(f', K)$.

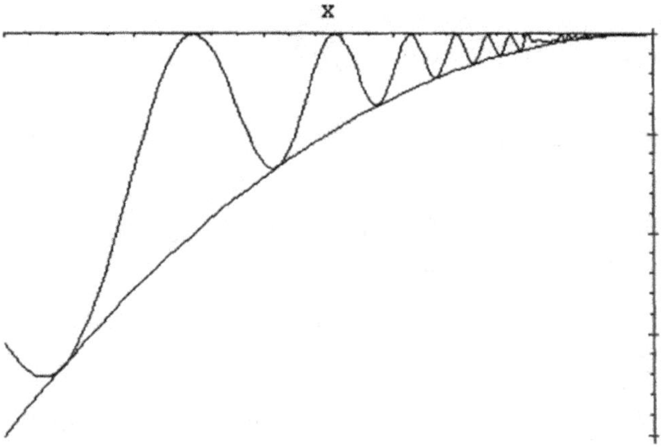

Figure 12.1. $f_1(x) = -x^2 sin\frac{1}{x} - x^2$.

Remark 3.1 For a vector valued function $f : K \subseteq \mathbb{R}^n \to \mathbb{R}^l$ one can define a level set as (Luc (1989)):

$$lev_c f = \{x \in K | f(x) \in c - C\},$$

where $c \in \mathbb{R}^l$. We observe that the previous example shows that Propositions 2.3 and 2.4 cannot be extended to $MVVI(f', K)$ with this definition of level set. In fact, if one considers $C = \mathbb{R}^2_+$ and for instance $c = (f_1(y_1), f_2(y_1))$, the corresponding level set is not convex.

Our idea, partially based on a technique proposed in Gong (2001) and applied also in Crespi (2002) for Stampacchia vector variational

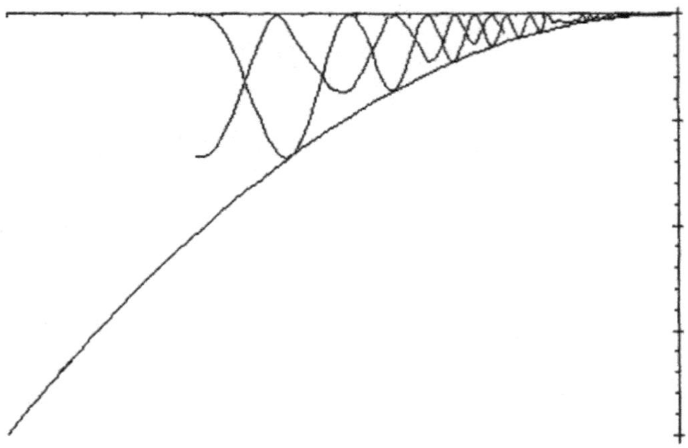

Figure 12.2. $f_1(x)$ and $f_2(x)$.

inequalities, is to consider a solution concept stronger than the one in Definition 3.2.

Definition 3.5 *A vector $x^* \in K$ is a (weak) solution of a convexified Minty vector variational inequality when:*

$$CMVVI(F,K) \qquad \text{conv}\,\Theta(x^*) \cap \left(-\text{int } C\right) = \emptyset,$$

where conv A *is the convex hull of a given set A.*

Remark 3.2 i) Clearly, if $l = 1$ Definition 3.5 collapses into Definition 2.2.

ii) If $l \geq 2$, it follows from the definitions that, if $x^* \in K$ solves *CMVVI(F, K)* then it solves also *MVVI(F,K)*.
The converse is not always true, as it is shown in the following example.

Example 3.2 *Let $l = 2$, $C = \mathbb{R}^2_+$, $F : \mathbb{R} \to \mathbb{R}^2$, with $F(x) = \begin{bmatrix} 1 \\ \frac{1}{x-1} \end{bmatrix}$ and $K = \left[-\frac{1}{2}, \frac{1}{2}\right]$. It is easy to check that $x^* = 0$ solves MVVI(F,K), since $\Omega(0) \cap \left(-\text{int } \mathbb{R}^2_+\right) = \emptyset$. However it is easy to see that $\text{conv}\Theta(0) \cap \left(-\text{int } \mathbb{R}^2_+\right) \neq \emptyset$.*

The following scalarization result plays a crucial role in the next proofs. We denote by C^* the positive polar cone of C, i.e.:

$$C^* := \left\{\lambda \in \mathbb{R}^l \mid \langle\lambda, c\rangle \geq 0, \ \forall c \in C\right\}.$$

Lemma 3.2 *A vector $x^* \in K$ solves CMVVI(F, K) if and only if there exists a nonzero vector $\lambda \in C^*$, such that x^* is a solution of the following scalar Minty variational inequality:*

$$MVI(\lambda F, K) \qquad \langle \lambda F(y), y - x^* \rangle \geq 0, \qquad \forall y \in K.$$

Proof. Let $x^* \in K$ solve $MVI(\lambda F, K)$ for some nonzero $\lambda \in C^*$. We have $\langle \lambda, w \rangle < 0$, $\forall w \in -\text{int } C$, while $\langle \lambda, w \rangle \geq 0$, $\forall w \in \Theta(x^*)$. It follows easily that:

$$\Theta(x^*) \subseteq \textbf{conv } \Theta(x^*) \subseteq \{w \in \mathbb{R}^l | \langle \lambda, w \rangle \geq 0\},$$

while:

$$-\text{int } C \subseteq \{w \in \mathbb{R}^l | \langle \lambda, w \rangle < 0\}$$

and so $\textbf{conv } \Theta(x^*) \cap -\text{int } C = \emptyset$.
Conversely, assume that $x^* \in K$ solves *CMVVI(F, K),* which means that $\textbf{conv } \Theta(x^*)$ and –int C are two disjoint convex sets. By classical separation arguments the thesis follows easily. □

Theorem 3.1 *Let $x^* \in K$ be a solution of $CMVVI(f', K)$. Then x^* is a solution of $VP(f, K)$.*

Proof. By Lemma 3.2 we know $x^* \in K$ solves $MVI(\lambda f', K)$ for some nonzero $\lambda \in C^*$ and, by Proposition 2.2 it follows that the scalar problem:

$$\min_{x \in K} \lambda f(x)$$

is also solved by x^*. By a classical scalarization result, (Luc (1995); Sawaragi et al (1985)) x^* solves $VP(f, K)$. □

Example 3.3 *Let $f : \mathbb{R} \to \mathbb{R}^2$ be defined as $f(x) = \begin{bmatrix} x \\ -x^2 \end{bmatrix}$, and K = [0,1]. Clearly f is not $\mathbb{R}^2_+$-**convex** and thus its Jacobian $f'(x) = \begin{bmatrix} 1 \\ -2x \end{bmatrix}$ is not $\mathbb{R}^2_+$-**monotone**. Consider the point $x^* = 0 \in K$. We have:*

$$\Theta(0) = \{w \in \mathbb{R}^2 \,|\, w = \begin{bmatrix} y \\ -2y^2 \end{bmatrix}, y \in K\}$$

and *hence* $\textbf{conv}\, \Theta(0) \cap (-\textbf{int}\, \mathbb{R}^2_+) = \emptyset$. *Consequently $x^* = 0$ solves $CMVVI(f', K)$ (and hence 0 solves $MVVI(f', K)$), and by Theorem 3.1 we can conclude $x^* = 0$ is a solution of the primitive problem*

$VP(f,K)$, as it can be easily seen. However Proposition 3.3 would not have allowed such a conclusion, since f is not $\mathbb{R}^2_+$-convex.

The converse of Theorem 3.1 can be stated under the assumption of C–convexity of f:

Theorem 3.2 *Let $f : \mathbb{R}^n \to \mathbb{R}^l$ be C-convex and differentiable. If $x^* \in K$ is a solution of $VP(f,K)$, then x^* solves $CMVVI(f',K)$*

Proof. By contradiction, assume x^* is efficient, but $\exists \omega \in \mathbf{conv}\,\Theta(x^*)$ such that $\omega \in (-\mathbf{int}\ C)$. By Caratheodory Theorem, each element of $\mathbf{conv}\ \Theta(x^*)$ can be written as a convex combination of at most $l+1$ points of $\Theta(x^*)$, that is:

$$\omega = \sum_{i=1}^{l'} \lambda_i \langle f'(y_i), y_i - x^* \rangle_l,$$

where $0 < l' \le l+1$, $y_i \in K$, $\lambda_i \ge 0$, $i = 1, \ldots, l'$ and $\sum_{i=1}^{l'} \lambda_i = 1$. Moreover, by the C–convexity of f, we have:

$$f(x^*) - f(y_i) \in \langle f'(y_i), x^* - y_i \rangle_l + C, \qquad \forall i = 1, \ldots, l'.$$

Since C is a convex cone, we obtain:

$$\sum_{i=1}^{l'} \lambda_i [f(x^*) - f(y_i)] = f(x^*) - \sum_{i=1}^{l'} \lambda_i f(y_i) \in \sum_{i=1}^{l'} \lambda_i \langle f'(y_i), x^* - y_i \rangle_l + C \subseteq$$

$$C + \text{int}\ C \subseteq \text{int}\ C.$$

Since K is convex, $\sum_{i=1}^{l'} \lambda_i y_i \in K$ and the C–convexity of f implies:

$$f(x^*) \in \sum_{i=1}^{l'} \lambda_i f(y_i) + \text{int}\ C \subseteq f\left(\sum_{i=1}^{l'} \lambda_i y_i\right) + C + \text{int}\ C \subseteq$$

$$f\left(\sum_{i=1}^{l'} \lambda_i y_i\right) + \text{int}\ C.$$

Hence we get the absurdo:

$$f\left(\sum_{i=1}^{l'} \lambda_i y_i\right) - f(x^*) \in -\text{int}\ C.$$

□

Remark 3.3 Theorems 3.1 and 3.2 actually reproduce for the minimization of vector valued functions f the known results for the scalar case (see Proposition 2.2).
Indeed a Minty type (vector) variational inequality is a sufficient condition for efficiency without assumptions on the differentiable objective functions. Necessity holds true as well, but under C–convexity assumption on f.

The last thing to check should be that any of the cases which fulfills Proposition 3.3, actually fulfills also Theorem 3.1. This would be the case if:

$$\begin{matrix} x^* \in MVVI(f', K) \\ f \;\; C - convex \end{matrix} \quad \Rightarrow \quad x^* \in CMVVI(f', K)$$

Corollary 3.1 *Let C be a polyhedral cone and let $f : \mathbb{R}^n \to \mathbb{R}^l$ be C-convex and differentiable. If x^* solves $MVVI(f', K)$, then x^* solves $CMVVI(f', K)$.*

Proof. Under the assumptions, Proposition 3.3 allows to conclude that x^* is efficient for f over K. Thus Theorem 3.2 implies the thesis. □

The following results extend Corollary 3.1 to any ordering cone C and any vector variational inequality, under additional hemicontinuity assumptions.

Theorem 3.3 *Let $F: \mathbb{R}^n \to \mathbb{R}^{l \times n}$ be hemicontinuous and C-monotone. Then any x^* which solves MVVI(F, K) is a solution of CMVVI(F, K).*

Proof. Let x^* solve *MVVI(F,K)*. Then it holds:

$$\Theta(x^*) \subseteq (-\text{int } C)^c.$$

If $w \in \text{conv}\,\Theta(x^*)$, by Caratheodory Theorem there exist an integer $l' \in (0, l+1]$, l' vectors $y_i \in K$ and scalars $\lambda_i \in [0,1]$ with $\sum_{i=1}^{l'} \lambda_i = 1$, such that:

$$\begin{aligned} w &= \sum_{i=1}^{l'} \lambda_i \langle F(y_i), y_i - x^* \rangle_l = \\ &= \sum_{i=1}^{l'} \lambda_i \langle F(y_i) - F(x^*), y_i - x^* \rangle_l + \sum_{i=1}^{l'} \lambda_i \langle F(x^*), y_i - x^* \rangle_l. \end{aligned}$$

Since f is C–monotone, we have:

$$\langle F(y_i) - F(x^*), y_i - x^* \rangle_l \in C, \qquad \forall i = 1, \ldots, l',$$

and since C is a convex cone:

$$\sum_{i=1}^{l'} \lambda_i \langle F(y_i) - F(x^*), y_i - x^* \rangle_l \in C, \qquad \forall i = 1, \ldots, l'.$$

Moreover, by the convexity of K, we have, $\forall y \in K$ and $t \in [0,1]$ that $z := ty - (1-t)x^* \in K$ and since x^* solves *MVVI(F,K)*, we get:

$$\langle F(z), z - x^* \rangle_l \notin -\text{int } C$$

$$i.e. \ \langle F(z), z - x^* \rangle_l \in (-\text{int } C)^c.$$

Since $(-\text{int } C)^c$ is a cone, we can conclude:

$$\langle F(z), y - x^* \rangle_l \in (-\text{int } C)^c,$$

simply noting that $z - x^* = t(y - x^*)$.
By the hemicontinuity of F, letting $t \downarrow 0$ in the previous inclusion, we get:

$$\langle F(x^*), y - x^* \rangle_l \in cl \left[(-\text{int } C)^c\right] = (-\text{int } C)^c.$$

Hence $\sum_{i=1}^{l'} \langle F(x^*), y_i - x^* \rangle = \langle F(x^*), (\sum_{i=1}^{l'} y_i) - x^* \rangle_l \notin -\text{int } C$ and so $w \notin (-\text{int } C)$. □

Remark 3.4 The function F in Example 3.2, which is not C–monotone, actually shows that monotonicity is necessary for Theorem 3.3 to hold true.

Remark 3.5 Combining Theorems 3.1, 3.2 and 3.3, one gets the extension of Proposition 3.3 to any cone C (convex, closed and with nonempty interior).

Theorem 3.3 allows to prove the following vector version of Minty Lemma:

Lemma 3.3 *Let F be hemicontinuous and monotone. Then x^* is a solution of CMVVI(F, K) if and only if it is a solution of VVI(F, K).*

Proof. Theorem 3.3 and Remark 3.2 (point *ii*) allow us to prove results just by passing through Lemma 3.1. □

4. Conclusions and further remarks

In this paper we focused on the relationships between Minty variational inequality and optimization both in scalar and vector case. We observed, in particular that the existing extension of MVI to the vector case does not allow to recover, without additional assumptions, the results holding in the scalar case, in particular with respect to the fact that MVI is a sufficient condition for optimality.
Having this in mind, we gave a stronger solution concept of vector MVI and we linked it to the weak solutions of a vector optimization problem.
Several steps ahead should be done on this topic. For instance one should try to give a characterization also of efficient solutions. However this looks to be a hard task which, at the moment, has no solution also for Stampacchia Vector Variational Inequality, as far as we know.
Moreover, dealing with vector optimization, proper efficiency has to be considered and more strict definitions of solution of a Minty vector variational inequality could be studied for the purpose of characterizing also this case.

References

Baiocchi, C. and Capelo, A. (1978), *Disequazioni variazionali e quasivariazionali. Applicazioni a problemi di frontiera libera,* Quaderni U. M. I. , Pitagora editrice, Bologna.

Chen, G.Y. and Cheng, G.M. (1987), Vector variational inequality and vector optimization, *Lecture notes in Economics and Mathematical Systems,* Vol. 285, Springer-Verlag, Berlin, pp. 408-416.

Crespi, G.P. (2002), Proper efficiency and vector variational inequalities, *Journal of Information and Optimization Sciences,* Vol. 23, No. 1, pp. 49-62.

Crespi, G.P., Ginchev, I. and Rocca, M., Existence of solutions and starshapedness in Minty variational inequalities, *Journal of Global Optimization* (to appear).

Dontchev A.L. and Zolezzi T. (1993), *Well–Posed Optimization Problems,* Springer, Berlin.

Giannessi, F. (1980), Theorems of the alternative, quadratic programs and complementarity problems, *Variational Inequalities and Complementarity Problems. Theory and applications* (R.W. Cottle, F. Giannessi, J.L. Lions eds.), Wiley, New York, pp. 151-186.

Giannessi, F. (1998), On Minty variational principle, *New Trends in Mathematical Programming* (F. Giannessi, S. Komlósi, T. Rapcsák eds.), Kluwer Academic Publishers, Boston, MA, pp. 93-99.

Gong, X.H. (2001), Efficiency and Henig Efficiency for Vector Equilibrium Problems, *Journal of Optimization Theory and Applications,* Vol. 108, No. 1, pp. 139-154.

Hadjisavvas, N. and Schaible, S. (1998), From scalar to vector equilibrium problems in the quasimonotone case, *Journal of Optimization Theory and Applications,* Vol. 96, No. 2, pp. 297-309.

Hirsch M.W. and Smale S., (1974), *Differential Equations, Dynamical Systems and Linear Algebra,* Academic Press, New York.

John R. (1998), Variational Inequalities and Pseudomonotone Functions: Some Characterizations, *Generalized Convexity, Generalized Monotonicity,* (J.P. Crouzeix, J.E. Martinez-Legaz, M. Volle eds.), Kluwer, Dordrecht, pp. 291-301.

John R. (2001), A note on Minty Variational Inequality and Generalized Monotonicity, *Generalized Convexity and Generalized Monotonicity* (N.Hadjisavvas, J.E. Martinez-Legaz, J.P. Penot eds.), Lecture notes in Economics and Mathematical Systems, Vol. 502, Springer, Berlin, pp. 240–246.

Karamardian, S. and Schaible, S. (1990), Seven kinds of monotone maps, *Journal of Optimization Theory and Applications,* Vol. 66, No. 1, pp. 37-46.

Kinderlehrer, D. and Stampacchia, G. (1980), *An introduction to variational inequalities and their applications,* Academic Press, New York.

Komlósi, S. (1998), On the Stampacchia and Minty Variational Inequalities, *Generalized Convexity and Optimization for Economic and Financial Decisions,* (G. Giorgi, F.A. Rossi eds.), Pitagora, Bologna.

Lee, G.M., Kim, D.S., Lee, B.S. and Yen, N.D. (1999), Vector Variational Inequalities as a tool for studying vector optimization Problems, *Nonlinear Analysis,* Vol. 84, pp. 745-765.

Luc, D.T. (1989), *Theory of Vector Optimization,* Springer Verlaag, Berlin.

Luc, D.T. and Swaminathan, S. (1993), A caracterization of convex functions, *Nonlinear Analysis,* Vol. 20, No. 6, pp. 697-701.

Luc, D.T. (1996), Hartman-Stampacchia's theorem for densely pseudomonotone Variational inequalities, *Internal Report,* Vietnam National Centre for Natural Science and Technology – Institute of Mathematics, Hanoi.

Minty, G.J. (1967), On the generalization of a direct method of the calculus of variations, *Bulletin of American Mathematical Society,* Vol. 73, pp. 314-321.

Nagurney A. (1993), *Network economics: A Variational inequality approach,* Kluwer Academic Publishers, Boston, MA.

Rockafellar R.T. (1967), Convex functions, monotone operators and variational inequalities, *Proceedings of the N.A.T.O. Advanced Study Institute,* pp. 35-65.

Sawaragi Y., Nakayama H. and Tanino T. (1985), *Theory of Multiobjective Optimization,* Academic Press, New York.

Stampacchia G. (1960), Formes bilinéaires coercitives sur les ensembles convexes, *C. R. Acad. Sciences de Paris,* t.258, 9 Groupe 1, pp. 4413-4416.

Chapter 13

SECOND ORDER OPTIMALITY CONDITIONS FOR NONSMOOTH MULTIOBJECTIVE OPTIMIZATION PROBLEMS

Giovanni P. Crespi*
Faculty of Economics
Université de la Vallée d'Aoste, Italy

Davide La Torre†
Department of Economics
University of Milan, Italy

Matteo Rocca‡
Department of Economics
University of Insubria, Italy

Abstract In this paper second-order necessary optimality conditions for nonsmooth vector optimization problems are given by smooth approximations. We extend to the vector case the approach introduced by Ermoliev, Norkin and Wets to define generalized derivatives for discontinuous functions as limit of the classical derivatives of regular functions.

Keywords: Vector optimization, Optimality conditions, Mollifiers, Taylor's Formula.

MSC2000: 90C29, 90C30, 26A24

---

*email:g.crespi@univda.it
†email:davide.latorre@unimi.it
‡email:mrocca@eco.uninsubria.it

1. Introduction

In this paper we extend to vector optimization the approach introduced by Ermoliev, Norkin and Wets to define generalized derivatives even for discontinuous functions, which often arise in applications (see Ermoliev et al (1995) for references on this point). To deal with such applications a number of approaches have been proposed to develop a subdifferential calculus for nonsmooth and even discontinuous functions. Among the many possibilities, let us remember the notions due to Clarke (1990), Michel et al (1984), in the context of Variational Analysis. The previous approaches are based on the introduction of first-order generalized derivatives. Extensions to higher-order derivatives have been provided for instance by Bonnans et al (1999), Cominetti and Correa (1990), Crespi et al (2002a), Ginchev and Guerraggio (1998), Guerraggio and Luc (2001), Guerraggio et al (2001), Hiriart-Hurruty (1977), Hiriart-Hurruty et al (1984), Klatte et al. (1988), La Torre and Rocca (2002), Luc (2002), Michel et al (1994), Penot (1998), Rockafellar (1989), Rockafellar (1988), Yang and Jeyakumar (1992), Yang (1993), Yang (1996), Wang (1991), Ward (1994). Most of these higher-order approaches assume that the functions involved are of class $C^{1,1}$, that is once differentiable with locally Lipschitz gradient, or at least of class C^1. Anyway, another possibility, concerning the differentiation of nonsmooth functions dates back to the 30's and is related to the theory of Sobolev spaces (Sobolev (1988)) and the concept of "distributional derivative" (Schwartz (1966)). These techniques are widely used in the theory of partial differential equations but have not been applied to deal with optimization problems involving nonsmooth functions, until the works of Craven (1986) and Ermoliev et al (1995). More specifically, the approach followed by Ermoliev, Norkin and Wets appeals to some of the results of the theory of distributions; they define a sequence of smooth functions $f_\epsilon : \mathbb{R}^n \to \mathbb{R}$, depending on a parameter $\epsilon > 0$ and converging to the given function $f : \mathbb{R}^n \to \mathbb{R}$ by sending ϵ to 0. The family of smooth functions f_ϵ is built by convolution of f with a "sufficiently regular" kernel; the result is the regularity of f_ϵ does not depend on the differentiability properties of f but only on the regularity of the kernel. So if the kernel is at least of class C^2, one can define first and second-order generalized derivatives as the cluster points of all possible values of first and second-order derivatives of f_ϵ. For more details one can see Ermoliev et al (1995). In this paper, section 2 recalls the notions of mollifier, of epi-convergence of a sequence of functions and some definitions introduced in Ermoliev et al (1995); section 3 is devoted to the introduction of second-order derivatives for scalar functions by means of mollified functions; sections 4 deal with second-order necessary optimality conditions for multiobjective optimization problems.

2. Preliminaries

To follow the approach presented in Craven (1986) and Ermoliev et al (1995), we first need to introduce the notion of mollifier (see e.g. Brezis (1963)).

Definition 2.1 *A sequence of* mollifiers *is any sequence of functions* $\psi_\epsilon : \mathbb{R}^m \to \mathbb{R}_+$, $\epsilon \downarrow 0$, *such that:*

i) $\text{supp}\ \psi_\epsilon := \{x \in \mathbb{R}^m \mid \psi_\epsilon(x) > 0\} \subseteq \rho_\epsilon \text{cl}\mathcal{B},\ \rho_\epsilon \downarrow 0,$

ii) $\displaystyle\int_{\mathbb{R}^m} \psi_\epsilon(x)dx = 1,$

where $\mathcal{B}$ *is the unit ball in* $\mathbb{R}^n$, $\text{cl}X$ *means the closure of the set X and* dx *denotes Lebesgue measure.*

Definition 2.2 (Brezis (1963)) *Given a locally integrable function* $f : \mathbb{R}^m \to \mathbb{R}$ *and a sequence of bounded mollifiers, define the functions* $f_\epsilon(x)$ *through the convolution:*

$$f_\epsilon(x) := \int_{\mathbb{R}^m} f(x - z)\psi_\epsilon(z)dz.$$

The sequence $f_\epsilon(x)$ *is said a sequence of* mollified functions.

In the following all the functions considered will be assumed to be locally integrable.

Remark 2.1 There is no loss of generality in considering $f : \mathbb{R}^m \to \mathbb{R}$. The results in this paper remain true also if f is defined on an open subset of $\mathbb{R}^m$.

Some properties of the mollified functions can be considered classical.

Theorem 2.1 (Brezis (1963)) *Let* $f \in C(\mathbb{R}^m)$. *Then* f_ϵ *converges continuosly to* f, *i.e.* $f_\epsilon(x_\epsilon) \to f(x)$ *for all* $x_\epsilon \to x$. *In fact* f_ϵ *converges uniformly to* f *on every compact subset of* $\mathbb{R}^m$ *as* $\epsilon \downarrow 0$.

The previous convergence property can be generalized.

Definition 2.3 (Rockafellar and Wets (1998)) *A sequence of functions* $f_n : \mathbb{R}^m \to \mathbb{R}$ epi-converges *to* $f : \mathbb{R}^m \to \mathbb{R}$ *at* x, *if:*

i) $\liminf_{n\to+\infty} f_n(x_n) \geq f(x)$ *for all* $x_n \to x$;

ii) $\lim_{n\to+\infty} f_n(x_n) = f(x)$ *for some sequence* $x_n \to x$.

The sequence f_n *epi–converges to* f *if this holds for all* $x \in \mathbb{R}^m$, *in which case we write* $f = e - \lim f_n$.

Remark 2.2 It can be easily checked that when f is the epi–limit of some sequence f_n then f is lower semicontinuous. Moreover if f_n converges continuously, then also epi–converges.

Definition 2.4 (Ermoliev et al (1995)) *A function* $f : \mathbb{R}^m \to \mathbb{R}$ *is said* strongly lower semicontinuous *(s.l.s.c.) at* x *if it is lower semicontinuous at* x *and there exists a sequence* $x_n \to x$ *with* f *continuous at* x_n *(for all* n*) such that* $f(x_n) \to f(x)$. *The function* f *is strongly lower semicontinuous if this holds at all* x.
The function f *is said* strongly upper semicontinuous *(s.u.s.c.) at* x *if it is upper semicontinuous at* x *and there exists a sequence* $x_n \to x$ *with* f *continuous at* x_n *(for all* n*) such that* $f(x_n) \to f(x)$. *The function* f *is strongly upper semicontinuous if this holds at all* x.

Proposition 2.1 *If* $f : \mathbb{R}^m \to \mathbb{R}$ *is s.l.s.c., then* $-f$ *is s.u.s.c.* .

Proof. It follows directly from the definitions. □

Theorem 2.2 (Ermoliev et al (1995)) *Let* $\varepsilon_n \downarrow 0$. *For any s.l.s.c. function* $f : \mathbb{R}^m \to \mathbb{R}$, *and any associated sequence* f_{ϵ_n} *of mollified functions we have* $f = e - \lim f_{\epsilon_n}$.

Remark 2.3 It can be seen that, according to Remark 2.2, Theorem 2.1 follows from Theorem 2.2.

Theorem 2.3 *Let* $\varepsilon_n \downarrow 0$. *For any s.u.s.c. function* $f : \mathbb{R}^m \to \mathbb{R}$ *and any associated sequence* f_{ϵ_n} *of mollified functions, we have for any* $x \in \mathbb{R}^m$:

i) $\limsup_{n\to+\infty} f_{\epsilon_n}(x_n) \leq f(x)$ *for any sequence* $x_n \to x$;

ii) $\lim_{n\to+\infty} f_{\epsilon_n}(x_n) = f(x)$ *for some sequence* $x_n \to x$.

Proof. Since f is s.u.s.c., we have $-f$ s.l.s.c. and thus Theorem 2.2 applies:

i) for any sequence $x_n \to x$, $\liminf_{n\to+\infty} \left(-f_{\epsilon_n}(x_n)\right) \geq -f(x)$, which implies:

$$\limsup_{n\to+\infty} f_{\epsilon_n}(x_n) = -\liminf_{n\to+\infty} \left(-f_{\epsilon_n}(x_n)\right) \leq f(x);$$

ii) for some sequence $x_n \to x$, $\lim_{n\to+\infty}\left(-f_{\epsilon_n}(x_n)\right) = -f(x)$, from which we conclude:

$$\lim_{n\to+\infty} f_{\epsilon_n}(x_n) = f(x).$$

□

The following Proposition plays a crucial role in the sequel.

Proposition 2.2 (Schwartz (1966); Sobolev (1988)) *Whenever the mollifiers ψ_ϵ are of class C^k, so are the associated mollified functions f_ε.*

By means of mollified functions it is possible to define generalized directional derivatives for a nonsmooth function f, which, under suitable regularity of f, coincide with Clarke's generalized derivative. Such an approach has been deepened by several authors (see e.g. Craven (1986) and Ermoliev et al (1995)) in the first–order case.

Definition 2.5 (Ermoliev et al (1995)) *Let $f : \mathbb{R}^m \to \mathbb{R}$, $\epsilon_n \downarrow 0$ as $n \to +\infty$ and consider the sequence f_{ϵ_n} of mollified functions with associated mollifiers $\psi_{\epsilon_n} \in C^1$. The* upper mollified derivative *of f at x_0 in the direction $d \in \mathbb{R}^m$, with respect to (w.r.t.) the mollifiers sequence ψ_{ϵ_n} is defined as:*

$$\overline{D}_\psi f(x_0, d) := \sup_{x_n \to x_0} \limsup_{n\to+\infty} \nabla f_{\epsilon_n}(x_n)^\top d.$$

Similarly, we might introduce the following.

Definition 2.6 *Let $f : \mathbb{R}^m \to \mathbb{R}$, $\epsilon_n \downarrow 0$ as $n \to +\infty$ and consider the sequence f_{ϵ_n} of mollified functions with associated mollifiers $\psi_{\epsilon_n} \in C^1$. The* lower mollified derivative *of f at x_0 in the direction $d \in \mathbb{R}^m$, w.r.t. the mollifiers sequence ψ_{ϵ_n} is defined as:*

$$\underline{D}_\psi f(x_0, d) := \inf_{x_n \to x_0} \liminf_{n\to+\infty} \nabla f_{\epsilon_n}(x_n)^\top d.$$

In Ermoliev et al (1995) it has been defined also a generalized gradient w.r.t. the mollifiers sequence ψ_{ϵ_n}, in the following way:

$$\partial_\psi f(x_0) := \left\{ L := \limsup_{n\to+\infty} \nabla f_{\epsilon_n}(x_n), \quad x_n \to x_0 \right\}$$

i.e. the set of cluster points of all possible sequences $\nabla f_{\epsilon_n}(x_n)$ such that $x_n \to x_0$. Clearly (see e.g. Ermoliev et al (1995)) for the above mentioned upper mollified derivative it holds:

$$\overline{\mathcal{D}}_{\psi} f(x_0; d) \geq \sup_{L \in \partial_{\psi} f(x_0)} L^{\top} d,$$

$$\underline{\mathcal{D}}_{\psi} f(x_0; d) \leq \inf_{L \in \partial_{\psi} f(x_0)} L^{\top} d.$$

This generalized gradient has been used in Craven (1986) and Ermoliev et al (1995) to prove first–order necessary optimality conditions for nonsmooth optimization. The equivalence with the well–known notions of Nonsmooth Analysis is contained in the following proposition.

Proposition 2.3 (Ermoliev et al (1995)) *Let $f : \mathbb{R}^m \to \mathbb{R}$ be locally Lipschitz at x; then $\partial_{\psi} f(x)$ coincides with Clarke's generalized gradient and $\overline{\mathcal{D}}_{\psi} f(x_0, d)$ coincides with Clarke's generalized derivative (Clarke (1990)).*

Remark 2.4 From the previous proposition and the well–known properties of Clarke's generalized gradient, we deduce that, if f and $\psi_{\varepsilon} \in \mathcal{C}^1$, then $\partial_{\psi} f(x) = \nabla f(x)$.

Properties of these generalized derivatives and their applications to optimization problems are investigated in Craven (1986); Ermoliev et al (1995). By the way, for the aim of our paper, we will need to point out the following proposition (contained in Ermoliev et al (1995)) of which we give an alternative proof.

Proposition 2.4 *Let $f : \mathbb{R}^m \to \mathbb{R}$ and $x \in \mathbb{R}^m$. Then:*

i) $\overline{\mathcal{D}}_{\psi} f(\cdot; d)$ is upper semicontinuous (u.s.c.) at x for all $d \in \mathbb{R}^m$;

ii) $\underline{\mathcal{D}}_{\psi} f(\cdot; d)$ is lower semicontinuous (l.s.c.) at x for all $d \in \mathbb{R}^m$.

Proof. We can prove only *i)*, since *ii)* follows with the same reasoning. Assume $d \in \mathbb{R}^m$ is fixed. First we note that upper semicontinuity is obvious if $\overline{\mathcal{D}}_{\psi} f(x; d) = +\infty$. Otherwise, for all $K > \overline{\mathcal{D}}_{\psi} f(x; d)$, there exist a neighbourhood $U(x)$ and an integer n_0 so that:

$$\nabla f_{\epsilon_n}(x')^{\top} d < K, \qquad \forall n > n_0, \ \ \forall x' \in U(x).$$

Therefore, for each $x' \in U(x)$, we have:

$$\overline{\mathcal{D}}_{\psi} f(x'; d) = \sup_{x_n \to x'} \limsup_{n \to +\infty} \nabla f_{\epsilon_n}(x_n)^{\top} d \leq K,$$

which shows that $\overline{\mathcal{D}}_{\psi} f(\cdot; d)$ is u.s.c. indeed. □

Furthermore, we point out the following property, which might be recalled from Ermoliev et al (1995) or Crespi et al (2003):

Proposition 2.5 $\overline{D}_\psi f(x;\cdot)$ and $\underline{D}_\psi f(x;\cdot)$ *are positively homogeneous functions. Furthermore, if* $\overline{D}_\psi f(x;\cdot)$ *(*$\underline{D}_\psi f(x;\cdot)$ *respectively) is finite then it is subadditive (resp. superadditive) and hence convex (resp. concave) as a function of the direction* d.

3. Second–order mollified derivatives

As suggested in Ermoliev et al (1995), by requiring some more regularity of the mollifiers, it is possible to construct also second–order necessary and sufficient conditions for optimization problems. To do this we introduce the following:

Definition 3.1 *Let* $f : \mathbb{R}^m \to \mathbb{R}$, $\epsilon_n \downarrow 0$ *and consider the sequence of mollified functions* f_{ϵ_n}, *obtained from a family of mollifiers* $\psi_{\epsilon_n} \in \mathcal{C}^2$. *We define the* second-order upper mollified derivative *of* f *at* x_0 *in the directions* d *and* v, *w.r.t. to the mollifiers sequence* ψ_{ϵ_n}, *as:*

$$\overline{D}^2_\psi f(x;d,v) := \sup_{x_n \to x} \limsup_{n \to +\infty} d^\top H f_{\epsilon_n}(x_n) v,$$

where $Hf_{\epsilon_n}(x)$ *is the Hessian matrix of the function* $f_{\epsilon_n} \in \mathcal{C}^2$ *at the point* x.

In a similar way we give the following (see e.g. Crespi et al (2003)):

Definition 3.2 *Let* $f : \mathbb{R}^m \to \mathbb{R}$, $\epsilon_n \downarrow 0$ *and consider the sequence of mollified functions* f_{ϵ_n}, *obtained from a family of mollifiers* $\psi_{\epsilon_n} \in \mathcal{C}^2$. *We define the* second–order lower mollified derivative *of* f *at* x_0 *in the directions* d *and* v, *w.r.t. the mollifiers sequence* ψ_{ϵ_n}, *as:*

$$\underline{D}^2_\psi f(x;d,v) := \inf_{x_n \to x} \liminf_{n \to +\infty} d^\top H f_{\epsilon_n}(x_n) v.$$

Proposition 3.1 *Let* $f : \mathbb{R}^m \to \mathbb{R}$ *and* $x \in \mathbb{R}^m$.

i) If $\lambda > 0$, *then:*

$$\overline{D}^2_\psi \lambda f(x;d,d) = \lambda \overline{D}^2_\psi f(x;d,d);$$
$$\underline{D}^2_\psi \lambda f(x;d,d) = \lambda \underline{D}^2_\psi f(x;d,d).$$

Moreover, if $\lambda < 0$ *we get:*

$$\overline{D}^2_\psi \lambda f(x;d,d) = \lambda \underline{D}^2_\psi f(x;d,d).$$

ii) The maps $(\mathbf{d}, v) \to \overline{D}^2_\psi f(x; \mathbf{d}, v)$ *and* $(\mathbf{d}, v) \to \underline{D}^2_\psi f(x; \mathbf{d}, v)$ *are symmetric (that is* $\overline{D}^2_\psi f(x; d, v) = \overline{D}^2_\psi f(x; v, d)$ *and* $\underline{D}^2_\psi f(x; d, v) = \underline{D}^2_\psi f(x; v, d)$*).*

iii) The functions $\overline{D}^2_\psi f(x; \mathbf{d}, \cdot)$ *and* $\underline{D}^2_\psi f(x; \mathbf{d}, \cdot)$ *are positively homogeneous, whenever* $\mathbf{d} \in \mathbb{R}^m$*.*

iv) If $\overline{D}^2_\psi f(x; \cdot, \cdot)$ *(*$\underline{D}^2_\psi f(x; \cdot, \cdot)$ *resp.) is finite, then it is sublinear (superlinear).*

v) $\overline{D}^2_\psi f(x; \mathbf{d}, -v) = -\underline{D}^2_\psi f(x; \mathbf{d}, v)$*.*

vi) $\overline{D}^2_\psi f(\cdot; \mathbf{d}, v)$ *is upper semicontinuous (u.s.c.) at* x *for every* $\mathbf{d}$, $v \in \mathbb{R}^m$*.*

vii) $\underline{D}^2_\psi f(\cdot; \mathbf{d}, v)$ *is lower semicontinuous (l.s.c.) at* x *for every* $\mathbf{d}, v \in \mathbb{R}^m$*.*

In the following we will set for simplicity:

$$\overline{D}^2_\psi f(x; d) := \overline{D}^2_\psi f(x; d, d)$$

and:

$$\underline{D}^2_\psi f(x; d) := \underline{D}^2_\psi f(x; d, d).$$

Remark 3.1 Clearly the previous derivatives may be infinity. A sufficient condition for these derivatives to be finite is to require $f \in C^{1,1}$ (that is once differentiable with locally Lipschitz partial derivatives). In fact, in this case the second-order mollified derivatives can be viewed as first-order mollified derivatives of a locally Lipschitz function and thus Proposition 2.3 applies.

Remark 3.2 It is important to underline that the previous derivatives are dependent on the specific family of mollifiers which we choose and also on the sequence ϵ_n. Practically, by changing one of this choices we might obtain different result for $\overline{D}^2_\psi f(x; d, v)$. However, the results which follow hold true *for any* mollifiers sequence (provided they are at least of class C^2) and any choice of ϵ_n. Moreover, by Proposition 4.10 in Ermoliev et al (1995), we have that, if $f \in C^1$, then for any choice of the sequence of mollifiers and of ϵ_n, $\overline{D}^2_\psi f(x; d, v)$ coincides with:

$$\limsup_{x' \to x,\ t \downarrow 0} \frac{\nabla f(x' + td)^\top v - \nabla f(x')^\top v}{t}.$$

Using these notions of derivatives, we shall introduce a Taylor's formula for strongly semicontinuous functions, as it is proven in Crespi et al (2002a):

Theorem 3.1 (Lagrange Theorem and Taylor's formula) *Let $f : \mathbb{R}^m \to \mathbb{R}$ be a s.l.s.c. (resp. s.u.s.c.) function and let $\epsilon_n \downarrow 0$, $t > 0$ and $d \in \mathbb{R}^m$.*

i) *If $\psi_{\epsilon_n} \in C^1$ is a sequence of mollifiers, there exists a point $\xi \in [x_0, x_0 + td]$ such that:*

$$f(x_0 + td) - f(x_0) \le t\overline{D}_\psi f(\xi; d).$$
$$(f(x_0 + td) - f(x_0) \ge t\underline{D}_\psi f(\xi; d))$$

ii) *If $\psi_{\epsilon_n} \in C^2$ is a sequence of mollifiers, there exists $\xi \in [x_0, x_0 + td]$ such that:*

$$f(x_0 + td) - f(x_0) \le t\overline{D}_\psi f(x_0; d) + \tfrac{t^2}{2}\overline{D}^2_\psi f(\xi; d)$$
$$(f(x_0 + td) - f(x_0) \ge t\underline{D}_\psi f(x_0; d) + \tfrac{t^2}{2}\underline{D}^2_\psi f(\xi; d))$$

assuming that the righthand sides are well defined, i.e. it does not happen the expression $+\infty - \infty$.

4. Second order optimality conditions

Given $f : \mathbb{R}^m \to \mathbb{R}^l$ and a subset $X \subseteq \mathbb{R}^m$ we now consider the following multiobjective optimization problem:

$$VP) \qquad \min_{x \in X} f(x)$$

where $a \le_{\mathbb{R}^l} b$ if and only if $b - a \in \mathbb{R}^l_+$. For this type of problem the notion of weak solution is recalled in the following definition.

Definition 4.1 $x_0 \in X$ *is a local weak solution of VP) if there exists a neighbourhood U of* x_0 *such that* $f(U \cap X) - f(x_0) \cap -\mathrm{int}\mathbb{R}^l_+ = \emptyset$.

In the sequel, the following definitions of first order set approximations will be useful.

Definition 4.2 *Let* $x_0 \in \mathrm{cl}\ X$, *where* cl X *is the closure of the set X. The following sets:*

- $WF(X, x_0) = \{d \in \mathbb{R}^m | \exists t_n \downarrow 0, x_0 + t_n d \in X\}$
- $T(X, x_0) = \{d \in \mathbb{R}^m | \exists t_n \downarrow 0, d_n \to d, x_0 + t_n d_n \in X\}$

are called, respectively, the cone of weak feasible directions and *the* contingent cone.

Theorem 4.1 *Assume that f_i, $i = 1, \dots, l$, are s.l.s.c. functions. Let $\epsilon_n \downarrow 0$, $\psi_{\epsilon_n} \in C^1$, and $x_0 \in X$ be a local weak solution of VP). Then the following system has no solution on the set $WF(X, x_0)$:*

$$\overline{\mathcal{D}}_{\psi} f_i(x_0; d) < 0, \quad i = 1, \dots, l,$$

that is:

$$\max_{i=1,\dots,l} \overline{\mathcal{D}}_{\psi} f_i(x_0; d) \geq 0, \quad \forall d \in WF(X, x_0).$$

Proof. First we claim that $\forall d \in WF(X, x_0)$, $\not\exists \alpha = \alpha(d) > 0$ such that $\overline{\mathcal{D}}_{\psi} f_i(x_0 + td; d) < 0$, $\forall t \in (0, \alpha)$, $i = 1, \dots, l$. In fact, if such an α would exist, the mean value theorem would imply:

$$f_i(x_0 + td) - f_i(x_0) \leq t\overline{\mathcal{D}}_{\psi} f_i(\xi_i; d) < 0,$$

where $\xi_i \in [x_0, x_0 + td]$, $i = 1, \dots, l$ which contradicts the fact that x_0 is a local solution of VP). Hence, for any fixed $d \in WF(X, x_0)$ one can find a sequence $t_n \downarrow 0$ such that for all n it holds:

$$\overline{\mathcal{D}}_{\psi} f_{i^*}(x_0 + t_n d; d) \geq 0,$$

for some given i^*. Recalling that the first-order upper mollified derivative is u.s.c. at x_0, we obtain that $\overline{\mathcal{D}}_{\psi} f_{i^*}(x_0; d) \geq 0$ and hence we get the thesis. □

Remark 4.1 If f_i are C^1 functions, this result coincides with the classical necessary optimality condition for C^1 functions.

Definition 4.3 *The* set of the descent directions *for f at x_0 is:*

$$D_{\leq}(f, x_0) = \{d \in \mathbb{R}^m : \overline{\mathcal{D}}_{\psi} f(x_0; d) \in \mathbb{R}^l_-\},$$

where $\overline{\mathcal{D}}_{\psi} f(x_0; d) := \{\overline{\mathcal{D}}_{\psi} f_1(x_0; d), \dots, \overline{\mathcal{D}}_{\psi} f_l(x_0; d)\}$.

Theorem 4.2 *Assume that f_i, $i = 1, \dots, l$, are s.l.s.c. functions, $\epsilon_n \downarrow 0$, $\psi_{\epsilon_n} \in C^2$. If $x_0 \in X$ is a local weak minimum point then*

$$\max_{i \in I(x_0; d)} \overline{\mathcal{D}}^2_{\psi} f_i(x_0; d) \geq 0$$

for all $d \in WF(X, x_0) \cap D_{\leq}(f, x_0)$, where

$$I(x_0; d) = \left\{i : \overline{\mathcal{D}}_{\psi} f_i(x_0; d) = 0, i = 1 \dots l\right\}.$$

Proof. If $\overline{\mathcal{D}}^2_\psi f_i(x_0;d) = +\infty$ for some $i \in I(x_0;d)$ then the thesis is trivial. Suppose ab absurdo that there exists $d \in WF(X,x_0) \cap D_{\leq}(f,x_0)$ such that $\overline{\mathcal{D}}^2_\psi f_i(x_0;d) < 0$ for all $i \in I(x_0;d)$. Since $d \in WF(X,x_0)$ then there exists $x_n \to x_0$, $x_n = x_0 + t_n d$ and $t_n \downarrow 0$. If $i \notin I(x_0;d)$ then, using the upper semicontinuity property of $\overline{\mathcal{D}}_\psi f_i$, we have:

$$f_i(x_n) - f_i(x_0) \leq \overline{\mathcal{D}}_\psi f_i(\xi_n;d) < 0.$$

for some $\xi_n \in [x_0, x_0 + t_n d]$ and for n sufficiently large. If $i \in I(x_0;d)$ using Taylor's formula and the upper semicontinuity property of $\overline{\mathcal{D}}^2_\psi f_i$, we obtain:

$$f_i(x_n) - f_i(x_0) \leq \frac{t_n^2}{2}\overline{\mathcal{D}}^2_\psi f_i(\xi_n;d) < 0$$

where $\xi_n \in [x_0, x_0 + t_n d]$ and n sufficiently large. This implies that x_0 is not a local weak minimum point. □

We now consider the vector optimization problem subject to inequality constraints:

$$VP1) \qquad \min f(x)$$

$$\text{s.t.} \quad f_i(x) \leq 0$$

where $f : \mathbb{R}^m \to \mathbb{R}^l$ and $f_i : \mathbb{R}^m \to \mathbb{R}$, $i = l+1, \ldots, l+r$. Let:

$$\tilde{D}_{\leq}(f,x_0) = \{d \in \mathbb{R}^m : \ \overline{\mathcal{D}}_\psi f_i(x_0;d) \leq 0, i = l+1 \ldots l+r\}.$$

Theorem 4.3 *Let f_i, $i = 1, \ldots, l+r$ be s.l.s.c. functions and $\epsilon_n \downarrow 0$, $\psi_{\epsilon_n} \in C^2$. If $x_0 \in \mathbb{R}^m$ is a local weak minimum point for the problem VP1) then for all $d \in D_{\leq}(f,x_0) \cap \tilde{D}_{\leq}(f,x_0)$ we have*

$$\max_{i \in I(x_0;d) \cup J(x_0;d)} \overline{\mathcal{D}}^2_\psi f_i(x_0;d) \geq 0$$

where $J(x_0;d) = \{i : \overline{\mathcal{D}}_\psi f_i(x_0;d) = 0, i = l+1 \ldots l+r\}$.

Proof. If there exist $d \in D_{\leq}(f,x_0) \cap \tilde{D}_{\leq}(f,x_0)$ and $i \in I(x_0;d) \cup J(x_0;d)$ such that $\overline{\mathcal{D}}^2_\psi f_i(x_0;d) = +\infty$ then the thesis is trivial. By contradiction, let $d \in D_{\leq}(f,x_0) \cap \tilde{D}_{\leq}(f,x_0)$ such that $\overline{\mathcal{D}}^2_\psi f_i(x_0;d) < 0$ for all $i \in I(x_0;d) \cup J(x_0;d)$. Then for all $i \in I(x_0;d)$, using the upper semicontinuity property of $\overline{\mathcal{D}}^2_\psi f_i$, we have:

$$f_i(x_0 + td) - f_i(x_0) \leq \frac{t^2}{2}\overline{\mathcal{D}}^2_\psi f_i(\xi;d) < 0,$$

where $\xi \in [x_0, x_0 + td]$ and t is small enough. If $i \notin I(x_0; d)$ and $i \leq l$, then:

$$f_i(x_0 + td) - f_i(x_0) \leq t\overline{\mathcal{D}}_{\psi} f_i(\xi; d) < 0,$$

for some $\xi \in [x_0, x_0 + td]$ and t sufficiently small. In a similar way, for all $i \in J(x_0; d)$, we have:

$$f_i(x_0 + td) \leq f_i(x_0) + \frac{t^2}{2}\overline{\mathcal{D}}^2_{\psi} f_i(\xi; d) < 0$$

and, for all $i \notin J(x_0; d)$, $i > l$, we obtain:

$$f_i(x_0 + td) \leq f_i(x_0) + t\overline{\mathcal{D}}_{\psi} f_i(\xi; d) < 0.$$

that is $x_0 + td$ is feasible for all t sufficiently small. □

5. Second order characterization of convex vector functions

In this section we give a characterization of convex vector functions by means of second–order mollified derivatives. We remember that a vector function $f : \mathbb{R}^m \to \mathbb{R}^l$ is $\mathbb{R}^l_+$**-convex** if and only if each component f_i, $i = 1 \ldots l$, is convex. The following results are classical:

Lemma 5.1 (Zygmund (1959)) *Let $f : \mathbb{R}^m \to \mathbb{R}$ be a continuous function. Then f is convex if and only if:*

$$\frac{f(x + td) - 2f(x) + f(x - td)}{t^2} \geq 0,$$

$\forall x, d \in \mathbb{R}^m$, $\forall t \in \mathbb{R}$.

Lemma 5.2 (Evans et al (1992)) *Let $f : \mathbb{R}^m \to \mathbb{R}$ be a continuous function. Then f is convex if and only if the mollified functions f_ε, obtained from a sequence of mollifiers ψ_ε, are convex for every $\varepsilon > 0$.*

Theorem 5.1 *Let $f : \mathbb{R}^m \to \mathbb{R}$ be a continuous function and let $\varepsilon_n \downarrow 0$ and $\psi_{\varepsilon_n} \in C^2$. A necessary and sufficient condition for f to be convex is that:*

$$\underline{\mathcal{D}}^2_{\psi} f(x; d) \geq 0, \quad \forall x \in \mathbb{R}^m, \quad \forall d \in \mathbb{R}^m.$$

Proof. Necessity. By definition:

$$\underline{\mathcal{D}}^2_{\psi} f(x; d) = \inf_{x_n \to x} \liminf_{n \to +\infty} d^\top H f_{\varepsilon_n}(x_n) d.$$

Recalling the previous lemma, from the convexity of the functions f_ε, we have:

$$d^\top H f_{\varepsilon_n}(x) d \geq 0, \quad \forall x, d \in \mathbb{R}^m, \quad \forall n$$

and the necessity follows.
Sufficiency. We can write for every n:

$$\frac{f_{\varepsilon_n}(x+td)-2f_{\varepsilon_n}(x)+f_{\varepsilon_n}(x-td)}{t^2}=$$

$$=\frac{1}{t^2}\left[t\nabla f_{\varepsilon_n}(x)^\top d+\frac{t^2}{2}d^\top Hf_{\varepsilon_n}(\xi_n)d-t\nabla f_{\varepsilon_n}(x)d+\frac{t^2}{2}d^\top Hf_{\varepsilon_n}(\xi_n')d\right]=$$

$$=\frac{1}{2}\left[d^\top Hf_{\varepsilon_n}(\xi_n)d+d^\top Hf_{\varepsilon_n}(\xi_n')d\right],$$

where $\xi_n\in(x,x+td)$, $\xi_n'\in(x,x-td)$. As $n\to+\infty$ we can assume that $\xi_n\to\xi\in[x,x+td]$ and $\xi_n'\to\xi'\in[x,x-td]$; then we obtain:

$$\begin{aligned}\frac{f(x+td)-2f(x)+f(x-td)}{t^2}&=\lim_{n\to+\infty}\frac{1}{2}\left[d^\top Hf_{\varepsilon_n}(\xi_n)d+d^\top Hf_{\varepsilon_n}(\xi_n')d\right]\\&\geq\liminf_{n\to+\infty}d^\top Hf_{\varepsilon_n}(\xi_n)d+\liminf_{n\to+\infty}d^\top Hf_{\varepsilon_n}(\xi_n')d\\&\geq\frac{1}{2}\underline{\mathcal{D}}_\psi^2f(\xi;d)+\frac{1}{2}\underline{\mathcal{D}}_\psi^2f(\xi';d)\geq0.\end{aligned}$$

□

Corollary 5.1 *Let* $f:\mathbb{R}^m\to\mathbb{R}^l$ *be a continuous function and let* $\varepsilon_n\downarrow0$ *and* $\psi_{\varepsilon_n}\in C^2$. *A necessary and sufficient condition for* f *to be* $\mathbb{R}_+^l$*-convex is that:*

$$\underline{\mathcal{D}}_\psi^2f_i(x;d)\geq0,\quad\forall x\in\mathbb{R}^m,\forall d\in\mathbb{R}^m,\forall i=1\dots l.$$

References

Aghezzaf, B., Hachimi, M. (1999), Second-order optimality conditions in multiobjective optimization problems, *Journal of Optimization Theory and Applications,* 102, 37-50.

Bigi, G.C., Castellani, M. (2000), Second-order optimality conditions for differentiable multiobjective problems, *RAIRO Operations Research,* 34, 411-426.

Bonnans, J.F., Cominetti, R., Shapiro, A. (1999), Second order optimality conditions based on parabolic second order tangent sets, *SIAM Journal on Optimization,* 9, 2, 466-492.

Brezis, H. (1963), *Analyse fonctionelle – Theorie et applications,* Masson editeur, Paris.

Clarke, F.H. (1990), *Optimization and nonsmooth analysis,* SIAM, Classics in applied mathematics, Philadelphia.

Cominetti, R., Correa, R. (1990), A generalized second-order derivative in nonsmooth optimization, *SIAM Journal on Control and Optimization,* 28, 789-809.

Craven, B.D. (1986), Nondifferentiable optimization by smooth approximations, *Journal of Optimization Theory and Applications,* 17, 1, 1, 3-17.

Craven, B.D. (1989), Nonsmooth multiobjective programming, *Numerical Functional Analysis and Optimization,* 10, 49-64.

Crespi, G.P., La Torre, D., Rocca, M. (2003), Second-order mollified derivatives and optimization, *Rendiconti del Circolo Matematico di Palermo,* Serie II, Tomo LII, 251-262.

Crespi, G.P., La Torre, D., Rocca, M. (2003a), Second-order mollified derivatives and second-order optimality conditions, *Journal of Nonlinear and Convex Analysis,* 4, 3, 437-454.

Ermoliev, Y.M., Norkin, V.I., Wets, R.J.B. (1995), The minimization of semicontinuous functions: mollifier subgradient, *SIAM Journal on Control and Optimization,* 33, 1, 149-167.

Evans L.C., Gariepy R.F. (1992): *Measure theory and fine properties of functions.* CRC Press, 1992.

Ginchev, I., Guerraggio, A. (1998), Second order optimality conditions in nonsmooth unconstrained optimization, *Pliska Studia Mathematica Bulgarica,* 12, 39-50.

Guerraggio, A., Luc, D.T. (2001), On optimality conditions for $C^{1,1}$ vector optimization problems, *Journal of Optimization Theory and Applications,* 109, 3, 615-629.

Guerraggio, A., Luc, D.T., Minh, N.B. (2001), Second-order optimality conditions for C^1 multiobjective programming problems, *Acta Mathematica Vietnamica,* 26, 3, 257-268.

Hiriart-Hurruty, J.B. (1977), Contributions á la programmation mathematique: deterministe et stocastique. *Doctoral thesis,* Univ. Clermont-Ferrand.

Hiriart-Hurruty, J.B., Strodiot, J.J., Hien Nguyen, V.(1984), Generalized Hessian matrix and second-order optimality conditions for problems with $C^{1,1}$ data., *Applied Mathematics and Optimization,* 11, 43-56.

Kanniappan, P. (1983), Necessary conditions for optimality of nondifferentiable convex multiobjective programming, *Journal of Optimization Theory and Applications,* 40, 167-174.

Jeyakumar, V., Luc, D.T. (1998), Approximate Jacobian matrices for nonsmooth continuous maps and C^1-**optimization,** *SIAM Journal on Control and Optimization,* 36, 5, 1815-1832.

Klatte, D., Tammer, K. (1988), On second-order sufficient optimality conditions for $C^{1,1}$ optimization problems, *Optimization,* 19, 169-179.

La Torre, D., Rocca, M. (2000), $C^{k,1}$ functions and Riemann derivatives. *Real Analysis Exchange,* 25, 2, 743-752.

La Torre, D., Rocca, M. (2002), $C^{1,1}$ functions and optimality conditions, *Journal Computational Analysis Application,* to appear.

La Torre, D., Rocca, M.(2002), A characterization of $C^{k,1}$ functions, *Real Analysis Exchange,* 27, 2, 515-534.

Luc, D.T. (1995), Taylor's formula for $C^{k,1}$ functions, *SIAM Journal on Optimization,* 5, 659-669.

Luc, D.T. (2002), A multiplier rule for multiobjective programming problems with continuous data, *SIAM Journal on Optimization,* 13, 1, 168-178.

Majumdar, A.A.K. (1997) Optimality conditions in differentiable multiobjective programming, *Journal of Optimization Theory and Applications,* 1997, 419-427.

Michel, P., Penot, J.P. (1994), Second-order moderate derivatives, *Nonlinear Analysis,* 22, 809-824.

Michel, P., Penot, J.P. (1984), Calcul sous-differential pour des fonctions lischitziennes an nonlipschitziennes, *Comptes Rendus de l'Academie des Sciences Paris,* 298, 269-272.

Minami, M. (1983), Weak Pareto-optimal necessary optimality conditions in a nondifferentiable multiobjective program on a Banach space, *Journal of Optimization Theory and Applications,* 41, 451-461.

Penot, J-P. (1998), Second-order conditions for optimization problems with constraints, *SIAM Journal on Control and Optimization,* 37, 1, 303-318.

Preda, V. (1992), On some sufficient optimality conditions in multiobjective differentiable programming, *Kybernetica,* 28, 263-270.

Rockafellar, R.T. (1989), Second-order optimality conditions in nonlinear programming obtained by way of epi-derivatives, *Mathematics of Operations Research,* 14, 3, 462-484.

Rockafellar, R.T. (1988), First- and second-order epi-differentiability in nonlinear programming, *Transactions of the American Mathematical Society,* 307, 1, 75-108.

Rockafellar, R.T., Wets, R.J-B. (1998), *Variational Analysis,* Springer Verlag.

Schwartz, L. (1966), *Théorie des distributions,* Hermann, Paris.

Sobolev, S.L. (1988), *Some applications of functional analysis in mathematical physics,* 3rd ed., Nauka, Moscow.

Yang, X.Q., Jeyakumar, V. (1992), Generalized second-order directional derivatives and optimization with $C^{1,1}$ functions., *Optimization,* 26, 165-185.

Yang, X.Q., Second-order conditions in $C^{1,1}$ optimization with applications, *Numerical Functional Analysis and Optimization,* 14, 621-632.

Yang, X.Q. (1996), On second-order directional derivatives, *Nonlinear Analysis,* 26, 1, 55-66.

Wang, S.Y. (1991), Second order necessary and sufficient conditions in multiobjective programming, *Numerical Functional Analysis and Optimization,* 12, 237-252.

Ward, D.E. (1993), Calculus for parabolic second-order derivatives, *Set-valued analysis,* 1, 213-246.

Zemin, L. (1996), The optimality conditions of differentiable vector optimization problems, *Journal of Mathematical Analysis and Applications,* 201, 35-43.

Zygmund, A. (1959): *Trigonometric series.* Cambridge, 1959.

Chapter 14

SECOND ORDER SUBDIFFERENTIALS CONSTRUCTED USING INTEGRAL CONVOLUTIONS SMOOTHING

Andrew Eberhard*
Dept of Mathematics and Statistics, RMIT University, Australia

Michael Nyblom
Dept of Mathematics and Statistics, RMIT University, Australia

Rajalingam Sivakumaran
Dept of Mathematics and Statistics, RMIT University, Australia

Abstract In this paper we demonstrate that second order subdifferentials constructed via the accumulation of local Hessian information provided by an integral convolution approximation of the function, provide useful information only for a limited class of nonsmooth functions. When local finiteness of associated second order directional derivative is demanded this forces the first order subdifferential to possess a local Lipschitz property. To enable the study of a broader classes of nonsmooth functions we show that a combination of the infimal and integral convolutions needs to be used when constructing approximating smooth functions.

Keywords: Second order Subdifferentials, integral convolution, infimal convolution.

MSC2000: 49J52, 26B09

*email:andy.eb@rmit.edu.au

1. Introduction

The use of the integral convolution smoothing in nonsmooth analysis has a long history. Its application to first order subdifferentials for Lipschitz functions was probably first explicitly made by Craven (1986) and Craven (1986) but such ideas were implicitly used in earlier work of Warga (1975), Warga (1976), Halkin (1976) and Halkin (1976). The most comprehensive treatment may be found in the later work of Ermoliev et al (1995) which is later refined by Rockafellar et al (1998). The first comprehensive treatment of its use in deriving second order results may be found in the thesis Nyblom (1998) on which part of this paper is based. More recently Crespi et al (2002) have investigated second order notions in conjunction with optimality conditions (see also Nyblom (1998) for results in this direction).

The theory of generalized functions and distributions arose out of a need to furnish a rigorous framework for the definition of such quantities as the well known Dirac **δ-function.** We will assume the standard theory of distributions as may be found in Gariepy et al (1995). One can generate a distribution on $\mathcal{D}(\Omega) := \{\phi \in C^\infty(\Omega) \mid \{x \mid \phi(x) \neq 0\} \subseteq U$ a compact set$\}$ using any locally integrable function f on $\mathbb{R}^n$ by means of the definition

$$\langle f, \phi \rangle = \int_\Omega f(x)\phi(x)\,dx,$$

Closely allied with this generalized function is the familiar regularization of f given by

$$f(\beta, x) := \int_{\mathbb{R}^n} f(u)H_\beta(x-u)\,du = \int_{\mathbb{R}^n} f(x-u)H_\beta(u)\,du, \qquad (14.1)$$

where $H_\beta(\cdot)$ is a density function (usually with compact support) with mean zero and variance β. One appealing feature of the function defined in (14.1) is that it is always smooth. This smoothing operation has proven useful in many areas of optimization theory, in recent times the potential for it's use in non-smooth optimization has been exploited by numerous authors. A generalized second order directional derivative is used in Crespi et al (2002) to derive optimality conditions.

The smoothing process (14.2) may also be viewed as an averaging process associated with a random variable $\varepsilon(\beta)$. That is we may write $f(\beta, x) = E(f(x - \varepsilon(\beta))$ where E is the expectation operator associated with the process ε with a density function H_β. Thus we average the function values of f around the base point x. When f is almost everywhere equal to an absolutely continuous function (for example if f is locally Lipschitz) then one may apply integration by parts to obtain

$$E\nabla f(x-\varepsilon(\beta)) = \int_{\mathbb{R}^n} \nabla f(u) H_\beta(x-u)\, du = -\int_{\mathbb{R}^n} f(x-u)\nabla H_\beta(u)\, du \tag{14.2}$$

and depending on specific assumptions on the density one can also show that (14.2) equals $\nabla f(\beta, x) = \nabla E f(x-\varepsilon(\beta))$. Outside of this context ∇f is not necessarily defined densely and so (14.2) is used as a definition to define (up to a set of zero measure) a locally integrable function Df which is referred to as the generalized gradient of f i.e. for all test functions $\phi \in \mathcal{D}(\Omega) \subseteq C^\infty(\Omega)$

$$\int_\Omega Df(u)\phi(x-u)\, du = -\int_\Omega f(x-u)\nabla\phi(u)\, du.$$

Inductively we may extend this definition to higher derivatives $D^k f$. In standard texts on generalized functions it is shown that when $f \in W^{1,p}(\mathbb{R}^n)$ (where $W^{k,p}(\mathbb{R}^n) := \{f \mid D^k f \in L^p(\mathbb{R}^n) \text{ exists}\}$) we have (for $\phi = H_\beta$) $\nabla f(\beta, x) = \nabla E f(x-\varepsilon(\beta)) = EDf(x-\varepsilon(\beta))$ and so $\nabla f(\beta, x)$ is once again a local averaging of a function (the generalized derivative of f). To move away from functional definitions a pointwise estimate may be obtained by taking all accumulation points

$$\limsup_{\beta \downarrow 0} EDf(x-\varepsilon(\beta)) := \partial^H f(x).$$

The question of whether this relates to any kind of subgradient information is the subject of many of the papers we have mentioned so far and they all compare $\partial^H f(x)$ with the Clarke subgradient $\partial^o f(x)$. It suffices to assume f is a locally Lipschitz, subdifferentially regular function in order to ensure $\{\nabla f(\beta, x)\}_{\beta>0}$ converges to $\nabla f(x)$ when it exists a x (see Rockafellar et al (1998)). Clearly we may extend this to this second order level but it is also clear that even for locally Lipschitz functions we have no direct connection (i.e. via integration by parts) to a Hessians $\nabla^2 f$ (even when these exist densely). To study the effectiveness of this approximation we need to compare the second–order subdifferentials constructed using integral convolution smoothings with some other kind of subhessian. This is the purpose of this paper. By doing so we may study how effective these constructions are in capturing the essential second–order information associated with the function.

The second order directional derivatives defined in Crespi et al (2002) appearance to be well defined for the (very) large class of all strongly lower semi–continuous functions. The assumption that such quantities are finite over a neighbourhood of directions often implies the underlying function is actually para–concave (i.e. a function f is para–concave when there existence of a $\lambda > 0$ such that the function $g(x) := f(x) - \frac{1}{2\lambda}\|x\|^2$ is finite and concave). See Theorem 5.1 of this paper for such a result. An example where such assumptions occur is Crespi et al (2002). Indeed we must fall back on convexity or concavity properties via Alexandrov's theorem in order to obtain results ensuring $\{\nabla^2 f(\beta, x)\}_{\beta>0}$ provides an approximation of $\nabla^2 f(x)$ when it exists at x. The natural class of functions to which such approximations work are the para-convex or para concave functions (i.e. $-f$ is para–convex when f is para–concave). In this paper we show that one strategy that can be used to avoid such a sever restriction to the class of para-concave functions is to apply the infimal convolution approximation prior to the integral convolution smoothing since the infimal convolution produces an initial approximation with a para–concave function (when f is minorized by a quadratic function). It is beyond the scope of this paper to apply these ideas but we refer the reader to the recent thesis Sivakumaran (2003) where these ideas have found application in to the study of the relationship between weak solutions and viscosity solutions of elliptic partial differential equations and in the earlier thesis Nyblom (1998) which studies second order optimality conditions.

2. Preliminaries

In the following we will assume the reader has a working knowledge of variational analysis, nonsmooth analysis and the associated notion of convergence of sets taken from set–valued analysis (see Rockafellar et al (1998)). One may always assume we are using Kuratowski–Painlevé convergence notions (see Rockafellar et al (1998)). We will make frequent use of Alexandrov's theorem a version of which may be found in Rockafellar et al (1998), Theorem 13.51 and Corollary 13.42.

Denoting by $\mathcal{S}(n)$ the set of all real $n \times n$ symmetric matrices and by $\mathbb{R}_+$ (respectively $\overline{\mathbb{R}}$) the real intervals $[0, +\infty)$ (respectively $(-\infty, +\infty]$). In the following we will always consider functions $f : \mathbb{R}^n \to \overline{\mathbb{R}}$ to be at least lower semi–continuous and proper and denote by $\langle \cdot, \cdot \rangle$ the inner product on $\mathbb{R}^n$. Denote by δ_C the indicator function of a set C ($\delta_C(x) = 0$ if $x \in C$ and $+\infty$ otherwise). When C is a convex set in a vector space X denote by $0^+C = \{x \in X \mid C + x \subseteq C\}$ the recession directions of C. Let $S(C, h) := \sup\{\langle x, h\rangle \mid x \in C\}$ be the support function of C.

Definition 2.1 *Let $\{f, f^v : \mathbb{R}^n \to \overline{\mathbb{R}}, v \in W\}$ be a family of proper extended-real-valued functions, where W is a neighbourhood of w (in some topological space). Then the upper epi-limit is defined to be*

$$\text{e-ls}_{v\to w} f^v(x) = \sup_{\delta>0} \limsup_{v\to w} \inf_{x'\in B_\delta(x)} f^v(x') (\equiv \limsup_{v\to w} \inf_{x'\to x} f^v(x'))$$

The lower epi-limit $\text{e-li}_{v\to w} f^v$ *is given by*

$$\text{e-li}_{v\to w} f^v(x) = \sup_{\delta>0} \liminf_{v\to w} \inf_{x'\in B_\delta(x)} f^v(x') (\equiv \liminf_{v\to w} \inf_{x'\to x} f^v(x')).$$

When these two functions are equal, the epi-limit function $\text{e-lim}_{v\to w} f^v$ *is said to exist. In this case the sequence* f^v *is said to epi-converge to* f.

Let us now state formally the main reason for our interest in epi-convergence via the following result (see Attouch, (1984)). We denote $\inf_O f^v = \inf\{f^v(x) \mid x \in O\}$.

Theorem 2.1 *Let* $\{f^v \mid \mathbb{R}^n \to \overline{\mathbb{R}}, v \in V\}$ *be a variational family of lower semi-continuous functions and* $f : \mathbb{R}^n \to \overline{\mathbb{R}}$. *If* $\{f^v \mid \mathbb{R}^n \to \overline{\mathbb{R}}, v \in V\}$ *epi–converges we have for all* $x^v \to x$ *and* $\varepsilon_v \downarrow 0$ *with* $f^v(x^v) \leq \inf f^v + \varepsilon_v$ *that* $x \in \operatorname{argmin} f$.

The issue of when the sum of two epi–convergent functions is also epi–convergent arises frequently.

Theorem 2.2 *Suppose* $\{f^v \mid \mathbb{R}^n \to \overline{\mathbb{R}}, v \in W\}$ *and* $\{g^v \mid \mathbb{R}^n \to \overline{\mathbb{R}}, v \in W\}$ *be a variational family of proper lower semi–continuous functions and* $f, g : \mathbb{R}^n \to \overline{\mathbb{R}}$, *then*

1 *If* $\{f^v\}$ *and* $\{g^v\}$ *are epi-lower semi–continuous with respect to* f *and* g *at* x *then* $\{f^v+g^v\}$ *is epi-lower semi–continuous with respect to* $f+g$ *at* x.

2 *Suppose* $\{f^v\}$ *and* $\{g^v\}$ *are epi-upper semi–continuous with respect to* f *and* g *for all* x. *Then* $\{f^v+g^v\}$ *is epi-upper semi–continuous with respect to* $f+g$ *when* g *is continuous and* $\{g^v\}$ *uniformly converges to* g *on bounded subsets.*

Proof. The first part may be found as Corollary 2.6 in Robinson (1987). Condition 2 is well known, see Beer (1993) Theorem 7.15 (specialized to $\mathbb{R}$). □

One can similarly define an alternate convergence concept based on the set convergence of the hypographs of f.

Let us now define a number of subderivative concepts arising in nonsmooth analysis. We denote $x_v \to^f x$ to mean $x_v \to x$ and $f(x_v) \to f(x)$.

Definition 2.2 *Let* $f : \mathbb{R}^n \to \overline{\mathbb{R}}$ *be lower semi–continuous,* $\bar{x} \in \operatorname{dom} f$ *and* $\bar{x}$, z, $h \in \mathbb{R}^n$.

1 *A vector $y \in \mathbb{R}^n$ is called a proximal sub-gradient to f at $\overline{x}$ if for some $c > 0$*

$$f(x) \geq f(\overline{x}) + \langle y, x - \overline{x} \rangle - \frac{c}{2}\|x - \overline{x}\|^2$$

in a neighbourhood of $\overline{x}$. The set of all proximal sub-gradients to f at $\overline{x}$ is denoted $\partial_p f(\overline{x})$.

2 *The basic subdifferential is given by*

$$\partial f(x) = \limsup_{x' \to^f x} \partial_p f(x').$$

3 *A function $f : X \to \overline{\mathbb{R}}$ is said to be twice sub-differentiable (or possess a subjet) at x if the following set is nonempty*

$$\partial^{2,-} f(x) = \{(\nabla\varphi(x), \nabla^2\varphi(x)) : f - \varphi \text{ has a local minimum } x \text{ for some } \varphi \in C^2(Re^n)\}.$$

4 *The limiting subjet of f at x is defined to be;*

$$\underline{\partial}^2 f(x) = \limsup_{u \to^f x} \partial^{2,-} f(u).$$

5 *The set $\underline{\partial}^2 f(x,p) = \{Q \in \mathcal{S}(n) \mid (p, Q) \in \underline{\partial}^2 f(x)\}$ is called limiting subhessians of f.*

Place $-\partial^{2,-}(-f)(x,-p) = \partial^{2,+} f(x,p)$ and $-\partial^{2,-}(-f)(x) = \partial^{2,+} f(x)$. It must be stressed that these quantities may not exist everywhere but $\partial^{2,-} f(x)$ is defined densely. We extend the above notation to write $(x',p') \to^f (x,p)$ to mean "$x' \to x$ ", "$p' \to p$ " and "$f(x') \to f(x)$". The following is easily proved and so the proof is omitted.

Lemma 2.1 *Let $f : \mathbb{R}^n \to \overline{\mathbb{R}}$ be lower semicontinuous near x and finite at x.*

$$\underline{\partial}^2 f(x,p) = \limsup_{(x',p') \to^f (x,p)} \underline{\partial}^2 f(x',p'). \tag{14.3}$$

As noted earlier $-\mathcal{P}(n) \subseteq 0^+\partial^{2,-} f(x,p)$ and $\mathcal{P}(n) \subseteq 0^+\partial^{2,+} f(x,p)$ and so consequently the limiting quantities also contain these directions of recession (if non-empty). When there exists a pair $(p, Q) \in \partial^{2,-} f(x) \cap \partial^{2,+} f(x)$ then f is twice differentiable at x. Thus it is useful to consider the following related concept. Denote by $S_2(f) = \{x \in \text{dom}\,(f) \mid \nabla^2 f(x)$ exists $\}$ and let $(x_n, \nabla f(x_n)) \to_{S_2(f)} (\bar{x}, p)$ mean $\{x_n\} \subseteq S_2(f)$, $x_n \to^f \bar{x}$ and $\nabla f(x_n) \to p$.

Definition 2.3 *Denote*

$$\overline{D}^2 f(\bar{x},p) = \{Q \in \mathcal{S}(n) \mid \exists\{x_n\} \text{ s.t. } Q \in \limsup_{(x_n,\nabla f(x_n))\to_{S_2(f)}(\bar{x},p)} \{\nabla^2 f(x_n)\}\}.$$

Recall that a function f is para–convex if $f(\cdot) + \frac{1}{2\lambda}\|\cdot\|^2$ is convex for some $\lambda > 0$ sufficiently small and f is para–concave if $-f$ is para–convex. When a function is either para–convex or para–concave (or both) we have (by Alexandrov's theorem) $S_2(f)$ dense in dom (f). If f is simultaneously para–convex and para–concave then f is $C^{1,1}(\mathbb{R}^n)$ (see Eberhard (2000) for a proof and earlier references). The next observation was first made in Penot and later used in Ioffe et al (1997).

Theorem 2.3 *If f is lower semicontinuous then when $\underline{\partial}^2 f(\bar{x},p) \neq \emptyset$ we have*

$$\overline{D}^2 f(\bar{x},p) - \mathcal{P}(n) \subseteq \underline{\partial}^2 f(\bar{x},p). \tag{14.4}$$

If we assume in addition that f is continuous and para–concave function around $\bar{x}$ then equality holds in (14.4).

Let $\langle Q, B\rangle := \operatorname{tr} B^t Q$ for Q, $B \in \mathcal{S}(n)$ be the Frobenius inner product and note that the inner product with a rank one matrix is $\langle Q, uu^t\rangle = u^t Q u$. The following is found in Ralph (1990).

Definition 2.4 *Denote by $\mathcal{M}(n,m)$ the real $n \times m$ matrices.*

1. *The rank one hull of a set $\mathcal{A} \subseteq \mathcal{M}(n,m)$ is given by*

   $$\mathcal{A}^1 = \{A \in \mathcal{M}(n,m) \mid \langle A, vu^t\rangle \le q(\mathcal{A})(u,v)\ \forall v \in \mathbb{R}^m, u \in \mathbb{R}^n\}.$$

   where $q(\mathcal{A})(u,v) := \sup\{\langle Q, vu^t\rangle \mid Q \in \mathcal{A}\}$.

2. *A set $\mathcal{A}$ is said to be a rank one representer if $\mathcal{A}^1 = \mathcal{A}$.*

3. *When $\mathcal{A} \subseteq \mathcal{S}(n)$ (the real symmetric matrices) we denote the symmetric rank one support by $q(\mathcal{A})(u) := \sup\{\langle Q, uu^t\rangle \mid Q \in \mathcal{A}\}$ and the symmetric rank one hull*

   $$\mathcal{A}^1_s = \{A \in \mathcal{S}(n) \mid \langle A, uu^t\rangle \le q(\mathcal{A})(u) \text{for all } u \in \mathbb{R}^n\}.$$

4. *When $-\mathcal{P}(n) \subseteq 0^+\mathcal{A} \subseteq \mathcal{S}(n)$ we define the symmetric rank one barrier cone as $b^1(\mathcal{A}) := \{u \in \mathbb{R}^n \mid q(\mathcal{A})(u) < \infty\}$.*

Remark 2.1 *If we restrict attention to the real symmetric matrices $\mathcal{S}(n)$ and sets $\mathcal{A}$ such that $-\mathcal{P}(n) \subseteq 0^+\mathcal{A}$ then $q(\mathcal{A})(u,v) = +\infty$ unless $v = u$. Thus in this case we only need consider the symmetric supports*

$q(\mathcal{A})(u)$ *and* $\mathcal{A}_s^1 = \mathcal{A}^1$. *Indeed we always have* $-\mathcal{P}(n) \subseteq 0^+\{A \in \mathcal{S}(n) \mid \langle A, uu^t\rangle \leq q(\mathcal{A})(u)$ *for all* $u \in I\!R^n\}$ *and* $q(\mathcal{A} - \mathcal{P}(n))(u) = q(\mathcal{A})(u)$.

In the first order case ($\Delta_1 f(x,t,h) := \frac{1}{t}(f(x+th) - f(x))$ and f lower semi–continuous) when the support of $\partial f(\bar{x})$ is finite we have

$$S(\partial f(\bar{x}), h) \;=\; f^{\uparrow}(\bar{x}, h) := \limsup_{\substack{x' \to_f x \\ t \downarrow 0}} \inf_{h' \to h} \Delta_1 f(x', t, h') \quad (14.5)$$

$$\text{and } S(\partial_p f(\bar{x}), h) \;\leq\; f'_-(\bar{x}, h) := \liminf_{t \downarrow 0} \inf_{h' \to h} \Delta_1 f(\bar{x}, t, h') \quad (14.6)$$

When f is locally Lipschitz the support function of the Clarke subgradient $\partial^o f(\bar{x})$ determines the convex set $\partial^o f(\bar{x})$ uniquely and

$$S(\partial^o f(\bar{x}), h) = f^o(\bar{x}, h) := \limsup_{\substack{x' \to_f x \\ t \downarrow 0}} \Delta_1 f(x', t, h').$$

Place $\Delta_2 f(\beta, x, t, u) = 2\frac{f(\beta, x+tu) - f(\beta, x) - t\langle \nabla f(\beta, x), u\rangle}{t^2}$. The lower second order epi–derivative at $\bar{x}$ with respect to z and h is given by

$$f''_-(\bar{x}, z, h) := \liminf_{\substack{t \downarrow 0 \\ h' \to h}} \Delta_2 f(\bar{x}, t, z, h') = e\text{–li}_{\,t\downarrow 0} \Delta_2 f(\bar{x}, t, z, h).$$

and if $f''_-(\bar{x}, p, h) < +\infty$ then $f'_-(\bar{x}, h) = \langle p, h\rangle$.

It was first observed in Eberhard et al (1998) that for subjets we have a similar relation at the second order level.

$$\begin{aligned} q\left(\partial^{2,-} f(\bar{x}, p)\right)(u) &= \liminf_{\substack{t \to 0 \\ h' \to h}} \left(\frac{2}{t^2}\right) \left(f(\bar{x} + th') - f(\bar{x}) - t\langle p, h'\rangle\right) \\ &:= f_s(\bar{x}, p, u) = \min\{f''_-(\bar{x}, p, -u), f''_-(\bar{x}, p, u)\}. \end{aligned} \quad (14.7)$$

Hence if we work with subjets we are in effect dealing with objects dual to the lower, symmetric, second-order epi-derivative. The subhessian is always a closed convex set of matrices while $\underline{\partial}^2 f(\bar{x}, p)$ may not be convex (just as $\partial_p f(\bar{x})$ is convex while $\partial f(\bar{x})$ often is not). The following was first observed in Eberhard et al (1998) In general we have (see Ioffe et al (1997))

$$\begin{aligned} & q\left(\underline{\partial}^2 f(\bar{x}, p)\right)(h) := \sup\{\langle Q, hh^t\rangle \mid Q \in \underline{\partial}^2 f(\bar{x}, p)\} \quad \text{for all } h. \\ & \leq f^{\uparrow\uparrow}(\bar{x}, p, h) = \limsup_{(x', p') \to_{S_p(f)} (\bar{x}, p),\; t \downarrow 0} \;\inf_{u' \to h} \Delta_2 f(x', t, p', u') \end{aligned} \quad (14.8)$$

(the so called second order circa derivative). Equality holds when f is "prox–regular and subdifferentially continuous" (see Rockafellar et al (1998) and Eberhard (2000) for details) and when f is finite and para–concave (see Nyblom (1998)).

3. The Mollifier Subjet

In this section we shall investigate a new second order subdifferential which we shall call a mollifier sub/super Hessian. It is similar in construction to the limiting sub/super Hessian, in that it consists of accumulation points of symmetric matrices, which in this case are formed by the Hessians of the integral convolution smoothing of f. Our main aim here is to interrelate these new concepts with those of the previous sections, by determining a hierarchy of containments.

To begin, let us introduce the class of mollifiers we will be working with. The definition that follows was suggested in Remark 3.14 of Ermoliev et al (1995). This paper contained a more restrictive assumption of bounded support on the mollifiers but many of the results concerning the epi-convergence of the family $\{f(\beta, x)\}_{\beta>0}$ readily extented to the case of unbounded supports if $H(\beta, x)$ satisfies Definition 3.1.

Definition 3.1 *We call a family* $\{H(\beta, \cdot) \mid \beta > 0\}$ *of real valued functions on* $\mathbb{R}^n$ *a* C^m***-admissible*** *mollifier family for a locally integrable function* $f : \mathbb{R}^n \to \overline{\mathbb{R}}$ *if:*

1 $\int_{\mathbb{R}^n} H(\beta, z)dz = 1$ *for all* $\beta > 0$,

2 For all $\delta > 0$ *we have* $\lim_{\beta\downarrow 0} \int_{\|z\|>\delta} |f(x-z)|H(\beta, z)dz = 0$, *uniformly in a neighbourhood of* x,

3 $\lim_{\beta\downarrow 0} \int_{\|z\|>\delta} H(\beta, z)dz = 0$ *for all* $\delta > 0$ *and*

4 $f(\beta, x) = \int_{\mathbb{R}^n} f(z)H(\beta, x-z)dz$ *is a* $C^m(\mathbb{R}^n)$ *smooth function of* x.

We will call the resulting smoothing $f(\beta, x)$ an averaged function. Epiconvergence of the integral convolution is implied by the following property.

Definition 3.2 *A function* $f : \mathbb{R}^n \to \overline{\mathbb{R}}$ *is strongly lower semi-continuous at* x, *if it is lower semi-continuous at* x *and there exists a sequence* $x^v \to x$ *with* f *continuous at* x^v *(for all* v*) such that* $f(x^v) \to f(x)$. *The function is said to be strongly lower semi-continuous if this holds at all* x.

The following observation were made in Ermoliev et al (1995) for densities with finite supports (and extended in Nyblom (1998) for C^m-**admissible** mollifier families).

Theorem 3.1 *Suppose that $\{f(\beta,\cdot)|\beta>0\}$ is a family of averaged functions associated with a C^m-admissible family $\{H(\beta,\cdot)|\beta>0\}$ for a function f.*

1 *Suppose that f is continuous then the averaged functions $\{f(\beta,\cdot)|\beta>0\}$ converge uniformly to f on every bounded set in $\mathbb{R}^n$ and so must converge continuously (i.e. $f(\beta,x^\beta)\to f(x)$ for all $x^\beta\to x$ and $\beta\downarrow 0$).*

2 *For every strongly lower semi-continuous function f the family $\{f(\beta,\cdot)|\beta>0\}$ of averaged functions epi-converge to f.*

We now introduce a modification on the concept of a mollifier subgradient found in Ermoliev et al (1995). We say $x'\to^f x$ if and only if both $x'\to x$ and $f(x')\to f(x)$. Similarly $x^\beta\to^{f^H} x$ if and only if both $x^\beta\to x$, $\beta\downarrow 0$ and $f(\beta,x)\to f(x)$ (which is necessary when f is not continuous).

Definition 3.3 *Suppose that $f:\mathbb{R}^n\to\overline{\mathbb{R}}$ is integrable. Denote by $\{f(\beta,\cdot)\,|\,\beta>0\}$ the family of averaged functions associated with a C^1-admissible family $\{H(\beta,\cdot)\,|\,\beta>0\}$.*

1 *The mollifier subgradient set of f at x is*

$$\partial^H f(x)=\limsup\{\nabla f(\beta,x^\beta)\,|\,\beta\downarrow 0 \text{ and } x^\beta\to^{f^H} x\}$$

2 *The singular mollifier subgradient set is given by*

$$\partial^{H,\infty} f(x):=\limsup\{\lambda_\beta\nabla f(\beta,x^\beta)\,|\,\beta\downarrow 0,\ \lambda_\beta\downarrow 0 \text{ and } x^\beta\to^{f^H} x\}.$$

Note that when f is continuous and $x_n\to x$ we are guaranteed the existence of a sequence $\beta_n\downarrow 0$ such that $f(\beta_n,x_n)\to f(x)$ and so $x_n\to^{f^H} x$. If $H(\beta,\cdot)$ is such that there exists a function $\rho(\beta)\downarrow 0$ as $\beta\downarrow 0$ with the property that $\operatorname{supp} H(\beta,\cdot):=\{x\mid H(\beta,x)\neq 0\}\subseteq B_{\rho(\beta)}(0)$, then for any locally Lipschitz function, the diameter of $\partial^H f(x)$ is no greater than twice the local Lipschitz constant of f. For such a class of mollifiers it is easily verified that $\partial^H f(x)$ satisfies the following inclusion

$$\partial^H(f+g)(x)\subseteq\partial^H f(x)+\partial^H g(x), \tag{14.9}$$

for locally Lipschitz functions. Furthermore for such mollifiers and locally integrable functions it was noted in Ermoliev et al (1995) that for all $h\in\mathbb{R}^n$

$$\limsup_{\beta\downarrow 0,x^\beta\to x}\langle\nabla f(\beta,x^\beta),h\rangle\le f^o(x;h), \tag{14.10}$$

Consequently as $f^o(x;h)$ corresponds to the support function of the set $\partial^o f(x)$ for Lipschitz f and we deduce

$$\partial^H f(x) \subseteq \partial^o f(x). \tag{14.11}$$

Unfortunately if f is merely lower semicontinuous $f^o(x;h)$ will not correspond to the support of $\partial f(x)$ unless $h \in \text{int}\,\{h : S(\partial f(x;h),h) < \infty\}$ (in general when the support is finite it coincides with the first order circa derivative $f^{\uparrow}(\bar{x};h)$). This motivates the definition of $\partial^{H,\infty} f(x)$. In Nyblom (1998) it is shown that when $\partial^H f(x) \neq \emptyset$, $\partial^H g(x) \neq \emptyset$ and in addition we assume that $\partial^{H,\infty} f(x) \cap (-\partial^{H,\infty} g(x)) = \emptyset$ then $\partial^H (f+g)(x) \subseteq \partial^H f(x) + \partial^H g(x)$. As $\partial^H f(x) \subseteq \partial^o f(x)$ for mollifiers having bounded support we have $\partial^H f(x) = \{\nabla f(x)\}$ and $\partial^{H,\infty} f(x) = \{0\}$ if x is a point of strict differentiability. Thus if g is a strictly differentiable function we have from (14.9) that $\partial^H (g+f)(x) = \partial^H g(x) + \nabla f(x)$. We begin now by introducing the mollifier sub/superhessian. First we need to characterize the rank one support of the mollifier subhessians (and hence that of the mollifier super Hessians). It is convenient to make the following general assumption.

Axiom 1 *Suppose that $f : \mathbb{R}^n \to \overline{\mathbb{R}}$ is strongly lower semi–continuous and quadratically minorized. Let $H(\beta,\cdot)$ be a mollifier with finite mean values*

$$m_i(\beta) = \int_{\mathbb{R}^n} y_i H(\beta,y)\,dy \to 0 \text{ for } i = 1,\ldots,n, \text{ as } \beta \downarrow 0 \tag{14.12}$$

and a finite covariance matrix with components $(i,j = 1,\ldots,n)$

$$V_{ij}(\beta) = \int_{\mathbb{R}^n} y_i y_j H(\beta,y)\,dy \to 0 \text{ as } \beta \downarrow 0. \tag{14.13}$$

Theorem 3.2 *Suppose that $f : \mathbb{R}^n \to \overline{\mathbb{R}}$ and Axiom 1 holds. Then*

1 $\overline{\text{co}}\,\partial f(\bar{x}) = \overline{\text{co}}\,\partial^H f(\bar{x})$ *and* $\overline{\text{co}}\,\partial^\infty f(\bar{x}) = \overline{\text{co}}\,\partial^{H,\infty} f(\bar{x})$ *and so*

2 $\partial f(\bar{x}) = \partial^H f(\bar{x})$ *and* $\partial^\infty f(\bar{x}) = \partial^{H,\infty} f(\bar{x})$ *when f is Clarke regular.*

Thus $\partial^o f(x) = \overline{\text{co}}\left(\partial^H f(x) + \partial^{H,\infty} f(x)\right)$ *and the convex closure may be omitted if f is subdifferentially regular.*

Proof. The parts 1 and 2 have essentially been proved in Rockafellar et al (1998) and leave this up to the reader as an exercise. □

Definition 3.4 *Suppose that $f : \mathbb{R}^n \to \overline{\mathbb{R}}$ is integrable. Denote by $\{f(\beta,\cdot) \mid \beta > 0\}$ the family of averaged functions associated with a C^2-**admissible** family$\{H(\beta,\cdot) \mid \beta > 0\}$.*

1 The mollifier sub–Hessian of f at x is given by;

$$\partial^{2,H^{\uparrow}} f(x,p) \quad := \quad \limsup_{\substack{x^{\beta} \to_{f}^{H} x \\ \nabla f(\beta,x^{\beta}) \to p}} \left(\{\nabla^2 f(\beta, x^{\beta})\} - \mathcal{P}(n) \right)$$

2 The mollifier subjet of f at x is given by;

$$\partial^{2,H^{\uparrow}} f(x) := \{(p,Q) \mid Q \in \partial^{2,H^{\uparrow}} f(x,p)\}.$$

Place $\partial^{2,H^{\uparrow}} f(x,p) = -\partial^{2,H^{\downarrow}}(-f)(x,-p)$ to be the super Hessians and similarly $\partial^{2,H^{\uparrow}} f(x) = -\partial^{2,H^{\downarrow}}(-f)(x)$ the superjet. The mollifier subgradient like the limiting Hessians are robust concepts in the following sense. The simple proof (based on diagonalization of a nested set of sequences) is omitted.

Lemma 3.1 *Suppose that $f : \mathbb{R}^n \to \overline{\mathbb{R}}$ is integrable then*

$$\limsup_{x' \to_f x} \partial^H f(x') \subseteq \partial^H f(x) \text{ and } \limsup_{x' \to_f x} \partial^{2,H^{\uparrow}} f(x') \subseteq \partial^{2,H^{\uparrow}} f(x).$$

We may now state the main result of this section. The proof is taken from Nyblom (1998) and since it has not appeared elsewhere is placed in Appendix A.

Theorem 3.3 *Suppose that $f : \mathbb{R}^n \to \overline{\mathbb{R}}$ is strongly lower semi–continuous and minorized by a quadratic function. Let $\{H(\beta,\cdot) \mid \beta > 0\}$ be a C^2-**admissible** mollifier family that satisfies Axiom 1. Then*

1 $\partial_p f(x) \subseteq \partial^H f(x)$;

2 $\partial^{\infty} f(x) \subseteq \partial^{H,\infty} f(x)$ and $\partial f(x) \subseteq \partial^H f(x)$;

3 If $p \in \partial_p f(x)$ we have $\partial^{2,-} f(x,p) \subseteq \partial^{2,H^{\uparrow}} f(x,p)$ and so

4 $\underline{\partial}^2 f(x,q) \subseteq \partial^{2,H^{\uparrow}} f(x,q)$ for $q \in \partial f(x)$.

Proof. See Appendix A for the proof. □

Corollary 3.1 *Assume the hypotheses of Theorem 3.3. Then*

1 $\partial^o f(x) \subseteq \overline{\text{co}}\,(\partial^H f(x) + \partial^{H,\infty} f(x))$.

2 *Whenever* $\partial^{2,H^{\uparrow}} f(x,p) \subseteq \overline{D}^2 f(\bar{x},p) - \mathcal{P}(n)$ *we have* $\partial^{2,H^{\uparrow}} f(x,p) = \underline{\partial}^2 f(x,p) = \overline{D}^2 f(\bar{x},p) - \mathcal{P}(n)$.

Proof. As $\partial^{\infty} f(x) \subseteq \partial^{H,\infty} f(x)$ we have $\partial^{o} f(x) = \overline{\text{co}}\,(\partial f(x) + \partial^{\infty} f(x)) \subseteq \overline{\text{co}}\,(\partial^{H} f(x) + \partial^{H,\infty} f(x))$. To see 2 we only need invoke Theorem 3.3 part 3 and Corollary 2.3. □

4. Rank–1 Supports and Para–Concavity

In this section we investigate different ways that one can generate a rank–1 support to the set of matrices $\partial^{2,H^{\uparrow}} f(x,p)$ for some $p \in \partial^{H} f(x)$. This can quite effectively be done when f is para–concave (i.e. $f - \frac{1}{2\lambda}\|\cdot\|^2$ is finite–concave for some $\lambda > 0$). We may then use the standard approximation of f with its infimal convolution

$$f_{\lambda}(x) := \inf_{y} \left\{ f(y) + \frac{1}{2\lambda}\|x-y\|^2 \right\} \tag{14.14}$$

to obtain a para–concave approximation. When the infimum is attained we denote by $P_{\lambda}(x)$ the set of all such minima of (14.14). It is well known that this technique leads to a finite approximating function whenever f is prox–bounded (i.e. $f \geq \gamma - \frac{r}{2}\|\cdot\|^2$ which is equivalent to $f + \frac{r}{2}\|\cdot\|^2$ being bounded below, see Rockafellar et al (1998)). The condition $\lambda < r^{-1}$ is sufficient for $f_{\lambda} > -\infty$ (and hence $P_{\lambda}(x) \neq \emptyset$).

Regarding the variational behavior of the rank one support we have the following which may be found as Corollary 3.3 in Eberhard (2000).

Proposition 4.1 *Let* $\{\mathcal{A}(v)\}_{v \in W}$ *be a family of non-empty rank one representers and W a neighbourhood of* w. *Suppose that* $\limsup_{v \to w} \mathcal{A}(v) = \mathcal{A}(w)$ *then*

$$\limsup_{v \to w} \inf_{(u,h) \to (\bar{u},\bar{h})} q\left(\mathcal{A}(v)\right)(u,h) = q\left(\mathcal{A}(v)\right)(\bar{u},\bar{h}) \tag{14.15}$$

Remark 4.1 *When* $-\mathcal{P}(n) \subseteq 0^{+}\mathcal{A}(v)$ *for all* $v \in W$ *then (14.15) may be interpreted as being*

$$\limsup_{v \to w} \inf_{h \to \bar{h}} q\left(\mathcal{A}(v)\right)(h) = q\left(\mathcal{A}(v)\right)(\bar{h}) \tag{14.16}$$

From this result, Theorem 2.3, Theorem 3.3, equation (14.7) (and definitions) we immediately obtain the following.

Theorem 4.1 *Suppose that* $f : \mathbb{R}^n \to \overline{\mathbb{R}}$ *is strongly lower semi–continuous and minorized by a quadratic function. Let* $\{H(\beta,\cdot) \mid \beta > 0\}$ *be a*

C^2**-admissible** *mollifier family that satisfies Axiom 1. Then*

$$q\left(\overline{D}^2 f(x,p)\right)(h) \leq q\left(\underline{\partial}^2 f(x,p)\right)(h) = \limsup_{(x',p')\to_{S_p(f)}(x,p)} \inf_{h'\to h} f_s''(x',p',h')$$

$$\leq q\left(\partial^{2,H^\dagger} f(x,p)\right)(h) = \limsup_{\substack{(x^\beta,\nabla f(\beta,x^\beta))\to^{f^H}(x,p)\\ \beta\downarrow 0}} \inf_{u\to h} \langle \nabla^2 f(\beta,x), uu^t\rangle.$$

Remark 4.2 *In Crespi et al (2002) a second order directional derivative defined by taking a sequence of mollifiers $\psi(\beta_n,\cdot)\in C^2(\mathbb{R}^n)$ and placing*

$$\overline{\mathcal{D}}_H^2 f(x;u) := \sup_{x_n\to x} \limsup_{n\to+\infty} u^t\nabla^2 f(\beta_n,x_n)u\ .$$

From the standpoint of this study the dependence of the definition of $\overline{\mathcal{D}}_H^2 f(x;u)$ on a given sequence $\{\beta_n\}_{n\geq 0}$ is troubling. In general the results may depend sensitively on this sequence $\{\beta_n\}_{n\geq 0}$ without any a-priori way of predetermining its choice. Thus all we can do is compare the worst outcome. Henceforth we will take $\underline{\mathcal{D}}_H^2 f(x;u) = -\overline{\mathcal{D}}_H^2(-f)(x;u)$ where

$$\begin{aligned}\overline{\mathcal{D}}_H^2 f(x;u) &:= \sup_{\substack{\{x_n\to x\}\\ \{\beta_n\to 0\}}} \limsup_{n\to\infty} u^t\nabla^2 f(x_n,\beta_n)u\\ &= \limsup_{\substack{x'\to x\\ \beta\downarrow 0}} u^t\nabla^2 f(x',\beta)u.\end{aligned}$$

We immediately have under the assumption of Theorem 4.1 that

$$q\left(\overline{D}^2 f(\bar{x},p)\right)(u) \leq q\left(\underline{\partial}^2 f(x,p)\right)(u) \leq q\left(\partial^{2,H^\dagger} f(x,p)\right)(u) \leq \overline{\mathcal{D}}_H^2 f(x;u).$$

The rank-1 support of the mollifier subjet can sometimes be viewed as a generalized directional derivative.

Lemma 4.1 *Suppose that $f:\mathbb{R}^n\to\overline{\mathbb{R}}$ is strongly lower semi-continuous and minorized by a quadratic function. Let $\{H(\beta,\cdot)\mid\beta>0\}$ be a* C^2**-admissible** *mollifier family that satisfies Axiom 1. Let $x\in S_1(f)$ be a point of strict differentiability of f. Then for $p=\nabla f(x)$ we have*

$$\begin{aligned}f^{2,H^\dagger}(x,p,h) &:= \limsup_{\substack{(x^\beta,\nabla f(\beta,x^\beta))\to^{f^H}(x,p)\\ \beta\downarrow 0,\ t\downarrow 0}} \inf_{u\to h} \Delta_2 f(\beta,x^\beta,t,u)\\ &= q\left(\partial^{2,H^\dagger} f(x,p)\right)(h).\end{aligned}$$

Proof. Observe that as x is a point of strict differentiability we have for any $x' \to x$ that $\nabla f(\beta, x') \to \nabla f(x) = p$ (since $\partial^o f(x) = \{\nabla f(x)\}$). Then it follows that $\limsup_{\substack{x^\beta \to_{f^H} x \\ \beta \downarrow 0}} \{\nabla^2 f(\beta, x^\beta) - \mathcal{P}(n)\} = \partial^{2,H^\uparrow} f(x,p)$. Next observe that by the mean value theorem

$$\begin{aligned}
\Delta_2 f(\beta, x^\beta, t, u) &= \left(\frac{2}{t^2}\right)\left(f(\beta, x^\beta + tu) - f(\beta, x^\beta) - t\langle \nabla f(\beta, x^\beta), u\rangle\right) \\
&= \langle \nabla^2 f(\beta, x^\beta + ts^\beta u), uu^t\rangle
\end{aligned}$$

for some $s^\beta \in (0,1)$. Thus

$$\begin{aligned}
f^{2,H^\uparrow}(x,p,h) := & \limsup_{\substack{(x^\beta, \nabla f(\beta, x^\beta)) \to^{f^H} (x,p) \\ \beta\downarrow 0,\ t \downarrow 0}} \inf_{u \to h} \Delta_2 f(\beta, x^\beta, t, u) \\
= & \limsup_{\substack{(x^\beta, \nabla f(\beta, x^\beta)) \to^{f^H} (x,p) \\ \beta\downarrow 0,\ t \downarrow 0}} \inf_{u \to h} \langle \nabla^2 f(\beta, x^\beta + ts^\beta u), uu^t\rangle \\
\leq & \limsup_{\substack{x^\beta \to_{f^H} x \\ \beta \downarrow 0}} \inf_{u \to h} \langle \nabla^2 f(\beta, x^\beta), uu^t\rangle = q\left(\partial^{2,H^\uparrow} f(x,p)\right)(h)
\end{aligned}$$

using again the fact that for $x^\beta + ts^\beta u \to x$ that $\nabla f(\beta, x^\beta + ts^\beta u) \to \nabla f(x) = p$, due to strict differentiability. □

Place

$$[f](H_\beta)(x) = \int f(\eta) H_\beta(x - \eta)\, dm(\eta)$$

where m is the Lebesgue measure. We note that if $T(g) = [f](g)$(for $g \in \mathcal{D}(U)$) then $D_v T(g) = (DT)(g)(v) = -[f](\nabla g \cdot v)$ and so $D^2 T(g) = [f](\nabla^2 g)$ is the second order generalized derivative of the distribution T generated by f. We now use the fact that the Radon–Nikodym derivative $\frac{d}{dm}\psi$ of a Radon measure ψ may be viewed as a classical limiting process. In the following we are going to assume that $B(x,\delta)$ is the usual open ball centered around x but using the box norm $\|x\| = \max_{i=1,\ldots,n} \{x_i\}$ and so $m(B_{r/2}(x)) = r^n$. Some texts refer to these as a cube (i.e. a regular interval around x).

Definition 4.1 *Let ψ be a real-valued set function on subsets of $\mathbb{R}^n$ place*

$$\begin{aligned}
\overrightarrow{\psi}'(x) &= \sup_{\delta \to 0} \left\{ \sup_{r \in (0,\delta)} \frac{\psi(B_{r/2}(x))}{m(B_{r/2}(x))} \right\} \quad \text{and} \\
\underline{\psi}'(x) &= \inf_{\delta \to 0} \left\{ \inf_{r \in (0,\delta)} \frac{\psi(B_{r/2}(x))}{m(B_{r/2}(x))} \right\}.
\end{aligned}$$

If $\overline{\psi}'(x) = \underline{\psi}'(x) := \psi'(x)$ exists, we say ψ is differentiable at c *with respect to* m.

In this definition the use of symmetric neighbourhoods of x is not necessary (see Gariepy et al (1995)). We note that if $\psi = \psi_1 + \psi_2$ and both ψ_1' and ψ_2' exist then $\psi' = \psi_1' + \psi_2'$ exists.

The variation of a set-function $\mu : \mathcal{B}(U) \to \overline{\mathrm{IR}}$, for $B \in \mathcal{B}_c(U) := \{B \in \mathcal{B}(U) : \overline{B}$ is compact with $\overline{B} \subseteq U\}$ (where $\mathcal{B}(U)$ denotes all Borel measurable sets in U) and $\mathcal{P}(B) := \{F \mid F$ is a finite partition of B into disjoint Borel measurable sets $\}$, is defined by:

$$|\mu|(B) := \sup\{\sum_{B_i \in F} |\mu(B_i)| \quad : \quad F \in \mathcal{P}(B)\}.$$

From Gariepy et al (1995) (see Theorem 7.12 extended to signed measures, and Lemma 7.11) we have the following result.

Theorem 4.2 *Suppose μ is (signed) Radon measure on an open subset U of IR^n. Let $\mu = \mu_c + \mu_s$ be its decomposition into its absolutely continuous part $\mu_c \prec m$ (i.e. $m(E) = 0$ implies $\mu_c(E) = 0$ for any measurable E) and singular part $\mu_s \perp m$ (i.e. there exists a measurable set E such that $m(E) = 0$ and $|\mu_s|(E^c) = 0$ where $|\mu_s|$ denotes the variation of μ). Finally let $\frac{d\mu_c}{dm} \in L^1_{loc}(U)$ denote the Radon–Nikodym derivative of μ_c with respect to m. Then*

$$|\mu_s|'(x) = 0 \text{ and } \frac{d\mu_c}{dm}(x) = \mu'(x) \,.$$

for almost all $x \in U$ (w.r.t. Lebesgue measure m).

Combining Theorem 4.2 with Theorem 5.1 of Dudley (1977) we immediately obtain the following result.

Proposition 4.2 *Suppose that $f : U \to \mathrm{IR}$ is convex and $\mu := D^2[f]$. Then for all $x \in U$ a.e. (with respect to Lebesgue measure m)*

$$\frac{d}{dm}\left(D^2[f]\right)_c(x) = \nabla^2 f(x) = \mu'(x)$$

where $\left(D^2[f]\right)_c = \mu_c$ is the absolutely continuous part of μ.

It was observed in Mignot (1976) that the points at which a maximal monotone operator (like ∂f) is differentiable with respect to its domain of existence (that is $S_1(f)$), in the Fréchet sense, is a set of full measure. Thus we may assume that on $S_1(f)$ we have Fréchet differentiability of

∇f. We note that it is well known that for convex (or concave) functions f that if $x \in S \subseteq S_1(f)$ (the set of points of Fréchet differentiability) then x is also a point of continuous differentiability i.e. f is strictly differentiable at all $x \in S_1(f)$. Recall $\mathcal{D}(A)$ denotes the set of all $C^\infty(\mathbb{R}^n)$ real valued functions with compact support A and $S_2(f)$ the points at which the Hessian of f exist.

Theorem 4.3 *Suppose $U \subseteq \mathbb{R}^n$ is convex and open, that f is a finite concave (or convex) function. Suppose also that (for all $\beta > 0$) $H_\beta \in \mathcal{D}(\mathbb{R}^n)$ and there exist constants $c_\beta < 1 < C_\beta$; both tending to unity as $\beta \downarrow 0$ and functions $V'(\beta) < V(\beta) \downarrow 0$ such that $\lim_{\beta \to 0} \frac{V'(\beta)}{V(\beta)} = 1$,*

$$c_\beta \frac{1}{V(\beta)^n} \delta_{B_{\frac{V'(\beta)}{2}}(0)}(u) \leq H_\beta(u) \leq C_\beta \frac{1}{V(\beta)^n} \delta_{B_{\frac{V(\beta)}{2}}(0)}(u). \tag{14.17}$$

Then on a set of full (Lebesgue)measure $S \subseteq S_2(f)$ and any $x^\beta (\in S) \to x \in S$ with $\|x^\beta - x\| < \frac{V'(\beta)}{2}$ as $\beta \downarrow 0$ we have

$$\begin{aligned} \lim_{\beta \downarrow 0} \langle D^2[f], H_\beta(x^\beta - \cdot)\rangle &= \lim_{\beta \downarrow 0} \langle [\nabla^2 f], (H_\beta(x^\beta - \cdot)\rangle \\ &= \int \nabla^2 f(u) H_\beta(x^\beta - u)\, du = \lim_{\beta \downarrow 0} \nabla^2 f(\beta, x^\beta) = \nabla^2 f(x), \end{aligned}$$

where $f(\beta, x) := [f](H_\beta(\cdot - x))$ (i.e. for all $x \in U$,

$$f(\beta, x) := \int_U f(u) H_\beta(x - u)\, du$$

for all β sufficiently small so $x + \operatorname{Supp} H_\beta \subseteq U$). If in addition we have $q\left(\underline{\partial}^2 f(x, \nabla f(x))\right)(u) \leq K\|u\|^2 < \infty$ for all u and $x \in S$ then $D^2[f] \prec m$, implying

$$\nabla^2 f(\beta, x) = [\nabla^2 f](H_\beta(x - \cdot))\,(x) = \int \nabla^2 f(u) H_\beta(x - u)\, du. \tag{14.18}$$

In particular this implies $D^2[f](H_\beta(x - \cdot)) = [\nabla^2 f](H_\beta(x - \cdot)) = \nabla^2 f(\beta, x)$ on S.

Proof. See Appendix A for the proof. □

Theorem 4.4 *Suppose that f is a finite concave function on a domain with interior. Suppose also that $H_\beta \in \mathcal{D}(U)$ satisfies condition (14.17). Then on a set of full Lebesgue measure $S \subseteq S_1(f)$ and $x \in S \subseteq S_1(f)$ with β sufficiently small we have*

$$\nabla^2 f(\beta, x) = \nabla^2[f](H_\beta(x - \cdot))(x) = [\nabla^2 f](H_\beta(x - \cdot))(x). \tag{14.19}$$

In particular this implies the second order distributional derivative of the distribution $[f]$ *generated by* f *is given by*

$$D^2[f](H_\beta(x-\cdot))(x) = [\nabla^2 f](H_\beta(x-\cdot))\,(x)\,.$$

Also for $x \in \mathbf{int\,dom}\, f$ *we have*

$$\langle \nabla^2 f(\beta, x), hh^t\rangle \leq \; [q(\overline{D}^2 f(x-\cdot))(h)](H_\beta(\cdot))\,(x) \tag{14.20}$$

where $\overline{D}^2 f(x) := \limsup_{x'(\in S_2(f))\to x} \nabla^2 f(x')$. *In particular when* $x \in S_1(f)$ *we have*

$$\left(\partial^{2,H^\uparrow} f(x, \nabla f(x))\right)^1 = \left(\overline{D}^2 f(x, \nabla f(x)) - \mathcal{P}(n)\right)^1 = \left(\underline{\partial}^2 f(x, \nabla f(x))\right)^1.$$

Proof. Apply Theorem 6.2.7 of Nyblom (1998) to deduce that since f is finite concave and $x \in S_1(f)$ we have $q\left(\underline{\partial}^2 f(x,p)\right)(u) = f^{\uparrow\uparrow}(x,p,u)$. Then use the concavity of f and the super-gradient inequality to deduce that $f^{\uparrow\uparrow}(x,p,u) \leq 0$ for all u and $x \in \mathbf{int\,dom}\, f$. Thus the first part (14.19) following immediately from Theorem 4.3. Next note that as $q\left(\underline{\partial}^2 f(x, \nabla f(x))\right)(h) = f^{\uparrow\uparrow}(x, \nabla f(x), h) \leq 0$ for all h and $x \in S \subseteq S_1(f)$ it follows from (14.4) that $q(\overline{D}^2 f(x, \nabla f(x)))(h) = q(\underline{\partial}^2 f(x, \nabla f(x)))(h)$ and we have $\langle \nabla^2 f(u), hh^t\rangle \leq 0$ for h and $u \in V \cap S$ for a sufficiently small neighbourhood V of any $x \in \mathbf{int\,dom}\, f$. Thus for all $x \in \mathbf{int\,dom}\, f$ we have

$$\langle [\nabla^2 f](H_\beta(x-\cdot))\,(x)\,, hh^t\rangle = \int \langle \nabla^2 f(u), hh^t\rangle H_\beta(x-u)\,dm \leq 0.$$

Thus we may apply Fatou's Lemma to $\langle \nabla^2(-f)(u), hh^t\rangle \geq 0$ for fixed $\beta > 0$ to obtain for any $x \in \mathbf{int\,dom}\, f$

$$\begin{aligned}
\langle \nabla^2(-f)(\beta, x), hh^t\rangle &= \lim_{x'(\in S)\to x} \int \langle \nabla^2(-f)(u), hh^t\rangle H_\beta(x'-u)\,dm \\
&\geq \int \langle \nabla^2(-f)(u), hh^t\rangle \left(\liminf_{x'(\in S)\to x} H_\beta(x'-u)\right) dm \\
&= \langle [\nabla^2(-f)](H_\beta(x-\cdot))\,(x)\,, hh^t\rangle \qquad (14.21)
\end{aligned}$$

where we have used linearity of the integral to obtain the last equality. Now observation that Fatou's Lemma also implies for any $k(x') \to h$ (as

$x' \to x$) to

$$\liminf_{x'(\in S)\to x} \sup_{k\in B_\delta(h)} \int \langle \nabla^2(-f)(u), kk^t\rangle H_\beta(x'-u)\,dm$$
$$\geq \liminf_{x'(\in S)\to x} \int \langle \nabla^2(-f)(x'-u), (k(x'))(k(x'))^t\rangle H_\beta(u)\,dm$$
$$\geq -\int \limsup_{x'(\in S)\to x} \langle \nabla^2 f(x'-u), (k(x'))(k(x'))^t\rangle H_\beta(u)\,dm$$

and so

$$\limsup_{x'(\in S)\to x} \inf_{k\in B_\delta(h)} \langle \nabla^2 f(\beta, x'), kk^t\rangle$$
$$\leq \int \limsup_{x'(\in S)\to x} \langle \nabla^2 f(x'-u), (k(x'))(k(x'))^t\rangle H_\beta(u)\,dm.$$

Thus we are able to write for all $\delta > 0$, using the monotone convergence theorem,

$$\limsup_{x'(\in S)\to x} \inf_{k\in B_\delta(h)} \langle \nabla^2 f(\beta, x'), kk^t\rangle$$
$$\leq \int \min_{\{k(x')\to h\}} \limsup_{x'(\in S)\to x} \langle \nabla^2 f(x'-u), (k'(x'))(k'(x'))^t\rangle H_\beta(u)\,dm.$$

Using the fact that $(x,h) \mapsto \langle \nabla^2 f(x,\beta), hh^t\rangle$ is uniformly continuous on bounded sets for all $\beta > 0$, on taking the supremum over $\delta > 0$ we have

$$\langle \nabla^2 f(\beta, x), hh^t\rangle = \text{e–ls}_{x'(\in S)\to x} \langle \nabla^2 f(\beta, x'), hh^t\rangle$$
$$\leq \int \text{e–ls}_{x'(\in S)\to x} \langle \nabla^2 f(x'-u), hh^t\rangle H_\beta(u)\,dm$$
$$= \int q(\overline{D}^2 f(x-u))(h) H_\beta(u)\,dm.$$

One may argue directly from (14.21) by bounding the integral by the rank–1 support of the convex hull of the Hessians $\{\nabla^2 f(x-u) \mid x-u \in \operatorname{supp} H_\beta \cap S_2(f)\}$ and then using (14.16) that

$$\text{e–ls}_{\substack{x^\beta \to^{f^H} x \\ \nabla f(\beta, x^\beta)\to p}} \langle \nabla^2 f(\beta, x^\beta), hh^t\rangle = q\left(\partial^{2,H^\dagger} f(x,p)\right)(h) \leq q\left(\overline{D}^2 f(x)\right)(h).$$

When $x \in S_1(f)$ as f is concave we have x a point of strict differentiability of f and so $\nabla f(\beta, x^\beta) \to \nabla f(x) = p$ for all $x^\beta \to^{f^H} x$ and also

$\overline{D}^2 f(x) = \bar{D}^2 f(x,p)$. Finally the above inequality between rank-1 supports implies $\left(\partial^{2,H^\dagger} f(x,p)\right)^1 \subseteq \left(\overline{D}^2 f(x,p) - \mathcal{P}(n)\right)^1$. Using Theorem 2.3 we have $\overline{D}^2 f(\bar{x},p) - \mathcal{P}(n) = \underline{\partial}^2 f(\bar{x},p)$ in full generality (when f is concave). Also Theorem 3.3 gives $\underline{\partial}^2 f(\bar{x},p) \subseteq \partial^{2,H^\dagger} f(x,p)$ resulting the following string of inclusions

$$\begin{aligned} \left(\underline{\partial}^2 f(\bar{x},p)\right)^1 &\subseteq \left(\partial^{2,H^\dagger} f(x,p)\right)^1 \\ &\subseteq \left(\bar{D}^2 f(x,p) - \mathcal{P}(n)\right)^1 = \left(\underline{\partial}^2 f(\bar{x},p)\right)^1 \end{aligned}$$

which establish equality. □

The following simple result may be found as Lemma 3.2.9 in Sivakumaran (2003).

Lemma 4.2 *For any function*

$$(f - \langle p,\cdot\rangle)_\lambda (x) = f_\lambda(x + \lambda p) - \langle p, x\rangle - \frac{\lambda}{2}\|p\|^2 . \qquad (14.22)$$

Thus $f(\bar{x}) + (\lambda/2)\|\bar{p}\|^2 = f_\lambda(\bar{x} + \lambda\bar{p})$ *implies* $(f - \langle \bar{p},\cdot\rangle)_\lambda (\bar{x}) = f(\bar{x}) - \langle \bar{p}, \bar{x}\rangle$, *or equivalently* $(\bar{p}, \frac{1}{2\lambda} I) \in \partial^{2,-} f(\bar{x})$.

The following is immediate from (14.16), Lemma 4.1 and Theorem 4.4.

Corollary 4.1 *Suppose that f is lower semi–continuous and quadratically minorized. Suppose also that $H_\beta \in \mathcal{D}(U)$ satisfies condition (14.17). Then if $x \in S_1(f)$ and $p = \nabla f(x)$*

$$\begin{aligned} f_\lambda^{2,H^\dagger}(x + \lambda p, p, h) &= q\left(\partial^{2,H^\dagger} f_\lambda(x + \lambda p, p)\right)(h) \\ &= q\left(\left(\underline{\partial}^2 f_\lambda(x + \lambda p, p)\right)^1\right)(h) \\ \text{and} \quad \left(\underline{\partial}^2 f(x,p)\right)^1 &= \limsup_{\lambda \downarrow 0} \left(\partial^{2,H^\dagger} f_\lambda(x + \lambda p, p)\right)^1 . \end{aligned} \qquad (14.23)$$

5. Restricting the Class of Mollifier in Constructions

At a "kink" in a function there will be a discontinuity in $\nabla f(x)$ which will inevitably result in an infinite "curvature" in certain directions. As a bridge between smooth and non-smooth analysis we investigate the limiting behaviour of:

$$h^T \nabla^2 E(f(x - \varepsilon(\beta)))h = \frac{d^2}{dt^2} E[f(x + th - \varepsilon(\beta)) \mid_{t=0} . \qquad (14.24)$$

but issues of finiteness arise leading to the need for the following concept.

Definition 5.1 *A function $f : \mathbb{R}^n \to \overline{\mathbb{R}}$ is said to be second order regular at $x \in S_1(f)$ with respect to h if and only if $u \to \langle \nabla f(u), h \rangle$ is locally radially Lipschitz at x with respect to $S_1(f)$.*

Clearly this is less restrictive than assuming the function is $C^{1,1}(\mathbb{R}^n)$. To handle densities with unbounded supports one needs the following restricted class of Lipschitz functions. Let $\operatorname{supp} H(\beta.\cdot) := \{u \mid H(\beta, u) \neq 0\}$ denote the support of the density H.

Definition 5.2 *Let $H(\beta, x)$ be a density function on $\mathbb{R}^n$ with a radially symmetric convex support* $\operatorname{supp} H(\beta.\cdot)$ *with* $\operatorname{int}\operatorname{supp} H(\beta.\cdot) \neq \emptyset$.

1 $H(x, \beta) \to \delta'(x)$ *the Dirac* δ-**function**, *as* $\beta \to 0$.

2 $E(\varepsilon(\beta)) = \int_{\mathbb{R}^n} uH(\beta, u)\, du = 0$

3 $Var(\varepsilon(\beta)) = \int_{\mathbb{R}^n} \|u\|^2 H(\beta, u)\, du = n\beta^2$.

4 *For all* $\epsilon \geq 0$ *and* $\eta \geq 0$ $\exists \hat{\eta} > 0$ *such that for* $0 < \beta \leq \hat{\beta}$ *we have*

$$P(\varepsilon(\beta) \notin B(0, \eta)) = \int_{B(0,\eta)^c} H(\beta, u)\, du < \epsilon$$

5 $E(\varepsilon(\beta)^k) < \infty \quad \forall k$

6 *For all integrable* $g(\cdot) : \mathbb{R}^n \to \mathbb{R}$ *and all* $\eta > 0$ *we have both*

$$\int_{B(0,\eta)} H(\beta, u) g(u)\, du = \int_{B(0,1)} H(\frac{\beta}{\eta}, u) g(\pm \eta u)\, du.$$

These properties are possessed by many useful densities such as the normal distribution and other useful distributions with finite support are constructed using:

$$H(\beta, u) = \frac{1}{V_n(\beta)} U\left(1 - \frac{\|u\|^2}{2\beta^2}\right) \tag{14.25}$$

where $V_n(\beta)$ is renormalization factor and $U \in C^\infty(\mathbb{R})$ with $U(t) = 0$ for $t < 0$.

Definition 5.3 *Given a density function $H(\beta, \cdot)$ with a support* $S_\beta := \operatorname{supp} H(\beta.\cdot)$ *put*

$$\begin{aligned} L^H_{app}(\mathbb{R}^n) \;=\; & \{f \in L_{loc}(\mathbb{R}^n) \mid \forall x,\ \exists D \in \mathbb{R}^+,\ \beta_0 > 0 \text{ and } k \in N^+ \text{ s.t.} \\ & \forall\, 0 < \beta < \beta_o \text{ and } \forall u \in x - S_\beta \\ & \text{we have } d^*(\partial^o f(u), 0) \leq D(1 + \|u\|^k)\}, \end{aligned}$$

where $L_{loc}(\mathbb{R}^n)$ *denotes the set of locally Lipschitz functions defined on* $\mathbb{R}^n$.

This class will at least contain all non-smooth functions arising as the supremum of finitely many smooth functions. We note that if $H(\beta,\cdot)$ is of bounded support we may force $\mathcal{S}_\beta \subseteq B_\varepsilon(0)$ for $\beta > 0$ small. Then $L^H_{app}(\mathbb{R}^n) = L_{loc}(\mathbb{R}^n)$ since on a bounded set there exists a Lipschitz constant applicable to the whole set and the range of the Clarke subgradient multi–function is locally contained in a ball of radius given by this local Lipschitz constant. For densities of unbounded support the functions of this class can be loosely described as those for which the local Lipschitz constant does not grow, as a function of locality u, faster than some polynomial in $\|u\|$.

The following result (see Nyblom (1998), Proposition 6.4.1) is only one of a number of similar results that can be proved.

Proposition 5.1 *Let* $\{H(\beta,x) \mid \beta > 0\}$ *be a* C^2*–admissible family of mollifiers with density with mean zero, variance* $\beta > 0$, $f \in L^H_{app}(\mathbb{R}^n)$.

1 *Let* $H(\beta,x)$ *be the normal density with mean zero, variance* $\beta > 0$ *and* $x \in S_1(f)$. *Suppose* f *is second–order regular at* x *with respect to* h, *where* $\|h\| \doteq 1$. *Then*

$$\limsup_{\beta \downarrow 0} \mid \hat{h}^T \nabla^2 E(f(x - \varepsilon(\beta)))h \mid \leq nK$$

for all $\hat{h} \in B_1(0)$ *and some* $K > 0$. *In fact K may be taken as the local radial Lipschitz constant of* $\langle \nabla f(x), h\rangle$ *on* $S_1(f)$ *around* x.

2 *Conversely suppose that for all* $u \in B_\delta(x) \cap S_1(f)$ *and for all* $\hat{h} \in B_1(0)$

$$\limsup_{\beta \downarrow 0} \mid \hat{h}^T \nabla^2 E(f(u - \varepsilon(\beta)))h \mid \leq nK. \tag{14.26}$$

Then $\langle \partial f(x), h\rangle := \{\langle z, h\rangle \mid z \in \partial f(x)\}$ *is single valued and locally Lipschitz on* $B_\delta(x)$. *If* f *is regular then* $\langle \nabla f(x), h\rangle$ *is locally Lipschitz relative to* $S_1(f) \cap B_\delta(x)$.

We now investigate the connection that mollifier subjets have to the second order directional derivative of R. Cominetti and R. Correa in Cominetti et al (1990). It turns out that for $C^{1,1}$ functions all second order concepts discussed generate the same rank one hull (in the symmetric sense).

Definition 5.4 *The generalized second–order directional derivative of a function* $f : \mathbb{R}^n \to \overline{\mathbb{R}}$ *at* $x \in \mathbb{R}^n$ *in the direction* $(u, v) \in \mathbb{R}^n \times \mathbb{R}^n$ *is defined by*

$$f^{\infty}(x,u,v) = \limsup_{y\to x, t,s\downarrow 0} \frac{f(y+su+tv) - f(y+su) - f(y+tv) + f(y)}{st}$$

and the generalized Hessian of f *at* x *as the point-to-set mapping* $\partial^2 f(x) : \mathbb{R}^n \to \mathcal{K}(\mathbb{R}^n)$ *(the convex subsets of* $\mathbb{R}^n$*) given by*

$$\partial^2 f(x)(u) = \{x' \in \mathbb{R}^n : \langle x', v\rangle \le f^{\infty}(x,u,v) \text{ for all } v \in \mathbb{R}^n\}.$$

For the rest of this section we will assume that $H(\beta,\cdot)$ is a mollifier which satisfies the assumptions (1) to (6) of Definition 5.2. Recall remark 4.2.

Proposition 5.2 *Let* f *be locally Lipschitz and suppose* $\hat{h}^T\nabla^2 E(f(x - \varepsilon(\beta)))h$ *is generated via convolution involving a density function* $H(\beta,u)$ *having bounded support. If the directions* $h, \hat{h}$ *are chosen such that* $f^{\infty}(x,\hat{h},h) > -\infty$ *then*

$$\limsup_{\substack{\beta\downarrow 0\\ x'\to x}} \hat{h}^T\nabla^2 E(f(x' - \varepsilon(\beta)))h \le f^{\infty}(x;\hat{h},h). \tag{14.27}$$

Proof. The result trivially holds if $f^{\infty}(x,\hat{h},h) = +\infty$ thus we assume $f^{\infty}(x,\hat{h},h) < +\infty$. By definition

$$\hat{h}^T\nabla^2 E(f(x' - \varepsilon(\beta)))h = \frac{d}{dt}\langle E\nabla f(x' + t\hat{h} - \varepsilon(\beta)), h\rangle \,|_{t=0}$$
$$= \lim_{t\downarrow 0}\langle \frac{E\nabla f(x' + t\hat{h} - \varepsilon(\beta)) - E\nabla f(x' - \varepsilon(\beta))}{t}, h\rangle.$$

Now $E\nabla f(x' + t\hat{h} - \varepsilon(\beta)) = \int_{(x'+t\hat{h}-S_1(f))} H(\beta,u)\nabla f(x' + t\hat{h} - u)\,du$ and

$$E(\nabla f(x' - \varepsilon(\beta))) = \int_{x'-S_1(f)} H(\beta,u)\nabla f(x' - u)\,du.$$

So for an arbitrary sequences $t_n \downarrow 0$ and $x_m \to x$ we have

$$\hat{h}^T\nabla^2 E(f(x_m - \varepsilon(\beta)))h$$
$$= \lim_{n\to\infty}\int_S H(\beta,u)\langle \frac{\nabla f(x_m + t_n\hat{h} - u) - \nabla f(x_m - u)}{t_n}, h\rangle\,du$$

where $S = \cap_{m=0}^{\infty} \cap_{n=0}^{\infty} \Big((x_m + t_n \hat{h} - S_1(f)) \cap (x_m - S_1(f)) \Big)$ is of full measure. Let $u \to_S 0$ mean that both $u \in S$ and $u \to 0$ then

$$\limsup_{\substack{m,n\to\infty \\ u\to_S 0}} \langle \frac{\nabla f(x_m + t_n\hat{h} - u) - \nabla f(x_m - u)}{t_n}, h \rangle$$
$$= \limsup_{\substack{m,n\to\infty \\ u\to_S 0}} \lim_{t'\to 0} \frac{1}{t'} [\Delta_{(x_m-u),t_n}(t') - \Delta_{(x_m-u),t_n}(0)]$$

where $\Delta_{(x_m-u),t_n}(t') = \frac{1}{t_n}[f(x_m + t_n\hat{h} + t'h - u) - f(x_m + t'h - u)]$. Thus

$$\limsup_{\substack{m,n\to\infty \\ u\to_S 0}} \langle \frac{\nabla f(x_m + t_n\hat{h} - u) - \nabla f(x_m - u)}{t_n}, h \rangle \leq f^{\infty}(x; \hat{h}, h).$$

Hence for m and n sufficiently large and $\varepsilon > 0$ we have for all $u \in S \cap N_\eta$ (an n-dimensional cube around the origin) and η small that

$$H(\beta, u) \langle \frac{\nabla f(x_m + t_n\hat{h} - u) - \nabla f(x_m - u)}{t_n}, h \rangle \leq H(\beta, u)(f^{\infty}(x; \hat{h}, h) + \varepsilon).$$

Integrating and using $\limsup_{\beta\downarrow 0} \int_{S\cap N_\eta} H(\beta, u)\, du \leq 1$ we obtain

$$\limsup_{\substack{m\to\infty \\ \beta\downarrow 0}} \hat{h}^t \nabla^2 E(f(x_m - \varepsilon(\beta)))h \leq f^{\infty}(x; \hat{h}, h) + \varepsilon$$
$$+ \limsup_{\substack{m\to\infty \\ \beta\downarrow 0}} \lim_{n\to\infty} \int_{S\cap N_\eta^c} H(\beta, u) \langle \frac{\nabla f(x_m + t_n\hat{h} - u) - \nabla f(x_m - u)}{t_n}, h \rangle\, du.$$

Placing $u = \beta u'$ in the last integral of the previous inequality and recalling property (6) of Definition 5.2 we obtain

$$\int_{S'\cap N_{\eta/\beta}^c} H(1, u) \langle \frac{\nabla f(x_m + t_n\hat{h} - \beta u) - \nabla f(x_m - \beta u)}{t_n}, h \rangle\, du \tag{14.28}$$

where $S' = S/\beta$. Since $H(1,u)$ has a bounded support S_H there exists $\beta_o > 0$ such that $(S' \cap N_{\eta/\beta}^c) \cap S_H = \emptyset$ for $\beta < \beta_o$. Thus for $\beta < \beta_o$ the integral in (14.28) is identically zero. Finally as ε arbitrary, we deduce

$$\limsup_{\substack{m\to\infty \\ \beta\downarrow 0}} \hat{h}^T \nabla^2 E(f(x_m - \varepsilon(\beta)))h \leq f^{\infty}(x; \hat{h}, h).$$

As $x_m \to x$ was arbitrary we have (14.27). □

Corollary 5.1 *Suppose that $f : \mathbb{R}^n \to \mathbb{R}$ is locally Lipschitz then if $p \in \partial f(x)$ we have*

$$\begin{aligned}&\underline{\partial}^2 f(x,p) \subseteq \partial^{2,H^\uparrow} f(x,p)\\ &\quad \subseteq \{Q \in \mathcal{S}(n) \mid \langle Q, hh^t\rangle \leq f^\infty(x,h,h) \text{for all } h\} := \left(\partial_s^2 f(x)\right)^1.\end{aligned}$$

If $f \in C^{1,1}(\mathbb{R}^n)$ then on taking the symmetric rank one hull in $\mathcal{S}(n)$ we have

$$\left(\underline{\partial}^2 f(x,\nabla f(x))\right)^1 = \left(\partial^{2,H^\uparrow} f(x,\nabla f(x))\right)^1 = \left(\partial_s^2 f(x)\right)^1.$$

Proof. By (14.16) we have for all h that

$$q(\partial^{2,H^\uparrow} f(x,p))(h) = \limsup_{\substack{\beta\downarrow 0\\ x^\beta \to_f^H x}} \inf_{u\to h} \langle \nabla^2 f(\beta, x^\beta), uu^t\rangle \leq f^\infty(x,h,h) \quad (14.29)$$

and so $\partial^{2,H^\uparrow} f(x,p) - \mathcal{P}(n) \subseteq \{Q \in \mathcal{S}(n) \mid \langle Q, hh^t\rangle \leq f^\infty(x,h,h) \text{ for } h\}$ while the other containment follow from Theorem 3.3.

In Páles et al (1996) (page 61) it was noted that if $f \in C^{1,1}(\mathbb{R}^n)$ we have $f^\infty(x,u,v) = \sup\{\langle A, uv^t\rangle \mid A \in \overline{D}^2 f(x,\nabla f(x)) \text{ for all } u,v \in \mathbb{R}^n\}$. By Corollary 2.3 and Theorem 3.3 we thus have

$$\begin{aligned}&\overline{D}^2 f(x,\nabla f(x)) - \mathcal{P}(n) \subseteq \partial^{2,H^\uparrow} f(x,\nabla f(x))\\ &\quad \subseteq \left(\partial_s^2 f(x)\right)^1 = \left(\overline{D}^2 f(x,\nabla f(x)) - \mathcal{P}(n)\right)^1.\end{aligned}$$

□

Appendix: A

Proof. (**of Theorem 3.3**) Take a $z \in \partial_p f(x)$ then there exists a $c > 0$ such that for all y in a neighbourhood of x we have

$$f(y) \geq f(x) + \langle z, y - x\rangle - \frac{c}{2}\|y - x\|^2.$$

Thus $(z, cI) \in \partial^{2,-} f(x)$. Hence we only need to demonstrate 3 and 1 will follow as a consequence of the first part of this proof. Thus we take $(p, X) \in \partial^{2,-} f(x)$ and $\phi(y)$ as promised in Proposition 6 of Eberhard et al (1998) where $\widetilde{c} \geq \|X\| + c$ and $\gamma - \frac{c}{2}\|x - y\|^2$ minorizes f. Thus for all $\lambda > 0$ there exists a $g_{\tilde{c}}(y) = -\varepsilon(\|x - y\|)\|x - y\|^2 \in C^2(\mathbb{R}^n)$ with $(\nabla g_{\tilde{c}}(x), \nabla^2 g_{\tilde{c}}(x)) = (0, 0)$, along with $\varepsilon(\|x - y\|) = \tilde{c}$ for $y \notin B_M(x)$ for which

$$\begin{aligned}\psi_\lambda(y) \;&:=\; f(y) - \phi(y) = f(y) - \langle p, y - x\rangle\\ &\qquad - \frac{1}{2}\langle X, (y - x)(y - x)^t\rangle + \lambda\|y - x\|^2 - g_{\tilde{c}}(y)\end{aligned}$$

has a strict global minimum at x. We note that for $y \notin B_M(x)$ we have $\psi_\lambda(y) = f(y) - \langle p, y - x\rangle - \frac{1}{2}\langle X, (y - x)(y - x)^t\rangle + (\lambda + \widetilde{c})\|y - x\|^2$. As f is minorized by

$\gamma - \frac{c}{2}\|x-y\|^2$ we have for $y \notin B_M(x)$ that $\psi_\lambda(y) \geq \gamma - \langle p, y-x\rangle + (\lambda + \tilde{c} - \frac{1}{2}(\|X\| + \frac{c}{2}))\,\|y-x\|^2 := h_\lambda(x)$ a strictly convex function. Thus $\mathbf{lev}_\alpha \psi_\lambda \subseteq \mathbf{lev}_\alpha h_\lambda$ which is bounded and convex for any α. Let $H(\beta,y)$ be a C^2 mollifier compatible with f which has expectation $\int_{\mathbb{R}^n} u_i H(\beta,u)\,du = m_i(\beta)$ and variance $\int_{\mathbb{R}^n} |u_i|^2 H(\beta,u)\,du = V_{ii}(\beta)$. As $\psi_\lambda - h_\lambda \geq 0$ we have on convolution $\psi_\lambda(\beta,y) - h_\lambda(\beta,y) \geq 0$ for all y and $\beta > 0$ and thus $\mathbf{lev}_\alpha \psi_\lambda(\beta,\cdot) \subseteq \mathbf{lev}_\alpha h_\lambda(\beta,\cdot)$. On convolution of $H(\beta,\cdot)$ with h_λ we get

$$\begin{aligned} h_\lambda(\beta,y) = \gamma - \langle p, y-x\rangle + \langle p, m(\beta)\rangle + (\lambda + \tilde{c} - \frac{1}{2}(\|X\| + \frac{c}{2}))(\|y-x\|^2 \\ -2\langle y-x, m(\beta)\rangle + \sum_{i=1}^{n} V_{ii}(\beta)) \end{aligned}$$

a quadratic strictly convex function and so the set $\mathbf{lev}_\alpha h_\lambda(\beta.\cdot)$ is bounded for all α and $\beta > 0$. Taking the integral convolution of ψ_λ we have

$$\begin{aligned} \psi_\lambda(\beta,y) &= f(\beta,y) - \langle p, y-x\rangle + \langle z, m(\beta)\rangle \\ &\quad -\frac{1}{2}\int_{\mathbb{R}^n} \langle X, (y-x-u)(y-x-u)^t\rangle H(\beta,u)\,du \\ &\qquad +\lambda \int_{\mathbb{R}^n} \|y-x-u\|^2\, H(\beta,u)\,du - g_{\tilde{c}}(\beta,y) \\ &= f(\beta,y) - \langle p, y-x\rangle + \langle z, m(\beta)\rangle - \frac{1}{2}\langle X, (y-x)(y-x)^t\rangle \\ &\quad +(y-x)^t(X + 2\lambda I)m(\beta) + \sum_{ij} X_{ij} V_{ij}(\beta) \\ &\quad +\lambda\left(\|y-x\|^2 + \sum_{i=1}^{n} V_{ii}(\beta)\right) - g_{\tilde{c}}(\beta,y) \end{aligned} \tag{14.A.1}$$

Since f is strongly lower semi–continuous by Theorem 3.1 we have $e\text{–}\lim_{\beta\downarrow 0} f(\beta,y) = f(y)$. As $g_{\tilde{c}}(y) \in C^2(\mathbb{R}^n)$ by the same theorem we have $g_{\tilde{c}}(\beta,y)$ converging uniformly on bounded sets and so epi–converges to g where $\mathbf{dom}\, g = \mathbb{R}^n$. Thus by the Theorem 2.2 we have $e\text{–}\lim_{\beta\downarrow 0} \psi_\lambda(\beta,y) = \psi_\lambda(y)$. As $\psi_\lambda(y)$ has a strict global minimum at x we have by Theorem 2.1 for any $x^\beta \in \mathbf{argmin}\, \psi_\lambda(\beta,\cdot)$ that $x^\beta \to x$. Now such minima are assured to exists since $y \mapsto \psi_\lambda(\beta,y)$ are continuous and for all $\beta > 0$ and $\mathbf{lev}_\alpha \psi_\lambda(\beta,\cdot)$ are bounded.

As $\psi_\lambda(\beta,\cdot)$ has a global minimum at x^β we have

$$\begin{aligned} \psi_\lambda(\beta,x^\beta) &\leq \psi_\lambda(\beta,y) \text{ for all } y \text{ and so} \\ \limsup_{\beta\downarrow 0} \psi_\lambda(\beta,x^\beta) &\leq \limsup_{\beta\downarrow 0} \psi_\lambda(\beta,y) \leq \mathrm{cl}_h \psi_\lambda(y) \text{ for all } y. \end{aligned}$$

As f is strongly lower semi–continuous we also have ψ_λ strongly lower semi–continuous and so there exists a sequence $y^k \to x$ with ψ_λ continuous at each y^k and $\psi_\lambda(y^k) \to \psi_\lambda(x)$. In particular for each such k we have $\mathrm{cl}_h \psi_\lambda(y^k) = \psi_\lambda(y^k)$ by the upper semi–continuity of ψ_λ at y^k. Thus

$$\limsup_{\beta\downarrow 0} \psi_\lambda(\beta,x^\beta) \leq \lim_{k\to\infty} \mathrm{cl}_h \psi_\lambda(y^k) = \psi_\lambda(x).$$

As ψ_λ has a strict local minimum at x it follows that

$$\begin{aligned}\psi_\lambda(x) &\leq \psi_\lambda(y) \text{ for all } y \text{ and so}\\ \psi_\lambda(x)\int_{\mathbb{R}^n} H(\beta,w)\,dw &\leq \int_{\mathbb{R}^n}\psi_\lambda(x^\beta - w)H(\beta,w)\,dw \text{ implying}\\ \psi_\lambda(x) &\leq \psi_\lambda(\beta,x^\beta) \quad \text{for all } \beta > 0.\end{aligned}$$

It follows that

$$\psi_\lambda(x) \leq \liminf_{\beta\downarrow 0}\psi_\lambda(\beta,x^\beta) \leq \limsup_{\beta\downarrow 0}\psi_\lambda(\beta,x^\beta) \leq \psi_\lambda(x)$$

and so $\psi_\lambda(\beta,x^\beta) \to \psi_\lambda(x)$. On inspection of (14.A.1) one can see that the only component of $\psi_\lambda(\beta,x^\beta)$ that does not converge uniformly on bounded sets is $f(\beta,y)$ and so it follows that $x^\beta \to^{f^H} x$. As each x^β is a local minimum of a smooth function we have $\nabla\psi_\lambda(\beta,x^\beta) = 0$ and $\nabla^2\psi_\lambda(\beta,x^\beta) \geq 0$ (i.e. positive semi–definite). The first order condition gives on application to (14.A.1)

$$\begin{aligned}0 &= \nabla\psi_\lambda(\beta,x^\beta) = \nabla f(\beta,x^\beta) - p - X(x^\beta - x)\\ &\qquad +(X+2\lambda I)m(\beta) + 2\lambda(x^\beta - x) - \nabla g_{\bar{c}}(\beta,x^\beta).\end{aligned}$$

For $y \notin B_M(0)$ we have $g_{\bar{c}}(y) = -\bar{c}\|y-x\|^2$ and so it follows from assumptions that for $t \leq \frac{M}{\|h\|}$ we have

$$\begin{aligned}&\lim_{\beta\downarrow 0}\int_{\|z\|>2M+\|x\|} |g_{\bar{c}}(x+th-z)|H(\beta,z)\,dz\\ &\qquad = \bar{c}\lim_{\beta\downarrow 0}\int_{\|z\|>2M+\|x\|}\|x+th-z\|^2 H(\beta,z)\,dz = 0,\\ \text{and}\quad &\lim_{\beta\downarrow 0}\int_{\|z\|>2M+\|x\|}\|\nabla g_{\bar{c}}(x-z)\|\,H(\beta,z)\,dz\\ &\qquad = -2\bar{c}\lim_{\beta\downarrow 0}\int_{\|z\|>2M+\|x\|}\|x-z\|\,H(\beta,z)\,dz = 0.\end{aligned}$$

As noted earlier $x^\beta \to x$ and as $g_{\bar{c}}(\cdot) \in C^2(\mathbb{R}^n)$ has $\nabla g_{\bar{c}}(x) = 0$ we may infer from (14.11) an the strict differentiability of g that $\lim_{\beta\downarrow 0}\nabla g_{\bar{c}}(\beta,x^\beta) = 0$ as $\partial^H g_{\bar{c}}(x) \subseteq \partial^o g_{\bar{c}}(x) = \{\nabla g_{\bar{c}}(x)\} = \{0\}$. Hence

$$\begin{aligned}\lim_{\beta\downarrow 0}\nabla f(\beta,x^\beta) &= \lim_{\beta\downarrow 0}(p + X(x^\beta - x) - (X+2\lambda I)m(\beta)\\ &\qquad -2\lambda(x^\beta - x) + \nabla g_{\bar{c}}(\beta,x^\beta)) = p,\end{aligned}$$

as $m(\beta) \to 0$ giving $p \in \partial^H f(x)$. Having established the inclusion $\partial_p f(x) \subseteq \partial^H f(x)$ the inclusion $\partial f(x) \subseteq \partial^H f(x)$ follows immediately from Lemma 3.1 and the robust nature of $\partial f(x)$.

Now suppose that $q \in \partial^\infty f(x)$. Then there exists $z_v \in \partial_p f(x_v)$, $x_v \to^f x$, $\lambda_v \downarrow 0$ such that $q = \lim_{v\to\infty}\lambda_v z_v$. For each v we may find a β_v and x^{β_v} such that $\|\nabla f(\beta_v,x^{\beta_v}) - z_v\| \leq \frac{1}{v\lambda_v}$ implying $\|\lambda_v\nabla f(\beta_v,x^{\beta_v}) - q\| \leq \frac{1}{v} + \|\lambda_v z_v - q\| \to 0$ as $v\to\infty$. Hence $\partial^\infty f(x) \subseteq \partial^{H,\infty} f(x)$.

From the second order condition $\nabla^2\psi_\lambda(\beta,x^\beta) \geq 0$ (in the order defined by the cone $\mathcal{P}(n)$) we have

$$\nabla^2\psi_\lambda(\beta,x^\beta) = \nabla^2 f(\beta,x^\beta) - X + 2\lambda - \nabla^2 g_{\bar{c}}(\beta,x^\beta) \geq 0. \tag{14.A.2}$$

As $g_{\bar{c}}(\beta, x^\beta)$ is $C^2(\mathbb{R}^n)$ and for $\|z\| > 2M + \|x\|$ we have $\nabla g_{\bar{c}}(z) = 2\bar{c}(z - x)$ it follows that we may apply (14.11) an the strict differentiability of ∇g once again. Applying this to each component of $\nabla g_{\bar{c}}(z)$ it follows that $\lim_{\beta \downarrow 0} \nabla^2 g_{\bar{c}}(\beta, x^\beta) = 0$ since $\partial^H(\nabla g_{\bar{c}}(z)_i) \subseteq \partial^o(\nabla g_{\bar{c}}(z)_i) = \{\nabla(\nabla g_{\bar{c}}(z)_i)\}$, corresponding to the i**th** row of $\nabla^2 g_{\bar{c}}(z)$. Then (14.A.2) gives $\liminf_{\beta \downarrow 0} \langle \nabla^2 f(\beta, x^\beta), uu^t \rangle \geq \langle X - 2\lambda, uu^t \rangle$ where λ is a fixed positive but arbitrary number. Hence $\liminf_{\beta \downarrow 0} \nabla^2 f(\beta, x^\beta) \geq X$ in the order determined by the cone $\mathcal{P}(n)$. Using Lemmas 2.1 and 3.1 we have

$$
\begin{aligned}
\partial^{2,\uparrow} f(x,p) &= \limsup_{(x',p') \to^f (x,p)} \partial^{2,-} f(x',p') \\
&\subseteq \limsup_{(x',p') \to^f (x,p)} \partial^{2,H^\uparrow} f(x',p') \subseteq \partial^{2,H^\uparrow} f(x,p)
\end{aligned}
$$

completing the proof. □

Proof. (**of Theorem 4.3**) We argue for f convex, the case for concave f being identical. As f is a convex function defined on a convex domain U with $\operatorname{int} U \neq \emptyset$ we have f is locally Lipschitz on int U. Now suppose that $\varphi \in \mathcal{D}(u)$ and hence has compact support (later we will place $\varphi(\cdot) = H(\beta, \cdot)$). Then $\left| f(u) \left(\frac{\varphi(x+th-u) - \varphi(x-u)}{t} \right) \right| \leq (L + f(x))(\nabla \varphi(x-u) + \varepsilon)$for t small and L the Lipschitz constant of f applicable to the compact support of φ. Then applying the Dominated Convergence Theorem we get $\nabla f(\beta, x) = \int f(u) \nabla \varphi(x-u)\, du$ existing. For $\varphi \in \mathcal{D}(U)$ we may argue in a similar way again to get $\nabla^2 f(\beta, x) = \int f(u) \nabla^2 \varphi(x-u)\, du$. By the properties of the convolution $f(\beta, x) = \int f(x-u) \varphi(u)\, du$ and as $\nabla f(x-u)$ exists a.e. we have, using the fact that Lipschitz functions are absolutely continuous, that $\nabla f(\beta, x) = \int \nabla f(x-u) \varphi(u)\, du$. As f is convex $S_2(f)$ is of full Lebesgue measure. As noted in Mignot (1976) relative to $S_2(f)$ we have ∇f Fréchet differentiable and so for (t, h, u) a.e. (and any $x \in S_2(f)$)

$$
\nabla f(x + th - u) - \nabla f(x - u) = t \nabla^2 f(x-u) h + o_u(t, h, x) \tag{14.A.3}
$$

where $\sup_{h \in B_1(0)} \|o_u(t,h,x)\| \leq o_u(t,x) \to 0$ as $t \downarrow 0$ for (for all $x \in S_2$) almost all $x \in U$ and almost all $u \in \mathbb{R}^n$ such that $x - u \in S_2(f)$. Thus noting that by Corollary 4.2 we have $\left(D^2[f]\right)_c(g) = \int \nabla^2 f(u) g(u)\, du$ for any bounded Borel measurable $g : U \to \mathbb{R}$ of compact support, then for any φ for which $\varphi_x = \varphi(x - \cdot) \in \mathcal{D}(U)$ and all h

$$
\begin{aligned}
D^2[f](\varphi_x)h &= \int f(u) \nabla^2 \varphi(x-u) h\, du \\
&= \left(\nabla_x^2 \int f(u) \varphi(x-u) du \right) h = \left(\nabla_x \int \varphi(u) \nabla f(x-u) du \right) h \\
&= \lim_{t \downarrow 0} \frac{1}{t} \int \left(\nabla f(x + th - u) - \nabla f(x-u) \right) \varphi(u)\, du \\
&= \int \nabla^2 f(x-u) h \varphi(u)\, du + \lim_{t \downarrow 0} \int \frac{o_u(t,h,x)}{t} \varphi(u)\, du \\
&= \left(\int \nabla^2 f(u) \varphi(x-u)\, du \right) h + \lim_{t \downarrow 0} \int \frac{o_u(t,h,x)}{t} \varphi(u)\, du \\
&= \left(D^2[f]\right)_c (\varphi_x) h + \lim_{t \downarrow 0} \int \frac{o_u(t,h,x)}{t} \varphi(u)\, du.
\end{aligned} \tag{14.A.4}
$$

Thus we have, on taking $\varphi = H_\beta$ (for β so small that $x + \operatorname{Supp} H_\beta \subseteq U$)

$$\nabla^2 f(\beta, x)h = \left(\nabla_x^2 \int f(u) H_\beta(x-u)\,du\right) h$$
$$= \int \nabla^2 f(u) h H_\beta(x-u)\,du + \lim_{t\downarrow 0} \int \frac{o_u(t,h,x)}{t} H_\beta(u)\,du \qquad (14.A.5)$$

As $D^2[f] = \left(D^2[f]\right)_c + \left(D^2[f]\right)_s$, we conclude from (14.A.4) that for all $x \in S_2(f)$

$$\langle \mu_s, H_\beta(x - \cdot)\rangle h := \langle \left(D^2[f]\right)_s, H_\beta(x-\cdot)\rangle h = \lim_{t\downarrow 0} \int \frac{o_u(t,h,x)}{t} \varphi_x(u)\,du. \qquad (14.A.6)$$

Let S be the set of full measure on which (14.A.6) holds (we know that $S \supseteq S_2(f)$). We have $|\mu_s|' \equiv 0$ a.e. in S and so for almost all $x \in S$ we obtain for any $x^\beta \in S$ such that $\|x^\beta - x\| < \frac{V'(\beta)}{2}$ as $\beta \downarrow 0$ that

$$\lim_{\beta\downarrow 0} \frac{|\mu_s|(B_{\frac{V(\beta)}{2}}(x^\beta))}{V(\beta)^n} = |\mu_s|'(x) = 0.$$

Indeed,

$$|\lim_{t\downarrow 0} \int \frac{o_u(t,h,x^\beta)}{t} H_\beta(u)\,du| = |\langle \mu_s, H_\beta(x^\beta - \cdot)\rangle|$$
$$= |\int_U H_\beta(x^\beta - u)\,d\mu_s(u)| \text{ from (14.A.6) with } x = x^\beta$$
$$\leq \int |H_\beta(x^\beta - u)|\,d|\mu_s|(u) \leq \frac{C_\beta |\mu_s| \left(B_{\frac{V(\beta)}{2}}(x^\beta)\right)}{V(\beta)^n}$$
$$\leq \frac{2^n C_\beta |\mu_s| \left(B_{V(\beta)}(x)\right)}{m\left(B_{V(\beta)}(x)\right)} \quad \to 2^n |\mu_s|'(x) = 0 \quad m\text{- a.e. in } S$$

$$\text{implying} \qquad \lim_{\beta\downarrow 0}\lim_{t\downarrow 0} \int_U \frac{o_u(t,h,x^\beta)}{t} H_\beta(u)\,du = 0.$$

Let L be the Lipschitz constant applicable to $u \in \overline{B}_\delta(x)$. Then noting that for β and t small we have $x' - u + th \in \overline{B}_\delta(x')$ if u is in the support of H_β and so for all $h \in B_1(0)$

$$\langle \nabla_{x'} \int f(u) H_\beta(x'-u)\,du, h\rangle = \langle \nabla_{x'} \int f(x'-u) H_\beta(u)\,du, h\rangle$$
$$= \lim_{t\downarrow 0} \int \left(\frac{f(x'-u+th) - f(x'-u)}{t}\right) H_\beta(u)\,du \leq L$$

Thus $\|\nabla f(\beta, x')\| \leq L$ and so for any $\beta \downarrow 0$ and $x^\beta \to x$ such that $\nabla f(\beta_n, x^\beta) \to z$ we have $z \in \partial f(x)$, since $\nabla f(\beta, x^\beta) = \int \nabla f(x^\beta - u) H_\beta(u)\,du \in \overline{\operatorname{co}}\{\nabla f(x) \mid x \in B_{\frac{\beta}{2}}(x^\beta)\}$, which itself converges to $\partial f(x)$ since $\partial f(x)$ coincides with the Clarke subgradient as f is convex and hence regular. Now as $x \in S_2(f)$ must be a point of strict differentiability we must have $\partial f(x) = \{\nabla f(x)\}$. Thus for any $\beta \downarrow 0$ and $x^\beta \to x$ we have $\nabla f(\beta, x^\beta) \to \nabla f(x)$. Thus when $\nabla^2 f(x^\beta - u^\beta) \to Q$ as $x^\beta \to x$ and $u^\beta \to 0$ we have $Q \in \overline{D}^2 f(x, \nabla f(x)) \subseteq \underline{\partial}^2 f(x, \nabla f(x))$ by Proposition 2.3.

Now take $\{x^\beta\}_{\beta>0}$ such that $\|x^\beta - x\| < \frac{V'(\beta)}{2}$. Then using (14.A.5) and (14.17) and the fact that $D^2[f]_c(\varphi_x) = \int \nabla^2 f(u) H_\beta(u)\, du$ we have for any i, j

$$
\begin{aligned}
&|\nabla^2 f(\beta, x^\beta)_{ij} - \langle ((D^2[f])_c)_{ij}, \frac{1}{V_\beta^n}\delta_{B_{\frac{V'_\beta}{2}}(0)}(x^\beta - \cdot)\rangle| \\
&= |\nabla^2 f(\beta, x^\beta)_{ij} - \langle (\mu_c)_{ij}, \frac{1}{V_\beta^n}\delta_{B_{\frac{V'_\beta}{2}}(0)}(x^\beta - \cdot)\rangle| \\
&\le |\nabla^2 f(\beta, x^\beta)_{ij} - \langle (\mu_c)_{ij}, H_\beta(x^\beta - \cdot)\rangle| + |\langle (\mu_c)_{ij}, H_\beta(x^\beta - \cdot) - \frac{1}{V_\beta^n}\delta_{B_{\frac{V'_\beta}{2}}(0)}(x^\beta - \cdot)\rangle| \\
&= |\lim_{t\downarrow 0}\int_U \frac{o_u(t, \underline{e}_j, x^\beta)}{t}\cdot \underline{e}_i H_\beta(u)\, du| + |\langle (\mu_c)_{ij}, H_\beta(x^\beta - \cdot) - \frac{1}{V_\beta^n}\delta_{B_{\frac{V'_\beta}{2}}(0)}(x^\beta - \cdot)\rangle|
\end{aligned}
$$

from (14.A.5) applied to x^β where $\{\underline{e}_1, \cdots, \underline{e}_n\}$ denotes the standard basis in $\mathbb{R}^n$. The first term $\to 0$ as $\beta \to 0$ as already argued earlier, so we focus on the second term.

$$
\begin{aligned}
&|\langle (\mu_c)_{ij}, H_\beta(x^\beta - \cdot) - \frac{1}{V_\beta^n}\delta_{B_{\frac{V'_\beta}{2}}(0)}(x^\beta - \cdot)\rangle| \\
&\le \int_{B_{\frac{V_\beta}{2}}(x^\beta)\setminus B_{\frac{V'_\beta}{2}}(x^\beta)} H_\beta\, d|(\mu_c)_{ij}| + (C_\beta - c_\beta)\frac{|(\mu_c)_{ij}|}{V_\beta^n}\left(B_{\frac{V'_\beta}{2}}(x^\beta)\right) \\
&\le C_\beta\left(\frac{|(\mu_c)_{ij}|}{V_\beta^n}\left(B_{\frac{V_\beta}{2}}(x^\beta)\right) - \frac{|(\mu_c)_{ij}|}{V_\beta^n}\left(B_{\frac{V'_\beta}{2}}(x^\beta)\right)\right) + (C_\beta - c_\beta)\frac{|(\mu_c)_{ij}|}{V_\beta^n}\left(B_{\frac{V'_\beta}{2}}(x^\beta)\right) \\
&= C_\beta\frac{|(\mu_c)_{ij}|}{V_\beta^n}\left(B_{\frac{V_\beta}{2}}(x^\beta)\right) - c_\beta\frac{|(\mu_c)_{ij}|}{(V'_\beta)^n}\left(B_{\frac{V'_\beta}{2}}(x^\beta)\right)\left(\frac{V'_\beta}{V_\beta}\right)^n \\
&\to |(\mu_c)_{ij}|'(x)\left(1 - \lim_\beta \left(\frac{V'_\beta}{V_\beta}\right)^n\right) = 0 \qquad \text{(14.A.7)}
\end{aligned}
$$

for m–almost all $x \in S$, since the $B_{\frac{V_\beta}{2}}(x^\beta)$, $B_{\frac{V'_\beta}{2}}(x^\beta)$ converge to x in the sense of the "regular differentiation basis", and

$$
|(\mu_c)_{ij}|'(x) = |(\nabla^2 f(x))_{ij}| \qquad \text{(14.A.8)}
$$

for almost all x (so the latter is finite and the limit in (14.A.7) is indeed zero). Now we shall verify (14.A.8). We already have (m–a.e.)

$$
\begin{aligned}
\frac{d}{dm}(\mu_c)_{ij} &= (\nabla^2 f)_{ij} \quad \text{on } U \quad \text{(a.e.)} \\
&= (\mu_{ij})' = (\mu_c)'_{ij}.
\end{aligned}
$$

Since $\mu_s \perp m$ so $\mu'_s \equiv 0$ a.e. By standard arguments,

$$
\frac{d}{dm}|(\mu_c)_{ij}| \overset{\text{a.e.}}{=} |\frac{d}{dm}(\mu_c)_{ij}| = |(\nabla^2 f)_{ij}| \quad \text{a.e. in } U
$$

and hence, on forming Hahn–Jordan decomposition ($(\mu_c)_{ij} = (\mu_c)^+_{ij} - (\mu_c)^-_{ij}$),

$$\begin{aligned}\frac{d}{dm}(\mu_c)^{\pm}_{ij} &= \frac{d}{dm}(\frac{1}{2}(|(\mu_c)_{ij}| \pm (\mu_c)_{ij}) = \frac{1}{2}\left(\frac{d}{dm}|(\mu_c)_{ij}| \pm \frac{d}{dm}(\mu_c)_{ij}\right) \\ &= \frac{1}{2}\left(|(\nabla^2 f)_{ij}| \pm (\nabla^2 f)_{ij}\right) \quad \text{a.e. in } U\end{aligned}$$

(from the m-a.e. uniqueness for Radon–Nikodym derivative)

Also, as $\mu_c^{\pm} << m$, have $\frac{d}{dm}(\mu_c)_{ij} \overset{\text{a.e.}}{\equiv} ((\mu_c)^{\pm}_{ij})'$ (from Theorem 4.2). So finally

$$\begin{aligned}|(\mu_c)_{ij}|' &= ((\mu_c)^+_{ij})'(x) + ((\mu_c)^-_{ij})'(x) = \frac{d}{dm}(\mu_c)^+_{ij}(x) + \frac{d}{dm}(\mu_c)^-_{ij}(x) \\ &= \frac{1}{2}\left(|(\nabla^2 f)_{ij}| + (\nabla^2 f)_{ij}\right) + \frac{1}{2}\left(|(\nabla^2 f)_{ij}| - (\nabla^2 f)_{ij}\right) = |\left(\nabla^2 f(x)\right)_{ij}|\end{aligned}$$

for almost all $x \in U$ (as required, giving (14.A.8)). Thus, on deletion of a m–null set from S, we have shown that $\nabla^2 f(\beta, x^\beta) \to \nabla^2 f(x)$ for $x \in S$ and $(x_\beta)_{\beta>0} \leq S$ such that $|x_\beta - x| < V'_{\beta/2}$ for all β.

When (for all $x \in S$) $q(\underline{\partial} f(x, \nabla f(x))(h) \leq K\|h\|^2 < \infty$ for all h it follows from the above observations that for any fixed $x \in S$, $\langle \nabla^2 f(x-u), hh^t \rangle \leq 2K\|h\|^2$ for almost all u in a sufficiently small neighbourhood V of zero. As f is finite and convex on a convex open domain it is regular and we may now apply Proposition 5.1 part 2 to deduce that $x \mapsto \langle \nabla f(x), h \rangle$ is locally Lipschitz for all h and hence $x \mapsto \nabla f(x)$ is locally Lipschitz on a neighbourhood of x. We have on this neighbourhood

$$\begin{aligned}h^t \frac{o_u(t,h,x)}{t} &\leq \frac{\|\nabla f(x+th-u) - \nabla f(x-u)\|}{t} + h^t \nabla^2 f(x-u)h \\ &\leq L\|h\| + 2K\|h\|^2.\end{aligned}$$

We may now apply the Dominated Convergence Theorem to deduce that for all $h \in B_1(0)$

$$\begin{aligned}h^t \mu_s(B_{\frac{V(\beta)}{2}}(x))h &= \lim_{t\downarrow 0} \int h^t \frac{o_u(t,h,x)}{t} H_\beta(x-u)\,du \\ &= \int_{B_{\frac{V(\beta)}{2}}(x)} h^t \left(\lim_{t\downarrow 0} \frac{o_u(t,h,x)}{t}\right) H_\beta(x-u)\,du = \int_{B_{\frac{V(\beta)}{2}}(x)} 0\,du = 0.\end{aligned}$$

As this holds for all h and β sufficiently small we have $\mu_s(A) = 0$ for all Borel sets $A \subseteq B_{\frac{V(\beta)}{2}}(x^\beta)$. By (14.A.6) it follows in a similar way that

$$h^t \langle (D^2[f])_s, H_\beta(x-\cdot)\rangle h = h^t \langle \mu_s, H_\beta(x-\cdot)\rangle h = \int h^t \left(\lim_{t\downarrow 0} \frac{o_u(t,h,x)}{t}\right) \varphi(u)\,du = 0$$

for β sufficiently small and all h giving $D^2[f]_s(H_\beta(x-\cdot)) = 0$. Since this holds for any $x \in S$ it follows that $h^t(D^2[f])_s h$ is the zero measure for any h, so $(D^2[f])_s = 0$ since it is symmetric. This completes the proof for case when f convex. If f concave, then argue as above with $-f$, for the same result. □

References

H. Attouch (1984) *Variational Convergencefor Functions and Operators,* Pitman Adv. Publ. Prog. Boston–London–Melbourne.

G. Beer (1993) *Topologies on Closed and Convex Sets,* mathematics and its applications Vol. 268, Kluwer Academic Publishers.

B. Craven (1986) Non-Differential Optimization by Smooth Approximations, *Optimization,* Vol. 17 no. 1, pp. 3-17.

B. Craven (1986) A Note on Non-Differentiable Symmetric Duality, *Journal of Australian Mathematical society Series B,* Vol. 28 no. 1, 30-35.

R. Cominetti and R. Correa (1990) A Generalized Second Order Derivative in Nonsmooth Optimization, *SIAM J. Control and Optimization,* Vol. 28, pp. 789-809.

G. Crespi, D. La Torre and M. Rocca, Mollified Derivative and Second–order Optimality Conditions, *Preprint communicated from the authors.*

R. M. Dudley (1977) On Second Derivatives of Convex Functions, *Math. Scand.,* Vol. 41, pp. 159-174.

A. Eberhard, M. Nyblom and D. Ralph (1998) Applying Generalised Convexity Notions to Jets, J.P Crouzeix et al. (eds), it Generalized Convexity, Generalized Monotonicity: Recent Results, Kluwer Academic Pub., pp. 111-157.

A. Eberhard and M. Nyblom (1998) Jets Generalized Convexity, Proximal Normality and Differences of Functions, *Non–Linear Analysis* Vol. 34, pp. 319-360.

A. Eberhard (2000) Prox–Regularity and Subjets, *Optimization and Related Topics,* Ed. A. Rubinov, Applied Optimization Volumes, Kluwer Academic Pub., pp. 237-313.

Y.M. Ermoliev, V.I. Norkin, R. J-B. Wets (1995) The Minimization of Semicontinuous Functions: Mollifier Subgradients, *SIAM J. Control and Optimization,* Vol. 33. No. 1, pp. 149-167.

R. F. Gariepy and W. P. Zoemer (1995) *Modern Real Analysis,* PWS Publishing Company, PWS Publishing Company, Boston Massachusetts.

H. Halkin (1976) Interior Mapping Theorem with Set–Valued Derivatives, *J. d'Analyse Mathèmatique,* Vol. 30, pp 200-207.

H. Halkin (1976) Mathematical Programming without Differentiability, *Calculus of Variations and Control Theory,* ed D. L. Russell, Academic Press, NY.

A. D. Ioffe (1989), On some Recent Developments in the Theory of Second Order Optimality Conditions, Optimization - fifth French-

German Conference, Castel Novel 1988, *Lecture Notes in Mathematics,* Vol. 405, Springer Verlag, pp. 55-68.

A.D. Ioffe and J-P. Penot (1987) Limiting Subhessians, Limiting Subjets and their Calculus, *Transactions of the American Mathematics Society,* Vol. 349, no. 2, pp 789–807.

F. Mignot (1976) Contrôle dans Inéquations Variationelles Elliptiques, *J. of Functional Analysis,* No. 22, pp. 130-185.

M. Nyblom (1998) Smooth Approximation and Generalized Convexity in Nonsmooth Analysis and Optimization, PhD thesis RMIT University.

Z. Páles and V. Zeidan (1996) Generalized Hessians for $C^{1,1}$ Functions in Infinite-Dimensional Normed Spaces, *Mathematical Programming,* Vol. 74, pp. 59-78.

J.-P. Penot (1994) Sub–Hessians, Super–Hessians and Conjugation, *Nonlinear Analysis, Theory Methods and Applications,* Vol. 23, no. 6, pp. 689–702.

D. Ralph (1990) *Rank-1 Support Functional and the Rank-1 Generalised Jacobian, Piecewise Linear Homeomorphisms,* Ph.D. Thesis, Computer Science Technical Reports #938, University of Wisconsin Madison.

S. M. Robinson, Local Epi-Continuity and Local Optimization, *Mathematical Programming,* Vol. 37, pp. 208-222.

R. T. Rockafellar and R. J-B. Wets (1998) *Variational Analysis,* Volume 317, A series of Comprehensive Studies in Mathematics, Springer.

R. Sivakumaran (2003) *A Study of the Viscosity and Weak Solutions to a Class of Boundary Valued Problems,* PhD Thesis, RMIT University.

J. Warga (1975) Necessary Conditions without Differentiability Assumptions in Optimal Control, *J. of Diff. Equ.,* Vol. 15, pp. 41-61.

J. Warga (1976) Derivative Containers, Inverse Functions and Controllability, *Calculus of Variations and Control Theory,* ed D. L. Russell, Academic Press, NY.

Chapter 15

APPLYING GLOBAL OPTIMIZATION TO A PROBLEM IN SHORT-TERM HYDROTHERMAL SCHEDULING

Albert Ferrer*

Departament de Matemàtica Aplicada I

Universitat Politècnica de Catalunya, Spain

Abstract A method for modeling a real constrained optimization problem as a reverse convex programming problem has been developed from a new procedure of representation of a polynomial function as a difference of convex polynomials. An adapted algorithm, which uses a combined method of outer approximation and prismatical subdivisions, has been implemented to solve this problem. The solution obtained with a local optimization package is also included and their results are compared.

Keywords: Canonical d.c. program, **ϵ-approximate** optimal solution, normal subdivision rule, prismatical and conical subdivision, outer approximation, semi-infinite program.

Mathematics Subject Classification (2000) 90C26, 90C30.

1. Introduction

The preparation of this paper has been motivated by the interest in applying global optimization procedures to problems in the real world which do not have any special structure but, whose solution has economic and technical implications. In this paper, we focus on the Short-Term Hydrothermal Coordination of Electricity Generation Problem (see Heredia et al (1995) for more details). Its importance stems from the economic and technical implications that the solution to this prob-

*email:alberto.ferrer@upc.es

lem has for electric utilities with a mixed (hydro an thermal) generation sistem. This kind of problem is very difficult to solve using global optimization algorithms because of the important role that the problem size plays in order to obtain satisfactory computational results. It suffices to see for instance Gurlitz et al (1991), where the authors have not found satisfactory results for a test reverse convex problem of dimension less than 10. Moreover, we need to know a representation of every nonlinear function, in the problem, as a difference of convex functions (d.c. functions) to transform it into an equivalent reverse convex program. Therefore, on attempting to solve programming problems without any special structure we have to develop new methods which are not needed for simpler problems with a special structure. In section 2 we describe the Short-Term Hydrothermal Coordination of Electricity Generation Problem. In section 3 we rewrite the problem as an equivalent reverse convex program by using the procedure described in Ferrer (2001) (to obtain a d.c. representation of a polynomial) and the properties of the d.c. functions (see Hiriart-Urruty (1985) and Horst et al (1990)). It should be stressed that several different transformations can be used to obtain an equivalent reverse convex program. The properties of the functions in the program are used to find a suitable complementary convex mathematical structure for the equivalent program. Section 4 is devoted to describing the algorithm and the basic operations where a prismatical subdivision process has been used to obtain an advantageous accommodation of the *Combined Outer Approximation and Cone Splitting Conical Algorithmfor Canonical D.C. Programming* (see Tuy (1998)). In section 5, by using the concept of *Least Deviation Decomposition* (see Luc et al (1999)), a semi-infinite programming problem has been formulated to calculate the optimal d.c. representation of a polynomial. In order to obtain more efficient implementations we have obtained the least deviation decomposition of each power hydrogeneration function (see (15.1)) following the algorithms described in Kaliski et al (1997) and Zhi-Quan et al (1999). These results are not explicity indicated in this paper and we only use them to obtain our computational results. In section 6 characteristics of generation systems and computational results are given. Finally, in section 7 conclusions are explained.

2. The problem

Given a short-term time period one wishes to find values for each time interval in the period so that we can satisfy the demand of electricity consumption for each time interval, a number of constraints are satisfied and the generation cost of thermal units is minimized. The model contains

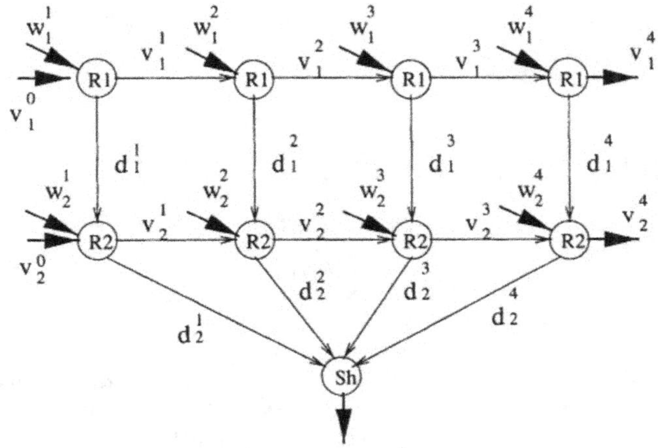

Figure 15.1. Four intervals and two reservoirs replicated hydronetwork

the replicated hydronetwork through which the temporary evolution of the reservoir system is represented. Figure 15.1 shows the network with only two reservoirs and where the time period has been subdivided into four intervals. We use j to indicate the jth reservoir, $j = 1 \ldots N_e$ and i to indicate the ith time interval, $i = 1 \ldots N_t$. It should be observed that

- the variables are the water discharges d_j^i from reservoir j over the ith interval and the volume stored v_j^i in reservoir j at the end of the ith time interval,
- in each time interval i, the water discharge from reservoir R_1 to reservoir R_2 establishes a link between the reservoirs,
- the volume stored at the end of the time interval i and the volume stored at the beginning of the time interval $i + 1$ are the same on each reservoir R_j, which establishes a link between each reservoir from the time interval i to $i + 1$,
- the volumes stored at the beginning and at the end of the time period are known (they are not variables). Acceptable forecasts for electricity consumption l^i and for natural water inflow w_j^i into the reservoirs of the hydrogeneration system at each interval must be available.

The main feature in this formulation is that the power hydrogeneration function h_j^i at the reservoir j over the ith interval can be approximated by a polynomial function of degree 4 in the variables v_j^{i-1}, v_j^i and d_j^i (see Heredia et al (1995)),

$$\begin{aligned} h_j^i(v_j^{i-1}, v_j^i, d_j^i) \;=\; & \mathsf{k}_j^i d_j^i [s_{vd} + \frac{s_{vl}}{2}(v_j^{i-1} + v_j^i) + \frac{s_{vq}}{3}(v_j^i - v_j^{i-1})^2 + \\ & + s_{vq} v_j^{i-1} v_j^i + \frac{s_{vc}}{4}((v_j^{i-1})^2 + (v_j^i)^2)(v_j^{i-1} + v_j^i) - \\ & - s_{dl} d_j^i - s_{dq}(d_j^i)^2] \qquad (15.1) \end{aligned}$$

where k_j^i (efficiency and unit conversion coefficient), s_{vd}, s_{vl}, s_{vq}, s_{vc}, s_{dl} and s_{dq} are technological coefficients which depend on each reservoir. The objective function, which will be minimized, is the generation cost of thermal units,

$$f(\ldots, v_j^{i-1}, v_j^i, d_j^i, \ldots) = \sum_{i=1}^{N_t} c_i \left(l^i - \sum_{j=1}^{N_e} h_j^i(v_j^{i-1}, v_j^i, d_j^i) \right). \qquad (15.2)$$

The linear constraints are the flow balance equations at all nodes of the network,

$$v_j^i - v_j^{i-1} - d_{j-1}^i + d_j^i = w_j^i \quad j = 1, ..., N_e, \; i = 1, ..., N_t. \qquad (15.3)$$

The nonlinear constraints are the thermal production with generation bounds,

$$\underline{g} \le l^i - \sum_{j=1}^{N_e} h_j^i(v_j^{i-1}, v_j^i, d_j^i) \le \overline{g} \quad i = 1, ..., N_t. \qquad (15.4)$$

There are positive bounds on all variables,

$$\underline{d_j} \le d_j^i \le \overline{d_j} \qquad j = 1, ..., N_e, i = 1, ..., N_t \qquad (15.5)$$

$$\underline{v_j} \le v_j^i \le \overline{v_j} \quad j = 1, ..., N_e, \; i = 1, ..., N_t - 1.$$

Hence, we can write

$$
\begin{array}{lll}
\text{minimize} & \sum_{i=1}^{N_t} c_i \left(l^i - \sum_{j=1}^{N_e} h_j^i(v_j^{i-1}, v_j^i, d_j^i) \right) & \\
\text{subject to} & \underline{g} \le l^i - \sum_{j=1}^{N_e} h_j^i(v_j^{i-1}, v_j^i, d_j^i) \le \overline{g} & i = 1, ..., N_t, \\
& v_j^i - v_j^{i-1} - d_{j-1}^i + d_j^i = w_j^i, & j = 1, ..., N_e, \\
& & i = 1, ..., N_t, \\
& \underline{d}_j \le d_j^i \le \overline{d}_j, & j = 1, ..., N_e, \\
& & i = 1, ..., N_t, \\
& \underline{v}_j \le v_j^i \le \overline{v}_j & j = 1, ..., N_e, \\
& & i = 1, ..., N_t - 1.
\end{array}
\tag{15.6}
$$

The problem has the following useful properties:

1 it is easy to generate problems of different sizes (Table 15.1) and instances with different degrees of nonconvexity which depend on the efficiency and unit conversion coefficient, whether the thermal units can satisfy all the demand of electricity during every time interval and the water inflows,

2 the objective function and the nonlinear constraints are polynomial functions,

3 the linear constraints are the flow balance equations at all nodes of a network.

3. The programming problem as an equivalent canonical d.c. program

A polynomial is a d.c. function on $\mathbb{R}^n$ because it has continuous derivatives of any order and we know that every function whose second partial derivatives are continuous on $\mathbb{R}^n$ is a d.c. function on $\mathbb{R}^n$. As we know how to construct a d.c. representation of the power hydrogeneration functions $h_j^i(v_j^{i-1}, v_j^i, d_j^i)$ (see Ferrer (2001)) then, we can obtain a d.c. representation of all functions within (15.6). Let

$$h_j^i(v_j^{i-1}, v_j^i, d_j^i) = f_j^i(v_j^{i-1}, v_j^i, d_j^i) - g_j^i(v_j^{i1}, v_j^i, d_j^i), \tag{15.7}$$

be a d.c. representation of the power hydrogeneration function, where $f_j^i(v_j^{i-1}, v_j^i, d_j^i)$ and $g_j^i(v_j^{i1}, v_j^i, d_j^i)$ are convex functions defined on a convex set which contains the feasible domain of the program (15.6). Then, by defining for all, $i = 1, ..., N_t$, the convex functions

$$F^i(\ldots, v_j^{i-1}, v_j^i, d_j^i, \ldots) = c^i \left(l^i + \sum_{j=1}^{N_e} g_j^i(v_j^{i-1}, v_j^i, d_j^i) \right) \tag{15.8}$$

and

$$G^i(\ldots, v_j^{i-1}, v_j^i, d_j^i, \ldots) = c^i \sum_{j=1}^{N_e} f_j^i(v_j^{i-1}, v_j^i, d_j^i), \tag{15.9}$$

and using these expressions to define

$$F(\ldots, v_j^{i-1}, v_j^i, d_j^i, \ldots) = \sum_{i=1}^{N_t} F^i(\ldots, v_j^{i-1}, v_j^i, d_j^i, \ldots) \tag{15.10}$$

and

$$G(\ldots, v_j^{i-1}, v_j^i, d_j^i, \ldots) = \sum_{i=1}^{N_t} G^i(\ldots, v_j^{i-1}, v_j^i, d_j^i, \ldots), \tag{15.11}$$

a d.c. representation of all functions within (15.6) can be obtained. By defining $n = N_e(2N_t - 1)$, $m = N_e N_t$ and $x = (\ldots, v_j^{i-1}, v_j^i, d_j^i, \ldots) \in \mathbb{R}^n$, and by expressing the linear constrains in the form $Ax = b$ the program (15.6) can be rewritten as the d.c. program

$$\begin{array}{ll} \text{minimize} & F(x) - G(x) \\ \text{subject to} & (G^i(x) + c^i \underline{g}) - F^i(x) \le 0 \quad i = 1, ..., N_t, \\ & F^i(x) - (G^i(x) + c^i \overline{g}) \le 0 \quad i = 1, ..., N_t, \\ & Ax = b, \\ & \underline{x} \le x \le \overline{x}, \end{array} \tag{15.12}$$

involving linear constraints of equality, where $A \in \mathbb{R}^{m*n}$, $b \in \mathbb{R}^m$ and $\underline{x}$, $\overline{x} \in \mathbb{R}^n$. The matrix A of the linear constraints in (15.12) can be written as $A = [B, N]$ where B is a non singular square matrix. Let y and z be the basic and non basic coordinates corresponding to the matrices B and N, respectively. Then, $x = (y, z)$ with $y = B^{-1}(b - Nz)$, so that it is possible to reduce the size of the d.c. program (15.12) by defining the functions $\varphi_1(z) = F(x)$, $\varphi_2(z) = G(x)$, $\varphi_1^i(z) = F^i(x)$ and $\varphi_2^i(z) = G^i(x)$. By using these functions in (15.12) we obtain an equivalent d.c. program of reduced size expressed by

$$\begin{array}{ll} \text{minimize} & \varphi_1(z) - \varphi_2(z) \\ \text{subject to} & (\varphi_2^i(z) + c^i \underline{g}) - \varphi_1^i(z) \le 0 \quad i = 1, ..., N_t, \\ & \varphi_1^i(z) - (\varphi_2^i(z) + c^i \overline{g}) \le 0 \quad i = 1, ..., N_t, \\ & \underline{b} \le Mz \le \overline{b}, \\ & \underline{z} \le z \le \overline{z}, \end{array} \tag{15.13}$$

where $M = B^{-1}N$, $\underline{b} = B^{-1}b - \overline{y}$ and $\overline{b} = B^{-1}b - \underline{y}$.

By adding the variable τ, the d.c. program (15.13) can be transformed into an equivalent d.c. program with a linear objective function

$$\begin{array}{ll} \textbf{minimize} & \tau \\ \textbf{subject to} & \varphi_1(z) - \varphi_2(z) - \tau \leq 0 \\ & (\varphi_2^i(z) + c^i\underline{g}) - \varphi_1^i(z) \leq 0 \quad i = 1, \dots, N_t, \\ & \varphi_1^i(z) - (\varphi_2^i(z) + c^i\overline{g}) \leq 0 \quad i = 1, \dots, N_t, \\ & \underline{b} \leq Mz \leq \overline{b}, \\ & \underline{z} \leq z \leq \overline{z}. \end{array} \tag{15.14}$$

The nonlinear constraints in (15.14) can be expressed using a single constraint by defining

$$r(z,\tau) = \max \left\{ \begin{array}{l} \varphi_1(z) - (\varphi_2(z) + \tau), \ (\varphi_2^i(z) + c^i\underline{g}) - \varphi_1^i(z), \\ \varphi_1^i(z) - (\varphi_2^i(z) + c^i\overline{g}), \ i = 1, \dots, N_t \end{array} \right\}$$

so that (15.14) can be written

$$\begin{array}{ll} \textbf{minimize} & \tau \\ \textbf{subject to} & r(z,\tau) \leq 0, \\ & \underline{b} \leq Mz \leq \overline{b}, \\ & \underline{z} \leq z \leq \overline{z}. \end{array} \tag{15.15}$$

From the properties of the d.c. functions (see Hiriart-Urruty (1985) and Horst et al (1990)), a d.c. representation of $r(z,\tau) = p(z,\tau) - q(z,\tau)$ can be obtained by using the convex functions

$$p(x,\tau) = \max \left\{ \begin{array}{l} 2\varphi_1(z) + \varphi_2(z) + N_t c^i \overline{g}, \\ \varphi_1(z) + 2\varphi_2(z) + \tau + (\varphi_2^i(z) - \varphi_1^i(z) + c^i\underline{g}) + N_t c^i \overline{g}, \\ i = 1, \dots, N_t, \\ \varphi_1(z) + 2\varphi_2(z) + \tau + (\varphi_1^i(z) - \varphi_2^i(z) - c^i\overline{g}) + N_t c^i \overline{g}, \\ i = 1, \dots, N_t \end{array} \right\}$$

and

$$q(z,\tau) = \varphi_1(z) + 2\varphi_2(z) + \tau + N_t c^i \overline{g}.$$

A more suitable d.c. representation of $r(z,\tau)$ can be obtained by defining the convex functions

$$p_0(z,\tau) = \max \left\{ \begin{array}{ll} 2\varphi_1(z) - \tau, & \\ \varphi_1(z) + \varphi_2(z) + (\varphi_2^i(z) - \varphi_1^i(z) + c^i\underline{g}) & i = 1, \dots, N_t, \\ \varphi_1(z) + \varphi_2(z) + (\varphi_1^i(z) - \varphi_2^i(z) - c^i\overline{g}) & i = 1, \dots, N_t \end{array} \right\}$$

and

$$q_0(z,\tau) = \varphi_1(z) + \varphi_2(z).$$

Then, we can write

$$p(z,\tau) = \varphi_2(z) + \tau + N_t c^i \overline{g} + p_0(z,t) \tag{15.16}$$

and

$$q(z,\tau) = \varphi_2(z) + \tau + N_t c^i \overline{g} + q_0(z,t), \tag{15.17}$$

so a new d.c. representation of $r(z,t)$ can be obtained

$$r(z,\tau) = p(z,\tau) - q(z,\tau) = p_0(z,\tau) - q_0(z,\tau). \tag{15.18}$$

By introducing a new variable t, the constraint $r(z,\tau) \leq 0$, can be replaced by an equivalent pair of convex and reverse convex constraints

$$\begin{array}{l} p_0(z,\tau) - t \leq 0, \\ q_0(z,\tau) - t \geq 0, \end{array} \tag{15.19}$$

respectively. Hence, by defining the closed convex sets

$$D = \{(z,\tau,t) : \ p_0(z,\tau) - t \leq 0, \ \underline{b} \leq Mz \leq \overline{b}, \ \underline{z} \leq z \leq \overline{z}\}$$

and

$$C = \{(z,\tau,t) : \ q_0(z,\tau) - t \leq 0\}$$

the d.c. program (15.15) is equivalent to the canonical d.c. program

$$\begin{array}{ll} \textbf{minimize} & \tau \\ \textbf{subject to} & (z,\tau,t) \in D \setminus \text{int } C. \end{array} \tag{15.20}$$

To solve the program (15.20) we need to find a vertex to the conical subdivisions by solving an initial convex program, and bound the closed convex sets of the resultant complementary mathematical convex structure.

3.1 A more advantageous equivalent reverse convex program

The pair of constraints (15.19) can be expressed by the $2N_t+1$ convex constraints

$$\begin{array}{ll} 2\varphi_1(z) - \tau - t \leq 0, & \\ \varphi_1(z) + \varphi_2(z) + (\varphi_2^i(z) - \varphi_1^i(z) + c^i \underline{g}) - t \leq 0 & i = 1, ..., N_t, \\ \varphi_1(z) + \varphi_2(z) + (\varphi_1^i(z) - \varphi_2^i(z) - c^i \overline{g}) - t \leq 0 & i = 1, ..., N_t \end{array} \tag{15.21}$$

and the reverse convex constraint

$$\varphi_1(z) + \varphi_2(z) - t \geq 0. \tag{15.22}$$

Hence, by defining the closed convex sets

$$\Omega = \left\{ (z,t): \begin{array}{l} \varphi_1(z) + \varphi_2(z) + (\varphi_2^i(z) - \varphi_1^i(z) + c^i \underline{g}) - t \leq 0, \\ \varphi_1(z) + \varphi_2(z) + (\varphi_1^i(z) - \varphi_2^i(z) - c^i \overline{g}) - t \leq 0, \\ i = 1, ..., N_t, \\ \underline{b} \leq Mz \leq \overline{b}, \\ \underline{z} \leq z \leq \overline{z} \end{array} \right\}$$

and

$$\Delta = \{(z,t): \ \varphi_1(z) + \varphi_2(z) - t \leq 0\},$$

and by using as objective function the convex function $2\varphi_1(z) - t$ a reverse convex program equivalent to the d.c. program (15.13) can be writen

$$\begin{array}{ll} \textbf{minimize} & 2\varphi_1(z) - t \\ \textbf{subject to} & (z,t) \in \Omega \setminus \text{int}\Delta, \end{array} \tag{15.23}$$

which is a more suitable transformation that allows us to use prismatical subdivisions and it is neither necessary to find an initial vertex by solving it nor bound the closed convex sets of the resultant complementary mathematical convex structure, as it is described in the next section.

4. Basic operations and the algorithm

Let D, C, C_ϵ and D_α be the closed and convex sets

$$\begin{array}{lll} D & := & \{(x,t) \in \mathbb{R}^n \times \mathbb{R}: \ Ax \leq b, \ g(x) - t \leq 0\}, \\ C & := & \{(x,t) \in \mathbb{R}^n \times \mathbb{R}: \ h(x) - t \leq 0\}, \\ C_\epsilon & := & \{(x,t) \in \mathbb{R}^n \times \mathbb{R}: \ h(x) - t + \epsilon \leq 0\}, \ \epsilon > 0, \\ D_\alpha & := & \{(x,t) \in D: \ f(x) - t \leq \alpha\}, \ \alpha \in \mathbb{R}, \end{array}$$

where A is a real $m \times n$ matrix, $b \in \mathbb{R}^m$ and $f(x)$, $g(x)$ and $h(x)$ are proper convex functions on $\mathbb{R}^n$. The notation cl *(F)* means the *closure* of the set F and ∂F denote the boundary of F. Notice that the sets D and C are not bounded but $D \setminus$ int C is a compact set when $Ax \leq b$ defines a polytope in $\mathbb{R}^n$. In this section we present some basic operations and a detailed description of the algorithm for solving the reverse convex programming problem of the form:

$$\text{minimize } \{f(x) - t : (x,t) \in D \setminus \text{int } C\}, \tag{15.24}$$

which has every global optimal solution in $D \cap \partial C$. Moreover, if the problem is *regular,* i.e., $D\backslash\text{int } C = \text{cl } (D\backslash C)$ then $(x^*, t^*) \in D\backslash\text{int } C$ is a global optimal solution if and only if $D_{\alpha^*} \subset C$ with $\alpha^* = f(x^*) - t^*$ (see Tuy (1998)). In the following we assume $D\backslash C \neq \emptyset$ and that for every feasible point (x_0, t_0) which verifies $t_0 = g(x_0) = h(x_0)$ we have

$$B(x_0, \epsilon) \cap \{x : Ax \leq b, g(x) < h(x)\} \neq \emptyset \quad \forall \epsilon > 0. \tag{15.25}$$

Lemma 4.1 *With the above-mentioned assumptions, the programming problem (15.24) is regular.*

Proof. We have $D\backslash\text{int } C \supset \text{cl } (D\backslash C)$ because cl $(D\backslash C)$ is the smallest closed set containing $D\backslash C$ and $D\backslash\text{int } C$ is a closed set in $\mathbb{R}^n \times \mathbb{R}$. On the other hand, let $(x^o, t^o) \in D\backslash\text{int } C$. Thus, there exists a sequence $\{(x_k, t_k),\ k = 1, 2, \dots\}$ of points of $D\backslash C$ that converge to (x^o, t^o). Indeed, three cases are possible:

1 $Ax_0 \leq b$ and $g(x_0) \leq t_0 < h(x_0)$. In this case $(x_0, t_0) \in D\backslash C$,

2 $Ax_0 \leq b$ and $g(x_0) < t_0 = h(x_0)$. In this case, the sequence

$$\{(x_0, t_0 - (t_0 - g(x_0))/2^k),\ k = 1, 2, \dots\}$$

of points of $D\backslash C$ converge to (x^o, t^o),

3 $Ax_0 \leq b$ and $g(x_0) = t_0 = h(x_0)$. By choosing

$$x_k \in B(x_0, 1/k) \cap \{x : Ax \leq b, g(x) < h(x)\} \neq \emptyset,$$

for every $k = 1, 2, \dots$, and taking $t_k = (h(x_k) + g(x_k))/2$ we can see that the sequence $\{(x_k, t_k),\ k = 1, 2, \dots\}$ of points of $D\backslash C$ converge to (x^o, t^o).

Hence, $(x^o, t^o) \in \text{cl } (D\backslash C)$ in $\mathbb{R}^n \times \mathbb{R}$ and $D\backslash\text{int } C \subset \text{cl } (D\backslash C))$, which proves the lemma. □

Define $\alpha^* := \inf\{f(x) - t : (x, t) \in D,\ h(x) - t + \epsilon > 0\}$. It is easily seen that the set $D_{\alpha^*}\backslash\text{int } C_\epsilon$ coincides with the set of the **ϵ-approximate** optimal solutions. The algorithm for solving the program (15.24), which we present in this section, is an adaptation of the *Combined OA/CS Conical Algorithm for CDC as* described in Tuy (1998), which responds to specific structure of this program. We introduce a branching process in which every partition set is a simplicial prism in $\mathbb{R}^{n+1}$, and the outer approximation process will be constructed by means of a sequence of polyhedrons generated through suitable piece linear functions. The algorithm has the advantage that it is not necessary to find any vertex for a conical subdivision process. This is substituted by a prismatical subdivision process.

4.1 Prismatical subdivision process

Let Z be an **n-simplex** in $\mathbb{R}^n$. The set

$$T(Z) := \{(x,\ t) \in \mathbb{R}^n \times \mathbb{R} :\ x \in Z\}$$

is called a *simplicial prism* of base Z. Every simplicial prism $T(Z)$ has $n+1$ edges that are parallel lines to the **t-axis.** Each edge pass through the $n+1$ vertices of Z. Then, every simplicial subdivision, $Z_1, \ldots, Z_r$, of the simplex Z via a point $z \in Z$ induces a *prismatical subdivision* of the prism $T(Z)$ in subprisms, $T(Z_1), \ldots, T(Z_r)$, via the parallel line to the **t-axis** through z (in this paper we are supposing that the simplicial subdivisions are proper, i.e., the point z doesn't coincide with any vertex of Z). A prismatical subdivision for $T(Z)$ is called a *bisection* of ratio ρ if it is induced by a bisection of ratio ρ of Z (see Tuy (1998)). A filter of simplices $Z_0 \supset Z_1 \supset \cdots \supset Z_k \supset \ldots$ induce a *filter of prisms,* $T_0 \supset T_1 \supset \cdots \supset T_k \supset \ldots$ with $T_k = T(Z_k)$, $k = 0, 1, 2, \ldots$. Also, every T_{k+1} is called a *child* of T_k. Moreover, a filter of prisms is said to be *exhaustive* if it is induced by an exhaustive filter of simplices, i.e., $T_\infty = \cap_{k=0}^{\infty} T_k$ is a parallel line to the **t-axis.** In that follows, the notation $[v^1, \ldots, v^{n+1}]$ means the simplex of vertices $v^1, \ldots, v^{n+1}$ and the notation $\mathbf{conv}\{v^1, \ldots, v^{n+1}\}$ is the *convex hull* of the set $\{v^1, \ldots, v^{n+1}\}$.

Proposition 4.1 *(Basic prismatical subdivision property) Let* $\{T_k,\ k = 0, 1, 2, \ldots\}$ *be a filter of prisms (with* $n+1$ *edges). Let* $q_k,\ k = 0, 1, 2, \ldots$ *be a point in the simplex spanned by the* $n+1$ *intersection points of the edges of* T_k *with* ∂C*. We assume that:*

1. *For infinitely many* k, T_{k+1} *is a child of* T_k *in a bisection of ratio* $\rho_k \geq \rho_0 > 0$.
2. *For all other* k, T_{k+1} *is a child of* T_k *in a subdivision via the parallel line to the* **t-axis** *through a point* $z_k \in Z_k$.

Then at least one accumulation point q *of the sequence* $\{q_k, k = 0, 1, 2, \ldots\}$ *satisfies* $q \in \partial C$.

Proof. Let Z_k be the simplex $Z_k = [v_k^1, \ldots, v_k^{n+1}]$ and let q_k^i be the point where the parallel line to the **t-axis** through the point v_k^i meets ∂C, so $q_k \in [q_k^1, \ldots, q_k^{n+1}]$. Let z_k be the point where the parallel line to the **t-axis** through the point q_k meets Z_k. At least one accumulation point of the sequence $\{z_k\}$ is a vertex of

$$Z_\infty := \cap_{k=0}^{\infty} Z_k = \mathbf{conv}\{v^1, \ldots, v^{n+1}\}$$

(see Tuy (1998) Theorem 5.1). Suppose $\{z_{k_s}\} \to v^1$. From $z_{k_i} = \sum_{i=1}^{n+1} \lambda_i^{k_s} v_{k_s}^i$ with $\sum_{i=1}^{n+1} \lambda_i^{k_s} = 1$ we have $(\lambda_1^{k_s}, \dots, \lambda_{n+1}^{k_s}) \to (1, 0, \dots, 0)$. On the other hand, from

$$q_{k_s} = \sum_{i=1}^{n+1} \lambda_i^{k_s} (v_{k_s}^i, h(v_{k_s}^i)) = (\sum_{i=1}^{n+1} \lambda_i^{k_s} v_{k_s}^i, \sum_{i=1}^{n+1} \lambda_i^{k_s} h(v_{k_s}^i))$$

we have

$$q_{k_s} = (z_{k_s}, \sum_{i=1}^{n+1} \lambda_i^{k_s} h(v_{k_s}^i)) \to q = (v^1, h(v^1)),$$

which proves that $q \in \partial C$. □

4.2 Outer approximation process

Let $\psi^\epsilon(x)$ be the convex proper function defined as

$$\psi^\epsilon(x) := \max \{h(x) + \epsilon, \ g(x)\}. \tag{15.26}$$

Lemma 4.2 *Consider $M := \{x_1, \dots, x_m\}$ a finite set of points in $\mathbb{R}^n$. Let p_i be a subgradient of the function $f(x)$ at the point $x_i \in M$ if $h(x_i) + \epsilon \geq g(x_i)$ or, else let p_i be a subgradient of the function $g(x)$ at point $x_i \in M$ if $h(x_i) < g(x_i)$. Thus, the function*

$$\psi_M^\epsilon(x) := max\{(x - x_i)^T p_i + \psi^\epsilon(x_i), \ i = 1, 2, \dots, m\} \tag{15.27}$$

satisfies the following properties:

1 $\psi_M^\epsilon(x)$ is piecewise linear and convex proper function on $\mathbb{R}^n$,

2 $P_M := \{(x, t) \in \mathbb{R}^n \times \mathbb{R} : \psi_M^\epsilon(x) - t \leq 0\}$ is a polyhedron.

If N is a finite set in $\mathbb{R}^n$ and $M \subseteq N$ then

3 $\psi_M^\epsilon(x) \leq \psi_N^\epsilon(x) \ \forall x \in \mathbb{R}^n$,

4 $P_M \supseteq P_N$.

Proof. Obvious. □

Let Z be the simplex of vertices $\{v^1, \dots, v^{n+1}\}$ and let

$$H := \{(x, t) \in \mathbb{R}^n \times \mathbb{R} : a^t x - t - \rho = 0\}$$

denote the uniquely defined hyperplane through the points $(v^i, h(v^i))$, $i = 1, 2, \dots, n+1$ with $a \in \mathbb{R}^n$ and $\rho \in \mathbb{R}$. Define the two closed halfspaces

$$H^+ := \{(x, t) \in \mathbb{R}^n \times \mathbb{R} : a^t x - t - \rho \leq 0\}$$

and

$$H^- := \{(x,t) \in \mathbb{R}^n \times \mathbb{R} : a^t x - t - \rho \geq 0\}.$$

Consider the filter of prisms $T_0 \supset T_1 \supset \cdots \supset T_k \supset \ldots$ where each $T_k := T(Z_k)$ is a prism which is induced by a proper subdivision of the simplex Z_{k-1} ($T_{k-1} := T(Z_{k-1})$) via a point $z_{k-1} \in Z_{k-1}$. Let

$$P_k = \{(x,\ t) \in \mathbb{R}^n \times \mathbb{R} : Ax \leq b,\ \psi^{\epsilon}_{M_k}(x) - t \leq 0\}$$

be the polyhedral generated from the set M_k, which contains the vertices of Z_0 and the points $z_0, z_1, \ldots, z_{k-1}$ generated in the subdivision process. In that follows, the function $\psi^{\epsilon}_{M_k}(x)$ will be denoted $\psi_k(x)$.

Lemma 4.3 *Let $(x(T_k), t(T_k))$ and $\mu(T_k)$ be the optimal solution and the optimal value of the linear program*

$$max\ \{a^t_k x - t - \rho :\ (x,t) \in T_k \cap P_k\}, \tag{15.28}$$

where $a^t_k x - t - \rho = 0$ is the hyperplane passing through the points $(v^i_k, h(v^i_k))$, $i = 1, 2, \ldots, n+1$ with $Z_k = [v^1_k, \ldots, v^{n+1}_k]$. Thus, the following assertions are true:

1 *if $\mu(T_k) > 0$ then $(x(T_k), t(T_k))$ doesn't lie on any edge of T_k,*

2 *if $\mu(T_k) \leq 0$ then $T_k \cap P_k \subset C$.*

Proof.

1 Suppose that the optimal solution of (15.28) lies on an edge of T_k. In this way, there exists a vertex v^i_k such that (v^i_k, t_i) satisfies $\mu(T_k) = a^t_k v^i_k - t_i - \rho > 0$. On the other hand, the vertex v^i_k satisfies that $\psi_k(v^i_k) \geq \psi(v^i_k) \geq h(v^i_k)$ so $\psi_k(v^i_k) - h(v^i_k) \geq 0$ and $a^t_k v^i_k - h(v^i_k) - \rho = 0$. Hence, $h(v^i_k) > t_i$ and $\psi_k(v^i_k) - t_i > \psi_k(v^i_k) - h(v^i_k) \geq 0$ which implies that $(v^i_k, t_i) \notin P_k$, which is a contradiction.

2 Let $(x^o, t^o) \in T_k \cap P_k$ be a feasible point of the linear program (15.28). Then, from the hypothesis in 2, we deduce that $a^t_k x^o - t^o - \rho \leq \mu(T_k) \leq 0$ so that $(x^o, t^o) \in T_k \cap P_k \cap H_+$. From $x^o \in Z_k$ we can write the expression $x^o = \sum_{i=1}^{n+1} \lambda^o_i v^i_k$ with $\sum_{i=1}^{n+1} \lambda^o_i = 1$. From the convexity of the function $h(x)$ we obtain the inequality $h(x^o) = h(\sum_{i=1}^{n+1} \lambda^o_i v^i_k) \leq \sum_{i=1}^{n+1} \lambda^o_i h(v^i_k)$. Finally, from the definition of the hyperplane H we know that each point $(v^i_k, h(v^i_k))$, $i = 1, \ldots, n+1$ verifies the equality $a^t_k v^i_k - h(v^i_k) - \rho = 0$ so that we have $h(v^i_k) = a^t_k v^i_k - \rho$. Hence, we can write

$$h(x^o) - t_o \leq \sum_{i=1}^{n+1} \lambda^o_i h(v^i_k) - t_o = \sum_{i=1}^{n+1} \lambda^o_i (a^t_k v^i_k - \rho) - t_o \leq$$

$$\leq a_k^t \sum_{i=1}^{n+1} \lambda_i^o v_k^i - \rho \sum_{i=1}^{n+1} \lambda_i - t_o = a^t x^o - \rho - t_o \leq \mu(T_k) \leq 0,$$

which proves that $(x^o, t^o) \in C$.

□

4.3 The algorithm

Initialization:
$\alpha_0 \leftarrow +\infty$; $\Gamma_0 \leftarrow \emptyset$;
Determine a simplex $Z_0 \supset \{x \in \mathbb{R}^n : Ax \leq b\}$,
its vertex set V_0 and the prism $T_0 = T(Z_0)$;
if $(\bar{x},\ \bar{t})$ is the best feasible solution available **then**
    $\alpha_0 \leftarrow f(\bar{x}) - \bar{t}$;
end if
$M_0 \leftarrow V_0 \cup \{\bar{x}\}$;
$P_0 \leftarrow \{(x,\ t) \in \mathbb{R}^n \times \mathbb{R} : Ax \leq b,\ \psi_0(x) - t \leq 0\}$;
Solve $\mu(T_0) = \max\{a_0^t x - t - \rho :\ (x,\ t) \in T_0 \cap P_0\}$,
with $(x(T_0), t(T_0))$ the optimal solution;
if $\mu(T_0) > 0$ **then**
    $\Phi_0 \leftarrow \{T_0\}$; $\Gamma_0 \leftarrow \{T_0\}$; $T_0^* \leftarrow T_0$; $Z_0^* \leftarrow Z_0$;
end if
stop $\leftarrow$ **false**; $k \leftarrow 0$;

while stop=false **do**
    if $\Gamma_k = \emptyset$ **then**
        if $\alpha_k = +\infty$ **then** the problem is infeasible;
        else $(\bar{x}, \bar{t})$ is an **ϵ-approximate** optimal solution; **end if**
        stop $\leftarrow$ **true**;
    else
        if $h(x(T)) - t(T) + \epsilon > 0$ for some $T \in \Gamma_k$ **then**
            $M_{k+1} \leftarrow M_k \cup \{x(T)\}$;
            $P_{k+1} \leftarrow \{(x,\ t) \in \mathbb{R}^n \times \mathbb{R} : Ax \leq b,\ \psi_{k+1}(x) - t \leq 0\}$;
            if $\psi(x(T)) = h(x(T))$ and $f(x(T)) - h(x(T)) < \alpha_k$ **then**
                $(\bar{x},\ \bar{t}) \leftarrow (x(T), h(x(T))$; $\alpha_{k+1} \leftarrow f(\bar{x}) - \bar{t}$;
            end if
        end if
        Split Z_k^* via the chosen normal rule to obtain a partition
        $T_{k_1}, \ldots, T_{k_r}$ of T_k^*; $\Psi_{k+1} \leftarrow \{T_{k_1}, \ldots, T_{k_r}\}$;
        For all prism $T_{k_i} \in \Psi_{k+1}$ solve the linear program:
        $\mu(T_{k_i}) = \max\{a_{k_i}^t x - t - \rho :\ (x,\ t) \in T_{k_i} \cap P_{k+1}\}$,
        with $(x(T_{k_i}), t(T_{k_i}))$ the optimal solution;

$\Phi_{k+1} \leftarrow (\Gamma_k \setminus \{T_k^*\}) \cup \Psi_{k+1}$; $\Gamma_{k+1} \leftarrow \{T \in \Phi_{k+1} : \mu(T) > 0\}$;
Choose $T_{k+1}^* \in$ **argmax** $\{\mu(T) : T \in \Gamma_{k+1}\}$;
end if
$k \leftarrow k + 1$;
end while

4.4 Convergence of the algorithm

Let $\sigma(x)$ be the convex proper function defined as

$$\sigma(x) := \max \{f(x) - \alpha^*, \ g(x)\}. \tag{15.29}$$

In that follows, each generated point $(x(T_k), t(T_k))$ in the algorithm will be denoted by (x_k, t_k). Thus, for each generated point (x_k, t_k) we can consider the cuts:

- $r_k(x,t) := p_k^T(x - x_k) - (t - \sigma(x_k))$ with p_k a subgradient of the function $\sigma(x)$ at the point x_k,
- $s_k(x,t) := p_k^T(x - x_k) - (t - \psi(x_k))$, as defined in Lemma 4.2.

Lemma 4.4 *Let $\{(x_k, t_k)\}$ be the sequence obtained in the algorithm by solving the linear problems (15.28). Thus, we have that $\psi(x_k) \leq \sigma(x_k)$ and the sequences $\{(x_k, t_k)\}$, $\{(x_k, \psi(x_k))\}$ and $\{(x_k, \sigma(x_k))\}$ are bounded.*

Proof. We have either $\psi(x_k) = g(x_k)$ or $\psi(x_k) = h(x_k) + \epsilon$. When $\psi(x_k) = g(x_k)$ then obviously $\psi(x_k) \leq \sigma(x_k)$. Otherwise $\bar{t_k} = \psi(x_k) = h(x_k) + \epsilon$ so $(x_k, \bar{t_k})$ is a feasible point and $f(x_k) - \bar{t_k} \geq \alpha^*$. Then, we can write $f(x_k) - \alpha^* \geq \bar{t_k} = h(x_k) + \epsilon$ and also $\psi(x_k) \leq \sigma(x_k)$.

On the other hand, the functions $\psi(x)$ and $\sigma(x)$ are continuous on the polytope defined by $Ax \leq b$ (which is a compact set). From $t_k < \psi(x_k) \leq \sigma(x_k)$ we can deduce that the sequences $\{(x_k, t_k)\}$, $\{(x_k, \psi(x_k))\}$ and $\{(x_k, \sigma(x_k))\}$ are bounded. □

Lemma 4.5 *The cuts $s_k(x,t)$ and $r_k(x,t)$ strictly separate each generated point (x_k, t_k) (of the sequence $\{(x_k, t_k)\}$ obtained in the algorithm) from D_{α^*}.*

Proof. Obviously, for all $(x,t) \in D_{\alpha^*}$ we have $\sigma(x) - t \leq 0$. Then, $r_k(x,t) = p_k^T(x - x_k) - (t - \bar{t_k}) \leq \sigma(x) - t \leq 0$ by using the convexity of the function $\sigma(x) - t$. On the other hand, let $\bar{t_k} = \psi(x_k)$ and $(x,t) \in D_{\alpha^*}$. Then, we can write

$$s_k(x,t) = p_k^T(x - x_k) - (t - \bar{t_k}) \leq \begin{cases} g(x) - t \leq 0 \\ \text{or} \\ f(x) - \alpha^* - t \leq 0. \end{cases} \tag{15.30}$$

Moreover, from $t_k < \psi(x_k) \leq \sigma(x_k)$ we obtain

$$r_k(x_k, t_k) = -(t_k - \sigma(x_k)) \geq -(t_k - \psi(x_k)) = s_k(x_k, t_k) > 0, \quad (15.31)$$

which proves the lemma. □

From lemma 4.4 we know that the sequence $\{(x_k, t_k)\}$ is bounded and that there exits a subsequence $\{k_i\}$ such that $\{(x_{k_i}, t_{k_i})\} \to (\bar{x}, \bar{t})$.

Lemma 4.6 *The following assertions are true:*

1 $r_{k_i}(x_{k_i}, t_{k_i}) = -(t_{k_i} - \sigma(x_{k_i})) \to 0$ *when* $i \to +\infty$ *and* $\sigma(\bar{x}) = \bar{t}$,

2 $s_{k_i}(x_{k_i}, t_{k_i}) = -(t_{k_i} - \psi(x_{k_i})) \to 0$ *when* $i \to +\infty$ *and* $\psi(\bar{x}) = \bar{t}$.

Proof. From (15.31) we have

$$\lim_{i \to +\infty} r_{k_i}(x_{k_i}, t_{k_i}) \geq 0. \quad (15.32)$$

On the other hand, if k is fixed we have $r_k(x_{k_i}, t_{k_i}) \leq 0$ for all $k_i > k$. Then, from $i \to +\infty$, we obtain $r_k(\bar{x}, \bar{t}) \leq 0$. Otherwise, for all k, we can write $r_k(x, t) = r_k(\bar{x}, \bar{t}) + p_k^T(x - \bar{x}) - (t - \bar{t})$. Hence, the relationship

$$r_{k_i}(x_{k_i}, t_{k_i}) \leq p_{k_i}^T(x_{k_i} - \bar{x}) - (t_{k_i} - \bar{t}) \quad (15.33)$$

can be obtained. Moreover, we know that $\{p_k\}$ is a bounded sequence (see Tuy (1998) Theorem 2.6). Then, letting $i \to +\infty$ in (15.33) we obtain

$$\lim_{i \to +\infty} r_{k_i}(x_{k_i}, t_{k_i}) \leq 0. \quad (15.34)$$

From (15.32) and (15.34) we can deduce that $r_{k_i}(x_{k_i}, t_{k_i}) \to 0$ and, as a direct consequence, we have $\sigma(\bar{x}) = \bar{t}$. The same proof holds true by using $s_{k_i}(x_{k_i}, t_{k_i})$ in place of $r_{k_i}(x_{k_i}, t_{k_i})$, which proves the lemma. □

From the preceding lemmas and by using the Proposition 4.1 we can enounce the following result.

Proposition 4.2 *The algorithm can only be infinite if* $\epsilon = 0$ *and in this case any accumulation point of the sequence* $\{(x_k, t_k)\}$ *is a global optimal solution for the program (15.24). Moreover, if* $\epsilon > 0$ *then the algorithm is finite and an* ϵ-*approximate optimal solution can be obtained.*

Proof. Let $(\bar{x}, \bar{t})$ be an accumulation point of the sequence $\{(x_k, t_k)\}$. From $h(x_k) - t_k > 0$ we obtain $h(\bar{x}) - \bar{t} \geq 0$. On the other hand, we know that $h(\bar{x}) - \bar{t} + \epsilon = 0$, which is a contradiction unless $\epsilon = 0$. In this case, every point $(x^o, t^o) \in D_{\alpha^*}$ satisfies $g(x^o) - t^o \leq 0$, $f(x^o) - t^o \leq \alpha^*$ and $Ax^o \leq b$. Suppose that $h(x^o) - t^o > 0$. This implies that $f(x^o) - t^o \leq \alpha^* < f(x^o) - t^o$, which is a contradiction. Thus, the point (x^o, t^o) must satisfy $h(x^o) - t^o + \epsilon \leq 0$ and therefore $(x^o, t^o) \in C$, i.e., $D_{\alpha^*} \subset C$. The optimality criterion, together with the regularity assumption, implies that $(\bar{x}, \bar{t})$ is a global optimal solution with α^* global optimal value. □

5. The least deviation problem

Let $\mathbb{R}_m[x_1, ..., x_n]$ and $H_k[x_1, ..., x_n]$ be the vector spaces of polynomials of degree less than or equal to m and of homogeneous polynomials of degree k, respectively. Both vector spaces are normed spaces using the norm of a polynomial $z(x) = \sum_{i \in I} a_i f_i(x)$ defined by $\|z(x)\| := (\sum_{i \in I} a_i^2)^{1/2}$ where $f_i(x), i \in I$ are the monomials of the usual base in $\mathbb{R}_m[x_1, ..., x_n]$ or the usual base in $H_k[x_1, ..., x_n]$. The notation $\|z(x)\|_k$ is used to indicate the norm in $H_k[x_1, ..., x_n]$. From the expression

$$\mathbb{R}_m[x_1, \ldots, x_n] = \bigoplus_{k=0}^{m} H_k[x_1, \ldots, x_n], \tag{15.35}$$

the following relationship between the norms can be deduced

$$\|z(x)\|^2 = \|z_0(x)\|_0^2 + \|z_1(x)\|_1^2 + \cdots + \|z_m(x)\|_m^2, \tag{15.36}$$

where $z_i(x) \in H_i[x_1, ..., x_n]$, $i = 0, 1, \ldots, m$ and $z(x) = \sum_{i=0}^{m} z_i(x)$. Let $\Lambda \subset \mathbb{R}^n$ be a closed convex set and let $K_m(\Lambda)$ and $K_k(\Lambda)$ be the nonempty closed convex cones of the polynomials in $\mathbb{R}_m[x_1, ..., x_n]$ and $H_k[x_1, ..., x_n]$, respectively which are convex on Λ. Denote $\boldsymbol{E} = \mathbb{R}_m[x_1, \ldots, x_n]$ and $K = K_m(\Lambda)$ or $\boldsymbol{E} = H_k[x_1, \ldots, x_n]$ and $K = K_k(\Lambda)$ because in the following all the properties to be deduced can be applied to both normed spaces. Let $D_\Lambda(z)$ be the set of all the d.c. representations of $z(x)$ on Λ, i.e.,

$$D_\Lambda(z) := \{(y_1, y_2) \in K \times K : \ z = y_1 - y_2 \text{ on } \Lambda\}, \tag{15.37}$$

which is a lower bounded ordered set by defining the relation

$$(\bar{y}_1, \bar{y}_2) \leq (y_1, y_2) \Leftrightarrow \|\bar{y}_1 + \bar{y}_2\| \leq \|y_1 + y_2\|, \tag{15.38}$$

so we can consider

$$\inf \{\|y_1 + y_2\| : (y_1, y_2) \in D_\Lambda(z)\} \geq 0. \tag{15.39}$$

On the other hand, the problem

$$\text{minimize } \{\|v\|^2 : v \in (-z + 2K) \cap (z + 2K)\}, \tag{15.40}$$

which we will refer to as *the minimal norm problem,* has a unique solution which is obtained by a unique point $v^* \in (-z+2K)\cap(z+2K)$ because, the feasible domain is a closed convex set and the function $\|v\|^2$ is strictly convex. The optimal solution $v^* = y_1^* + y_2^*$ give us an optimal d.c. represetation for $z = y_1^* - y_2^*$ and moreover allows us to substitute the expression (15.39) by

$$\|y_1^* + y_2^*\| = \min\{\|y_1 + y_2\| : (y_1, y_2) \in D_\Lambda(z)\} \geq 0, \tag{15.41}$$

which is called *the least deviation problem* (see Luc et al (1999)), and the pair (y_1^*, y_2^*) is called the *least deviation decomposition (LDD)* of $z(x)$ on Λ.

5.1 The equivalent semi-infinite minimal norm problem

A peculiarity of the the minimal norm problem (15.40) is that it can be transformed into a semi-infinite quadratic programming problem with linear constraints. The Hessian $\nabla^2(v \pm z)(x) = \sum_{i\in I}(v_i \pm a_i)\nabla^2 f_i(x)$, of the sum and difference of the polynomials $z(x) = \sum_{i\in I} a_i f_i(x)$ and $v(x) = \sum_{i\in I} v_i f_i(x)$, is a semidefinite positive matrix because $(v \pm z)(x)$ must be a convex polynomial. Hence, we can write

$$\lambda^t \nabla^2(v \pm z)(x)\lambda \geq 0 \quad \forall\lambda \in S^n,\ \forall x \in \Lambda \subset \mathbb{R}^n, \tag{15.42}$$

where $S^n = \{x \in \mathbb{R}^n : \ \|x\| = 1\}$, or its equivalent

$$\lambda^t \sum_{i\in I}(v_i \pm a_i)\nabla^2 f_i(x)\lambda \geq 0 \quad \forall\lambda \in S^n\ ,\forall x \in \Lambda \subset \mathbb{R}^n. \tag{15.43}$$

By substituting the set constraints of (15.40) by the equivalent set constraints (15.42), the problem (15.40) can be transformed into the equivalent semi-infinite quadratic programming problem

$$\begin{array}{ll} \textbf{minimize} & \|v\|^2 \\ \textbf{subject to} & \lambda^t \sum_{i\in I}(v_i \pm a_i)\nabla^2 f_i(x)\lambda \geq 0, \\ & \forall\lambda \in S^n\ ,\forall x \in \Lambda \subset \mathbb{R}^n, \end{array} \tag{15.44}$$

which depend on a family of parameters x and λ. Usually, Λ will be a convex compact set in the form $\Lambda = \prod_{i=1}^n [r_i, t_i]$.

The relationship (15.35) between $\mathbb{R}_m[x_1, ..., x_n]$ and $H_k[x_1, ..., x_n], k = 0, 1, \ldots, m$ can sometimes be used to simplify the calculus of the *LDD* of a polynomial $z(x) \in \mathbb{R}_m[x_1, ..., x_n]$. Consider $y_1^*(x) = \sum_{i=0}^m y_{1,i}^*(x)$ and $y_2^*(x) = \sum_{i=0}^m y_{2,i}^*(x)$ where $y_{1,i}^*(x)$ and $y_{2,i}^*(x)$ are polynomials in $H_i[x_1, ..., x_n], i = 0, 1, \ldots, m$.

Proposition 5.1 *(The decomposition property) Let* $\Lambda \subset \mathbb{R}^n$ *be a closed convex set and let* $z(x) = \sum_{k=0}^{m} z_k(x)$ *be a polynomial with* $z_k(x) \in H_k[x_1, ..., x_n]$. *Consider* $(y_{1,i}^o(x),\ y_{2,i}^o(x))$ *the LDD of* $z_i(x)$. *Then, the pair* $(y_1^o(x),\ y_2^o(x))$ *where* $y_1^o(x) = \sum_{i=0}^{m} y_{1,i}^o(x)$ *and* $y_2^o(x) = \sum_{i=0}^{m} y_{2,i}^o(x)$ *is the LDD of* $z(x)$ *on* Λ *when* $y_{1,i}^*(x)$ *and* $y_{2,i}^*(x)$ *are polynomials in* $K_i(\Lambda), i = 0, 1, \ldots, m$ *(which is not always true).*

Proof. Let $(y_1^*(x),\ y_2^*(x))$ be the *LDD* of the given polynomial $z(x)$. Thus, we know that the polynomial $v^*(x) = y_1^*(x) + y_2^*(x)$ solves the minimal norm program. Hence, we can write the inequality

$$\| v^o(x) \|^2 \geq \| v^*(x) \|^2, \tag{15.45}$$

where $v^o(x) = y_1^o(x) + y_2^o(x)$. On the other hand, we can consider the polynomials $y_1^*(x) = \sum_{i=0}^{m} y_{1,i}^*(x)$ and $y_2^*(x) = \sum_{i=0}^{m} y_{2,i}^*(x)$ where $y_{1,i}^*(x)$ and $y_{2,i}^*(x)$ are polynomials in $K_i(\Lambda), i = 0, 1, \ldots, m$. Thus, each $v_i^*(x) = y_{1,i}^*(x) + y_{2,i}^*(x), i = 0, 1, \ldots, m$ satisfies the inequality

$$\| v_i^*(x) \|_i^2 \geq \| v_i^o(x) \|_i^2$$

because the pair $(y_{1,i}^o(x),\ y_{2,i}^o(x))$ is the LDD of $z_i(x)$. Then, we can write

$$\| v^*(x) \|^2 = \sum_{i=0}^{m} \| v_i^*(x) \|_i^2 \geq \sum_{i=0}^{m} \| v_i^o(x) \|_i^2 = \| v^o(x) \|^2 . \tag{15.46}$$

From (15.45) and (15.46) we deduce that $\| v^*(x) \|^2 = \| v^o(x) \|^2$, which proves the proposition. □

We have obtained the least deviation decomposition of each power hydrogeneration function at each reservoir by using the algorithms described in the articles Kaliski et al (1997) and Zhi-Quan et al (1999). These results are not explicity indicated in this paper and we only use them to obtain our computational results.

6. Characteristics of generation systems and computational results

The characteristics of the generation systems can be found in Table 15.1. The names of the problems in Table 15.1 have the expression **c***n***e***m***i** and the names of the problem instances in Table 15.2 have the expression **c***n***e***m***iXYZ** where n, m, X, Y and Z mean:

- n (one digit) is the number of nodes,
- m (two digits) is the number of time intervals,

- $X = v$ when we know thatk^i_j in (15.1) depends on water discharges, or else it is a constant and then $X = k$,
- $Y = 1$ when thermal units satisfy the entire demand for electricity in every time interval, or else this is not possible and then $Y = 0$.
- $Z = b$ when we solve the problem instance using the optimal d.c. representation of the the power hydrogeneration functions, or else $Z = a$.

We use MINOS 5.5 to solve all problem instances and, also to check all gradients of the functions in the reverse convex program. The number maximum of iterations allowed in the global optimization algorithm has been of 5000 and the precision $\epsilon = 0.005$. In Table 15.2 *Iter* indicates the number of iterations required; *Sdv* indicates the number maximum of subdivisions that have been simultaneously active; *Fsb* indicates the total number of feasible points computed; *MINOS* indicates the optimal value at the solution obtained by MINOS; *Obj. Val* indicates the optimal value at the ϵ-approximate optimal solution obtained by the global optimization algorithm; *CPU time* is the CPU time in seconds. To solve all problems we have used a computer SUN ULTRA 2 with 256 Mb of main memory and 2 CPU of 200 MHz, SPCint95 7.88, SPCfp95 14.70. Moreover, to compare different speeds of solution, problems number 17 and 18 in Table 15.2 have been solved with a computer Compaq AlphaServer HPC320: 8 nodes ES40 (4 EV68, 833 MHz, 64 KB/8 MB), 20 GB of main memory, 1.128 GB on disk and top speed of 53,31 Gflop/s, connected with Memory Channel II de 100 MB/s.

Table 15.1. **Characteristics of Generation Systems**

Problem	*Nodes* N_e	*Intervals* N_t	*Dimension* $N_e(2N_t - 1)$	*Linear const.* N_eN_t	*Nonlinear const.* N_t
c2e02i	2	2	6	4	2
c2e03i	2	3	10	6	3
c4e03i	4	3	20	12	3

7. Conclusions

The instances with constant coefficient of efficiency and unit conversion seem to work well and we can find exact values for the optimal objective function. Otherwise, instances with a variable coefficient of efficiency have worse optimal values for the objective function but all solutions are very near to the solution founded by MINOS. Of course, this is not an ideal situation but it is not as bad as we might suppose.

Table 15.2. **Results and CPU requirements of problems solved**

Num	*Instance*	*Iter*	*Sdv*	*Fsb*	*MINOS*	*Obj. Val*	*CPU time*
1	c2e02ik0a	13	7	10	114.348	114.348	1.00
2	c2e02ik0b	11	6	4		114.347	0.77
3	c2e02iv0a	277	66	151	*infeas.*	93.036	89.66
4	c2e02iv0b	21	9	7		93.266	3.44
5	c2e02ik1a	12	7	10	114.348	114.348	0.88
6	c2e02ik1b	10	5	7		114.347	0.67
7	c2e02iv1a	14	9	12	93.090	93.093	1.11
8	c2e02iv1b	13	8	10		93.092	0.95
9	c2e03ik0a	1062	199	767	262.796	262.795	2037.04
10	c2e03ik0b	1030	189	738		262.296	1409.44
11	c2e03iv0a	1077	175	152	222.400	250.016	1951.55
12	c2e03iv0b	904	133	296		262.498	1433.19
13	c2e03ik1a	1223	191	632	262.796	262.796	1972.71
14	c2e03ik1b	1117	199	624		262.796	1810.76
15	c2e03iv1a	1079	175	192	222.399	250.015	1454.01
16	c2e03iv1b	1056	170	136		250.015	1420.34
17	c4e03ik0a	5000	1242	2605	374.040	374.068	115359.00
							23064.80
18	c4e03ik0b	5000	1225	2665		374.067	70771.50
							13105.80

Working alone MINOS can not find any solution for the problem *c2e02iv0Z* (which is related with the problem instances number 3 and 4). MINOS gives the problem as infeasible. On the other hand, when MINOS parts from the first feasible point found by using our global optimization procedure then MINOS give a solution which has the same optimal value as the optimal value calculated by using our algorithm.

From a computational standpoint and on observing the table 15.2, the efficency of using the optimal d.c. representation of the power hydrogeneration functions is obvious. In all instances where we have used them, the algorithm has obtained better CPU time and has carried out less iterations than problem instances where they have not been used (this difference must be outstanding for the problem instances number 3 and 4). Note that the optimal d.c. representation of the power hydrogeneration functions give us a more efficient d.c. representation of the functions in (15.6) but, they are not the optimal d.c. representation of these functions which would have required the solution of a very hard semi-infinite programming problem.

When the size of the problems increase then they become more and more difficult to solve. The size of the problem instances is a very serious limitation. We can observe from the instances number 17 and 18 that the CPU time can be reduced to one fifth by using the Compaq AlphaServer HPC320 computer. Obviously, the world of global optimization is the world of the high computers but I am sure that there exist a lot of available mathematical results (such as the concept of Least Deviation Decomposition) which could be used in order to obtain more efficient implementations for problems both with and without any specific structure.

Acknowledgments

We gladly thank the *CESCA,* The Supercomputing Center of Catalonia, for providing us with access to their computer Compaq AlphaServer *HPC*320.

References

Ferrer A. (2001), Representation of a polynomial function as a difference of convex polynomials, with an application, *Lectures Notes in Economics and Mathematical Systems,* Vol. 502, pp. 189-207.

Gurlitz T.R. and Jacobsen S.E. (1991), On the use of cuts in reverse convex programs, *Journal of Optimization Theory and Applications,* Vol. 68, pp. 257-274.

Heredia F.J. and Nabona N. (1995), Optimum short-term hydrothermal scheduling with spinning reverse through network flows, *IEEE Trans. on Power Systems,* Vol. 10(3), pp. 1642-1651.

Hiriart-Urruty J.B. (1985), Generalized differentiability, duality and optimization for problems dealing with differences of convex functions, *Lectures Notes in Economics and Mathematical Systems,* Vol. 256, pp. 27-38.

Horst R. and Tuy H. (1990), *Global optimization. Deterministic approaches,* Springer-Verlag, Heidelberg.

Horst R., Pardalos P.M. and Thoai Ng.V. (1995), *Introduction to global optimization,* Kluwer Academic Publishers, Dordrecht.

Horst R., Phong T.Q., Thoai Ng.V. and de Vries J. (1991), On solving a d.c. programming problem by a sequence of linear programs, *Annals of Operations Research,* Vol. 25, pp. 1-18.

Kaliski J., Haglin D., Roos C. and Terlaky T. (1997), Logarithmic barrier decomposition methods for semi-infinite programming, *Int. Trans. Oper. Res.,* Vol. 4(4), pp. 285-303.

Luc D.T., Martinez-Legaz J.E. and Seeger A. (1999), Least deviation decomposition with respect to a pair of convex sets, *Journal of Convex Analysis,* Vol. 6(1), pp. 115-140.

Tuy H (1998), *Convex analysis and global optimization,* Kluwer Academic Publishers, Dordrecht.

Zhi-Quan L., Roos C. and Terlaky T. (1999), Complexity analysis of logarithmic barrier decomposition methods for semi-infinite linear programming, *Applied Numerical Mathematics,* Vol. 29, pp. 379-394.

Chapter 16

ε-OPTIMALITY FOR NONSMOOTH PROGRAMMING ON A HILBERT SPACE

Misha G. Govil*
Department of Mathematics
Shri Ram College of Commerce
University of Delhi, India

Aparna Mehra
Department of Mathematics
Indian Institute of Technology, Delhi, India

Abstract Lagrange multiplier rules characterizing **ε-optimality/ε-efficiency** for nonsmooth programming problems on a real Hilbert space are established in terms of the limiting subgradients.

Keywords: Nonlinear programming ; limiting subgradient ; variational principle ; approximate solution ; Lagrange multiplier rule.

MSC2000: 90C29, 90C30

1. Introduction

The calculus results of nonsmooth analysis are frequently used to derive optimality conditions for nondifferentiable optimization problems. The most significant contribution in this direction was made by Clarke in 1983. He developed Lagrange multiplier rule for nondifferentiable scalar-valued Lipschitz programming problem by replacing the usual gradient of the function by Clarke's generalized gradient. Motivated by the work of Clarke (1983), Hamel (2001) extended the Lagrange multiplier rule to

---

*Corresponding Author. Email: mishagovil@yahoo.com

nondifferentiable scalar-valued programming problem on a real Banach space by using the notion of **ε-optimal** solution.

The notion of **ε-optimal** solution seems to be particularly useful for the class of optimization problems which otherwise have no optimal solutions. For this reason several authors (Loridan (1982), Liu (1991), Hamel (2001)) have turned their attention to develop **ε-optimality** conditions for mathematical programs. In all these works, Ekeland's variational principle (see Ekeland (1974)) is used as a basic tool to derive the main results. However, one notable limitation of this principle is that the perturbed objective function is not differentiable even though the original objective function is differentiable. Borwein and Preiss (1987) provided a smooth version of the variational principle in which the perturbed function is obtained by adding a smooth convex function to the original objective function.

In this paper, the variational principle of Borwein and Preiss (1987) is used to derive the Lagrange multiplier rule to characterize **ε-optimality** for nondifferentiable programming problems on a real Hilbert space in terms of limiting subgradient of the functions.

The paper is organized as follows. In Section 2, we present some definitions and results that are required in the subsequent sections. **ε-optimality** conditions for nonsmooth scalar-valued programming problem are derived in Section 3 while Section 4 is devoted to characterize **ε-efficiency** for nonsmooth multiobjective programming problem.

2. Preliminaries

2.1 Proximal Analysis

Let X be a real Hilbert space and S be a nonempty subset of X. Let x be a point not lying in S, and s be a point in S which is closest to x. The vector $x - s$ is a proximal normal direction to S at s and any nonnegative multiple of such a vector is called a proximal normal to S at s. The proximal normal cone to S at s, denoted by $N_S^P(s)$, is given by

$$N_S^P(s) = \{\xi : \xi = t(x - s),\ t \geq 0\}.$$

Let f be a lower semi continuous (l.s.c.) function from X to $R \cup \{+\infty\}$. A vector $\xi \in X$ is called a proximal subgradient of f at $x \in \mathbf{dom} f = \{x \in X : f(x) < +\infty\}$, if

$$(\xi, -1) \in N^P_{\mathrm{epi}\, f}\ (x, f(x)),$$

where $\mathrm{epi}\, f = \{(x, v) \in \mathbf{dom} f \times R : f(x) \leq v\}$ is an epigraph of f.

The set of all such ξ, denoted by $\partial_P f(x)$, is referred to as the proximal subdifferential of f at x. The proximal subdifferential usually has only a fuzzy calculus and, in general, does not satisfy the sum - rule as desired, that is,

$$\partial_P(f+g)(x) \not\subseteq \partial_P f(x) + \partial_P g(x) .$$

So, in some sense the proximal subdifferential is inadequate for the purpose of developing necessary optimality conditions. This subdifferential is therefore enlarged to the smallest adequate closed subdifferential, called the limiting subdifferential. In this context the limiting cone to S at s is given by

$$N_S^L(s) = \{w - \lim \xi_i : \xi_i \in N_S^P(s_i),\ s_i \to s\}$$

where $s_i \in S$ for all i and w—$\lim \xi_i$ is the weak limit of the sequence ξ_i. A vector ξ is called a limiting subgradient of f at x if and only if $(\xi, -1) \in N_{\mathrm{epi}\, f}^L(x, f(x))$. The set of all such limiting subgradients is called the limiting subdifferential, denoted by $\partial_L f(x)$, that is

$$\partial_L f(x) = \{w - \lim \xi_i : \xi_i \in \partial_P f(x_i),\ x_i \to x,\ f(x_i) \to f(x)\} .$$

Theorem 2.1 ((Sum Rule) Mordukhovich (1984)) *If one of f_1, f_2 is Lipschitz in a neighbourhood of x, then*

$$\partial_L(f_1 + f_2)(x) \subseteq \partial_L f_1(x) + \partial_L f_2(x) .$$

Lemma 2.1 (Clarke et al (1998)) *Let $f : X \to R \cup \{+\infty\}$ be l.s.c. function on X. If f has a local minimum at $x \in X$ then $0 \in \partial_P f(x) \subset \partial_L f(x)$.*

Remark 2.1 *The limiting normal cone and sub-differential are introduced in the paper of B.S. Mordukhovich (1976) for finite dimensional spaces. These concepts were extended to Banach space by A.Y. Kruger and B.S. Mordukhovich (1980). The normal cone and the subgradient for locally Lipschitz functions constructed by B.S. Mordukhovich and Y. Shao (1996) in an arbitrary Asplund space coincide respectively with the cone of limiting proximal normals and limiting subgradients of Clarke et al. (1998) in the Hilbert space setting.*

2.2 Variational Principle and its Applications

The following minimization rule due to Borwein and Preiss (1987), extensively studied by Clarke et al ((1997), Chapter 1, Theorem 4.2) will be used as a principle tool in proving the main results of the paper.

Theorem 2.2 *Let $f : X \to R \cup \{+\infty\}$ be l.s.c., bounded below function on a real Hilbert space X and let $\varepsilon > 0$. Suppose that x_ε is a point in X satisfying $f(x_\varepsilon) < \operatorname*{Inf}_{x \in X} f(x) + \varepsilon$. Then, for any $\lambda > 0$ there exist points y_ε and z_ε with*

(i') $\|y_\varepsilon - z_\varepsilon\| < \lambda, \quad \|z_\varepsilon - x_\varepsilon\| < \lambda$

(ii') $f(y_\varepsilon) \leq f(x_\varepsilon) < \operatorname*{Inf}_{x \in X} f(x) + \varepsilon$

(iii') *y_ε is the unique minimum of the function* $x \mapsto f(x) + \frac{\varepsilon}{\lambda^2}\|x - z_\varepsilon\|^2$.

Remark 2.2 *If y_ε is the unique minimum of the function $x \mapsto f(x) + \frac{\varepsilon}{\lambda^2}\|x - z_\varepsilon\|^2$ then it follows from Lemma 2.1 that*

$$0 \in \partial_L \mathbf{f}(\mathbf{y}_\varepsilon) + \frac{2\varepsilon}{\lambda^2}(\mathbf{y}_\varepsilon - \mathbf{z}_\varepsilon)$$

which implies, there exists $\tau \in \partial_L \mathbf{f}(\mathbf{y}_\varepsilon)$ with $\|\tau\| < \frac{2\varepsilon}{\lambda}$.

Consider the following constrained programming problem

$$\text{(CP)} \qquad \begin{array}{ll} \text{Inf} & f(x) \\ \text{subject to} & x \in C \subset X, \end{array}$$

where f is l.s.c. bounded below function on a nonempty subset C of a real Hilbert space X.

Definition 2.1 *$x_\varepsilon \in C$ is called an **ε-optimal** solution of (CP) if $f(x_\varepsilon) \leq \operatorname*{Inf}_{x \in C} f(x) + \varepsilon$.*

*The problem (CP) is equivalent to unconstrained problem (UCP) in the sense that **ε-optimal** solution of (CP) is equivalent to the **ε-optimal** solution of (UCP), where*

$$\text{(UCP)} \qquad \operatorname*{Inf}_{x \in X} f(x) + I_C(x)$$

and

$$I_C(x) = \begin{cases} 0 & \text{if } x \in C \\ +\infty & \text{if } x \notin C \end{cases} \qquad \text{is an indicator function of } C.$$

Theorem 2.3 *If x_ε is an **ε-optimal** solution of (CP) then there exist $y_\varepsilon \in X$ and $z_\varepsilon \in X$ such that for all $\lambda > 0$*

(i) $\|y_\varepsilon - z_\varepsilon\| < \lambda, \quad \|z_\varepsilon - x_\varepsilon\| < \lambda$

(ii) $f(y_\varepsilon) + I_C(y_\varepsilon) < \operatorname*{Inf}_{x \in X}(f(x) + I_C(x)) + \varepsilon \leq \operatorname*{Inf}_{x \in C}(f(x) + I_C(x)) + \varepsilon$

(iii) y_ε *is the unique minimum of the function* $x \mapsto f(x) + I_C(x) +$ $\frac{\varepsilon}{\lambda^2}\|x - z_\varepsilon\|^2$.

3. ε-Optimality for Scalar-Valued Problem

Consider a problem

$$\text{(P)} \qquad \begin{array}{rl} \text{Inf} & f(x) \\ \text{subject to} & g(x) \leq 0,\ h(x) = 0,\ x \in C\,. \end{array}$$

Let $Q = \{x \in X : g(x) \leq 0,\ h(x) = 0,\ x \in C\}$ be the feasible set of (P), $f : X \to R \cup \{+\infty\}$ be l.s.c., bounded below on Q, $g : X \to R^m$, $h : X \to R^p$ and C be a nonempty closed subset of a real Hilbert space X. f, g, h are locally Lipschitz functions, except possibly one, on X.

Definition 3.1 (Clarke et al (1998)) *The problem (P) is said to satisfy Growth Hypothesis if the set* $\{x \in C : f(x) \leq r,\ g(x) \leq s,\ \|h(x)\| \leq t\}$ *is bounded for each* $(r, s, t) \in R \times R^m \times R$.

Definition 3.2 *A point* x *is called normal, if*

$$\gamma \geq 0,\ \sum_{j=1}^{m} \gamma_j\, g_j(x) = 0,$$

$$0 \in \partial_L \left(\sum_{j=1}^{m} \gamma_j\, g_j + \sum_{k=1}^{p} s_k h_k \right)(x) \quad \Longrightarrow \quad \gamma = 0, s = 0.$$

In the following Theorem, we present a Lagrange multiplier rule of Fritz-John type that characterizes ε-**optimal** solution of (P).

Theorem 3.1 *Let* $x_\varepsilon \in X$ *be an* ε-**optimal** *solution of (P). Then there exist* $y_\varepsilon \in X$, $\tau \in X^*$ *and multipliers* $(\lambda_0(\varepsilon),\ \mu(\varepsilon),\ s(\varepsilon)) \in R \times R^m \times R^p$ *such that for all* $\lambda > 0$, *we have*

(a) $\|y_\varepsilon - x_\varepsilon\| < 2\lambda$

(b) $f(y_\varepsilon) < \operatorname*{Inf}_{x \in Q} f(x) + \varepsilon$

(c) $\tau \in \partial_L(\lambda_0(\varepsilon)f)(y_\varepsilon) + \sum_{j=1}^{m} \partial_L(\mu_j(\varepsilon)g_j)(y_\varepsilon) + \sum_{k=1}^{p} \partial_L(s_k(\varepsilon)h_k)(y_\varepsilon)$
$+N_C^L(y_\varepsilon)$

(d) $\|\tau\| < \frac{2\lambda_0(\varepsilon)\varepsilon}{\lambda}, \sum_{j=1}^{m} \mu_j(\varepsilon)g_j(y_\varepsilon) = 0,\ \lambda_0(\varepsilon) \geq 0,\ \mu(\varepsilon) \geq 0,$
$(\lambda_0(\varepsilon), \mu(\varepsilon), s(\varepsilon)) \neq 0$.

Proof. Since x_ε is an ε-**optimal** solution of (P) hence x_ε is also an ε-**optimal** solution of unconstrained problem

$$\operatorname*{Inf}_{x \in X} \ (f(x) + I_Q(x)).$$

Thus, by Theorem 2.2 there exist $y_\varepsilon \in X$, $z_\varepsilon \in X$ such that for all $\lambda > 0$, we have

(a′) $\|y_\varepsilon - z_\varepsilon\| < \lambda$, $\|z_\varepsilon - x_\varepsilon\| < \lambda$ which implies $\|y_\varepsilon - x_\varepsilon\| < 2\lambda$.

(b′) $f(y_\varepsilon) + I_Q(y_\varepsilon) < \operatorname*{Inf}_{x \in X}(f(x) + I_Q(x)) + \varepsilon \leq \operatorname*{Inf}_{x \in Q} f(x) + \varepsilon$ which implies $y_\varepsilon \in Q$.

(c′) y_ε is the unique minimum of the function
$x \mapsto f(x) + I_Q(x) + \frac{\varepsilon}{\lambda^2}\|x - z_\varepsilon\|^2$, which is equivalent to

$$f(y_\varepsilon) + I_Q(y_\varepsilon) + \frac{\varepsilon}{\lambda^2}\|y_\varepsilon - z_\varepsilon\|^2$$
$$\leq f(x) + I_Q(x) + \frac{\varepsilon}{\lambda^2}\|x - z_\varepsilon\|^2, \text{ for all } x \in X.$$

As $y_\varepsilon \in Q$ and $Q \subset X$, the above inequality implies

$$f(y_\varepsilon) + \frac{\varepsilon}{\lambda^2}\|y_\varepsilon - z_\varepsilon\|^2 \leq f(x) + \frac{\varepsilon}{\lambda^2}\|x - z_\varepsilon\|^2, \quad \text{for all } x \in Q.$$

That is, y_ε is the unique optimal solution of the problem

$$(\mathrm{P})_{z_\varepsilon} \qquad \text{Min} \ \ f(x) + I_C(x) + \frac{\varepsilon}{\lambda^2}\|x - z_\varepsilon\|^2$$
$$\text{subject to} \quad g(x) \leq 0,\ h(x) = 0.$$

Clearly, $(\mathrm{P})_{z_\varepsilon}$ also satisfies the Growth Hypothesis conditions. So by the necessary optimality conditions of Clarke et al (see Clarke et al (1998),

Chapter 3) there exist scalars $\lambda_0(\varepsilon) \geqq 0$, $\mu_j(\varepsilon) \geqq 0$, $j = 1, \ldots, m$, $s_k(\varepsilon)$, $k = 1, \ldots, p$ such that

$$0 \in \partial_L \left(\lambda_0(\varepsilon)\left(f(\cdot) + I_C(\cdot) + \frac{\varepsilon}{\lambda^2}\|\cdot - z_\varepsilon\|^2\right) + \sum_{j=1}^{m} \mu_j(\varepsilon) g_j(\cdot) + \sum_{k=1}^{p} s_k(\varepsilon) h_k(\cdot) \right)(y_\varepsilon)$$

and $\sum_{j=1}^{m} \mu_j(\varepsilon) g_j(y_\varepsilon) = 0$. Using Theorem 2.1, it follows that there exists $\tau \in X^*$, such that

$$\tau \in \partial_L(\lambda_0(\varepsilon) f)(y_\varepsilon) + \sum_{j=1}^{m} \partial_L(\mu_j(\varepsilon) g_j)(y_\varepsilon) + \sum_{k=1}^{p} \partial_L(s_k(\varepsilon) h_k)(y_\varepsilon) + N_C^L(y_\varepsilon)$$

with $\|\tau\| < \dfrac{2\lambda_0(\varepsilon)\varepsilon}{\lambda}$. This completes the proof. □

Corollary 3.1 *Let x_ε be an ε-optimal solution of (P) and let (P) satisfy the Growth Hypothesis. If y_ε is normal to the problem $(P)_{z_\varepsilon}$, that is, if $\mu(\varepsilon) \geq 0$, $\sum_{j=1}^{m} \mu_j(\varepsilon) g_j(y_\varepsilon) = 0$,*

$$0 \in \partial_L \left(\sum_{j=1}^{m} \mu_j(\varepsilon) g_j(\cdot) + \sum_{k=1}^{p} s_k(\varepsilon) h_k(\cdot) \right)(y_\varepsilon)$$

implies $\mu(\varepsilon) = 0$, $s(\varepsilon) = 0$, then $\lambda_0(\varepsilon) \neq 0$. So without loss of generality we can take $\lambda_0(\varepsilon) = 1$.

Remark 3.1 *Although the necessary optimality conditions developed above for the problem (P) follow from Theorem 4.2 (b) of Mordukhovich and Wang ((2002), pp. 635-636) yet the approach used in our paper to establish the said result is different and is based on the work of Clarke et al. (1998). Mordukhovich and Wang (2002) used exact penalization technique by adding an indicator function of the feasible set of (P) in the objective function thus converting the constrained problem (P) into an unconstrained problem. However, note that the indicator function is never differentiable at all the boundary points of the feasible set. Subsequently, we are obliged to deal with a nonsmooth minimization problem*

(Problem (5.12), pp. 635, Mordukhovich and Wang (2002)) even if the original problem (P) has smooth data. In our paper, we have followed the Value Function Analysis technique of Clarke et al. (1998) to convert constrained problem (P) into an unconstrained problem. The necessary conditions are then obtained by using the variational principle of Borwein and Preiss (1987) and the result of Clarke et al. (1998, pp. 110, Chapter 3). The main advantage of this approach is that if the original programming problem is differentiable the perturbed problem remains differentiable.

4. Multiobjective Optimization

In this section, we study the following nonsmooth multiobjective programming problem

$$\text{(MP)} \qquad \text{Inf} \quad (f_1(x), \ldots, f_r(x)) \quad \text{subject to} \quad x \in Q,$$

where $Q = \{x \in X : g(x) \leq 0, h(x) = 0, x \in C\}$.
The vector $\varepsilon = (\varepsilon_1, \ldots, \varepsilon_r)$, $\varepsilon_i > 0$, $1 \leq i \leq r$, is the permissible error vector and $\gamma = \sum_{i=1}^{r} \varepsilon_i$.

Definition 4.1 (Loridan (1982)) *$x_\varepsilon \in Q$ is said to be an **ε-efficient** solution of (MP) if there does not exist any $x \in Q$ such that*

$$f(x) \leq f(x_\varepsilon) - \varepsilon,$$

that is, there does not exist any $x \in Q$ such that

$$\begin{aligned} f_i(x) &\leqq f_i(x_\varepsilon) - \varepsilon_i, && 1 \leqq i \leqq r, \\ f_t(x) &< f_t(x_\varepsilon) - \varepsilon_t, && \text{for some } t. \end{aligned}$$

Assumption (A1). *We assume that for any $\varepsilon \in R^r_{++}$ the set of **ε-efficient** solutions of (MP) is nonempty.*
Following scalar-valued problem is associated with (MP)

$$\text{(SP)} \qquad \text{Inf} \quad \sum_{i=1}^{r} f_i(x) \quad \text{subject to} \quad x \in F_\varepsilon, \text{ where } F_\varepsilon = Q \cap (S(\varepsilon) \cup \{x_\varepsilon\})$$

and $S(\varepsilon) = \{x \in X : f_i(x) \leqq f_i(x_\varepsilon) - \varepsilon_i, \text{ for all } 1 \leqq i \leqq r\}$.

Lemma 4.1 *$x_\varepsilon \in Q$ is an **ε-efficient** solution of (MP) if and only if x_ε is a **γ-optimal** solution of (SP).*

Proof follows immediately from Definitions 4.1 and 2.1, and by the nature of the set $S(\varepsilon)$.

The problem (SP) is said to satisfy Growth Hypothesis if the set $\{x \in C : \sum_{i=1}^{r} f_i(x) \leq \zeta,\ g(x) \leq \vartheta,\ \|h(x)\| \leq t,\ f(x) - f(x_\varepsilon) + \varepsilon \leq \omega\}$ is bounded for each $(\zeta, \vartheta, t, \omega) \in R \times R^m \times R^p \times R^r$.

In the next theorem, we establish a Lagrange multiplier rule of Frtiz-John type characterizing ε-**efficiency** for (MP) under the above stated Growth Hypothesis.

Theorem 4.1 *Let x_ε be an ε-efficient solution of (MP) and let the Growth of Hypothesis for (SP) hold. Then there exist $y_\varepsilon \in Q$, $\tau \in X^*$ and multipliers $(\lambda_0(\varepsilon),\ \alpha(\varepsilon),\ \mu(\varepsilon),\ s(\varepsilon)) \in R_+ \times R^r \times R^m \times R^p$ such that for all $\lambda > 0$, we have*

1. $\|y_\varepsilon - x_\varepsilon\| < 2\lambda$

2. $\displaystyle\sum_{i=1}^{r} f_i(y_\varepsilon) < \operatorname*{Inf}_{x \in Q \cap S_\varepsilon} \sum_{i=1}^{r} f_i(x) + \gamma$

3. $$\tau \in \sum_{i=1}^{r} \partial_L((\lambda_0(\varepsilon) + \alpha_i(\varepsilon)) f_i)(y_\varepsilon) + \sum_{j=1}^{m} \partial_L(\mu_j(\varepsilon) g_j)(y_\varepsilon) + \sum_{k=1}^{p} \partial_L(s_k(\varepsilon) h_k)(y_\varepsilon) + N_C^L(y_\varepsilon)$$

4. $\displaystyle\|\tau\| < \frac{2\lambda_0(\varepsilon)\gamma}{\lambda}$

5. $\displaystyle\sum_{j=1}^{m} \mu_j(\varepsilon) g_j(y_\varepsilon) = 0$

6. $(\lambda_0(\varepsilon),\ \alpha(\varepsilon),\ \mu(\varepsilon),\ s(\varepsilon)) \neq 0,\ \lambda_0(\varepsilon) \geqq 0,\ \alpha(\varepsilon) \geqq 0,\ \mu(\varepsilon) \geqq 0.$

The proof easily follows by using the Lemma 4.1 and Theorem 3.1.

Corollary 4.1 *If in the above theorem y_ε is normal to the problem*

$$(\mathrm{SP})_{z_\varepsilon} \qquad \operatorname{Inf} \sum_{i=1}^{r} f_i(x) + I_C(x) + \frac{\gamma}{\lambda^2}\|x - z_\varepsilon\|^2$$

$$\text{subject to } \ x \in Q \cap S(\varepsilon) \cup \{x_\varepsilon\}.$$

Then $\lambda_0(\varepsilon) + \alpha_i(\varepsilon) \neq 0$ for at least one i, $1 \leqq i \leqq r$. So, without loss of generality, we can take

$$\sum_{i=1}^{r} (\lambda_0(\varepsilon) + \alpha_i(\varepsilon)) = 1.$$

5. Conclusions

In this paper, we have developed Lagrangian necessary **ε-optimality** conditions for nonsmooth programming problems on a real Hilbert space. These results, unlike those of Liu (1996) and Loridan (1982), do not require any convexity hypothesis on the functions. Moreover, the main difference between this work and the earlier work of Hamel (2001) is that a small generalized gradient, namely limiting subgradient, is used instead of Clarke's generalized gradient.

The Value Function Analysis (VFA) technique of Clarke et al. (1998) and Borwein and Preiss (1987) smooth variational principle are used to derive the main results. The VFA technique and the latter variational principle are significant due to the computational advantage over penalty function technique and Ekeland's variational principle respectively as the perturbed problem remains differentiable if the original constrained problem is so. Thus by following this approach the differentiability is maintained.

Although we have derived the necessary optimality conditions with nonsmooth data, however, in lieu of the above argument, algorithms can be designed for finding the approximate solutions of the differentiable constrained problem using the fact that the equivalent intermediate problems are differentiable.

Acknowledgement

Authors are thankful to the referee for suggesting new references and for the useful comments. Authors are also thankful to Dr. (Mrs.) S.K. Suneja, Department of Mathematics, Miranda House, University of Delhi, Delhi, India for her inspiration throughout the preparation of this paper.

References

Borwein, J. M. and Preiss, D. (1987), A smooth variational principle with applications to subdifferentiability and to differentiability of convex functions, *Trans. Amer. Math. Soc.,* Vol. 303, pp. 517-527.

Clarke, F. H. (1983), *Optimization and Nonsmooth Analysis,* Wiley, New York.

Clarke, F. H., Ledyavev, Y. S., Stern, R. J. and Wolenski, P. R. (1998), *Nonsmooth Analysis and Control Theory,* Springer, Berlin.

Ekeland, I. (1974), On the variational principle, *J. Math. Anal. Appl.,* Vol. 47, pp. 324-353.

Hamel, A. (2001), An **ε-Lagrange** multiplier rule for a mathematical programming on Banach spaces, *Optimization,* Vol. 49, pp. 137-149.

Kruger, A.Y. and Mordukhovich, B.S. (1980), Extremal points and the Euler equation in nonsmooth optimization, *Dokl. Akad. Nauk BSSR,* Vol. 24, pp. 684-687.

Liu, J. C. (1991), **ε-duality** theorem of nondifferentiable nonconvex multiobjective programming, *J. Optim. Th. Appl.,* Vol. 69 , pp. 153-167.

Liu, J. C. (1996), **ε-Pareto** optimality for nondifferentiable multiobjective programming via penalty function, *J. Math. Anal. Appl.,* Vol. 198, pp. 248-261.

Loridan, P. (1982), Necessary conditions for **ε-optimality,** *Math. Prog. Study,* Vol. 19, pp. 140-152.

Mordukhovich, B.S. (1976), Maximum principle in the problem of time optimal control with nonsmooth constraints, *J. Appl. Math. Mech.,* Vol. 40, pp. 960-969.

Mordukhovich, B.S. (1984), Nonsmooth analysis with nonconvex generalized differentials and adjoint mappings, *Dokl. Akad. Nauk BSSR,* Vol. 28, pp. 976-979.

Mordukhovich, B.S. and Shao, Y. (1996), Nonsmooth sequential analysis in Asplund spaces, *Trans. Amer. Math. Soc.,* Vol. 348, pp. 1235-1280.

Mordukhovich, B.S. and Wang, B. (2003), Necessary suboptimality and optimality conditions via variational principles, *SIAM J. Control Optim.,* Vol. 41, pp. 623-640.

Chapter 17

IDENTIFICATION OF HIDDEN CONVEX MINIMIZATION PROBLEMS

Duan Li*
Department of Systems Engineering and Engineering Management
The Chinese University of Hong Kong, Hong Kong

Zhiyou Wu
Department of Mathematics and Computer Science
Chongqing Normal University, P. R. China

Heung Wing Joseph Lee
Department of Applied Mathematics
The Hong Kong Polytechnic University, Hong Kong

Xinmin Yang
Department of Mathematics and Computer Science
Chongqing Normal University, P. R. China

Liansheng Zhang
Department of Mathematics, Shanghai University, P. R. China

Abstract If a nonconvex minimization problem can be converted into an equivalent convex minimization problem, the primal nonconvex minimization problem is called a hidden convex minimization problem. Sufficient conditions are developed in this paper to identify such hidden convex minimization problems. Hidden convex minimization problems possess the same desirable property as convex minimization problems: Any lo-

* Corresponding author. Email: dli@se.cuhk.edu.hk

cal minimum is also a global minimum. Identification of hidden convex minimization problem extends the reach of global optimization.

Keywords: Convex programming, nonconvex optimization, global optimization, convexification.

MSC2000: 90C25, 90C26, 90C30, 90C46

1. Introduction

We consider in this paper the following mathematical programming problem:

$$
\begin{array}{llll}
(P) & \min & g_0(x) & \\
& \text{subject to} & g_i(x) \leq b_i, & i = 1, \cdots, m \\
& & x \in X &
\end{array}
\tag{17.1}
$$

where $g_i : R^n \to R$, $i = 0, 1, \cdots, m$, are second-order differentiable functions and

$$
X = \{x \in R^n \mid l_i \leq x_i \leq u_i, i = 1, \cdots, n\}. \tag{17.2}
$$

Convexity is a key assumption in achieving a global optimality of (P) and in designing efficient solution schemes. When g_i, $i = 0, 1, \ldots, m$, are all convex functions, problem (P) is a convex programming problem whose local minimum is also a global minimum. Interesting research topics are to investigate (i) the existence of a certain class of nonconvex programming problems that also possess the desirable property that any local minimum is also a global minimum, and (ii) identification schemes to determine such a class of nonconvex programming problems.

The concept of hidden convexity was recently introduced in Li et al. (2003). If a nonconvex minimization problem (P) can be converted into an equivalent convex minimization problem, the primal nonconvex minimization problem (P) is called a hidden convex minimization problem. General sufficient conditions are derived in Li et al. (2003) to identify hidden convex minimization problems. Study of hidden convex minimization problems extends the reach of global optimization to a class of seemingly nonconvex minimization problems. The purpose of this paper is to reinforce the results in Li et al. (2003) via a different approach. Specifically, a variable transformation is adopted to derive a sufficient condition for identifying if a nonconvex optimization problem is hidden convex.

2. Variable Transformation for Convexification

Let function $h \in C^2$ be defined on X in (17.2). If h is convex on X, it is well known (see Avriel (1976)) that its convexity is preserved in a functional transformation $F(h(x))$ if $F : R \to R$ is convex and increasing. An inverse, and more difficult, question is: Given that h is nonconvex, do there exist a functional transformation F: $R \to R$ and a variable transformation $x = t(y)$ such that $F(h(t(y)))$ is a convex function of y on Y $(= t^{-1}(X))$? An answer to this question is given in Li et al. (2001) and Sun et al. (2001) in the context of monotone global optimization. As discussed in Horst (1984), if a nonconvex function h can be convexified by a functional transformation F, i.e., $F(h(x))$ is convex, then the primal function h must be quasiconvex. This section will propose a variable transformation for convexification in the context of hidden convex functions. Specifically, the following question will be answered: Given that h is nonconvex, under which situations the proposed variable transformation $x = t(y)$ will yield a convex function $h(t(y))$ of y on Y $(= t^{-1}(X))$?

Definition 2.1 *A function $h : R^n \to R$ is increasing (decreasing) on X with respect to x_i if*

$$h(x_1, \cdots, x_{i-1}, x_i^1, x_{i+1}, \cdots, x_n) \leq (\geq)\ h(x_1, \cdots, x_{i-1}, x_i^2, x_{i+1}, \cdots, x_n)$$

for any $x_i^1 < x_i^2$, where

$$x_i^1, x_i^2 \in X_i = \{x_i \mid (x_1, \cdots, x_{i-1}, x_i, x_{i+1}, \cdots, x_n) \in X\};$$

A function $h : R^n \to R$ is strictly increasing (decreasing) on X with respect to x_i if

$$h(x_1, \cdots, x_{i-1}, x_i^1, x_{i+1}, \cdots, x_n) < (>)\ h(x_1, \cdots, x_{i-1}, x_i^2, x_{i+1}, \cdots, x_n)$$

for any $x_i^1 < x_i^2$, where x_i^1, $x_i^2 \in X_i$.

Definition 2.2 *A function h: $R^n \to R$ is said to be monotone on its domain X if for every x_i, $i = 1, \cdots, n$, h is either increasing or decreasing; A function h defined on X is said to be strictly monotone if for every x_i, $i = 1, \cdots, n$, h is either strictly increasing or strictly decreasing.*

Define the following separable variable transformation

$$x = t_p(y) = (-\frac{1}{p_1} \ln y_1, \cdots, -\frac{1}{p_n} \ln y_n)^T, \tag{17.3}$$

where $p = (p_1, \cdots, p_n)$ is a vector of nonzero parameters.

Consider now the following transformed function on Y_p:

$$h_p(y) = h\Big(t_p(y)\Big), \tag{17.4}$$

where

$$\begin{aligned} Y_p &= t^{-1}X \\ &= \{y \in R^n \left| \begin{cases} e^{-p_i u_i} \le y_i \le e^{-p_i l_i}, & p_i > 0 \\ e^{-p_i l_i} \le y_i \le e^{-p_i u_i}, & p_i < 0 \end{cases} \right., i = 1, \cdots, n\}. \end{aligned} \tag{17.5}$$

If there exists a parameter vector p such that $h_p(y)$ defined in (17.4) is convex, then the primal function $h(x)$ is called a hidden convex function.

Denote by M_i and m_i upper and lower bounds of $\dfrac{\partial h(x)}{\partial x_i}$ over X, $i = 1, \ldots, n$, respectively, i.e.,

$$M_i \ \ge \ \max_{x \in X} \frac{\partial h(x)}{\partial x_i} \tag{17.6}$$

$$m_i \ \le \ \min_{x \in X} \frac{\partial h(x)}{\partial x_i}. \tag{17.7}$$

Denote by λ_0 a lower bound of the minimum eigenvalue of the Hessian of h over X, i.e.,

$$\lambda_0 \le \min_{x \in X} \lambda(x) = \min_{x \in X, d \in S^n} d^T H(x) d, \tag{17.8}$$

where S^n is the unit sphere in R^n, $H(z)$ is the Hessian of h at z and $\lambda(x)$ is the minimum eigenvalue of $H(x)$.

Let $\min_{x \in \emptyset} x = +\infty$, $\max_{x \in \emptyset} x = -\infty$ for a purpose of convenience. Let

$$A_i^0 \ = \ \{p_i > 0 \mid p_i m_i \ge -\lambda_0\} \cup \{p_i < 0 \mid p_i M_i \ge -\lambda_0\} \tag{17.9}$$

$$\bar{A}_i^0 \ = \ \{p_i > 0 \mid p_i m_i > -\lambda_0\} \cup \{p_i < 0 \mid p_i M_i > -\lambda_0\}. \tag{17.10}$$

Theorem 2.1 *Assume $h \in C^2(X)$. If for all $i = 1, \cdots, n$, $A_i^0 \neq \emptyset$, then $h_p(y)$ is a convex function on Y_p when $p_i \in A_i^0$, $i = 1, \ldots, n$. Furthermore, if for all $i = 1, \cdots, n$, $\bar{A}_i^0 \neq \emptyset$, then $h_p(y)$ is strictly convex on Y_p when $p_i \in \bar{A}_i^0$, $i = 1, \ldots, n$.*

Proof. By (17.3) and (17.4), we have

$$\frac{\partial h_p(y)}{\partial y_i} = -\frac{1}{p_i y_i} \frac{\partial h(x)}{\partial x_i} \tag{17.11}$$

Taking derivatives of (17.11) further yields the following,

$$\frac{\partial^2 h_p(y)}{\partial y_i^2} = \frac{1}{p_i^2 y_i^2}\frac{\partial^2 h(x)}{\partial x_i^2} + \frac{1}{p_i y_i^2}\frac{\partial h(x)}{\partial x_i} \tag{17.12}$$

$$= \frac{1}{p_i^2 y_i^2}\Big(\frac{\partial^2 h(x)}{\partial x_i^2} + p_i\frac{\partial h(x)}{\partial x_i}\Big) \tag{17.13}$$

$$\frac{\partial^2 h_p(y)}{\partial y_i \partial y_j} = \frac{1}{p_i p_j y_i y_j}\frac{\partial^2 h(x)}{\partial x_i \partial x_j}. \tag{17.14}$$

Let

$$A(x) = \operatorname{diag}\Big(\frac{1}{p_1 y_1}, \cdots, \frac{1}{p_n y_n}\Big) \tag{17.15}$$

$$B(x) = \operatorname{diag}\Big(p_1\frac{\partial h(x)}{\partial x_1}, \cdots, p_n\frac{\partial h(x)}{\partial x_n}\Big). \tag{17.16}$$

Then the Hessian of $h_p(y)$, $H_p(y)$, can be expressed by

$$H_p(y) = A(x)\Big(H(x) + B(x)\Big)A(x).$$

Since $A(x)$ is nonsingular, it is clear that $H_p(y)$ is positive definite if and only if $H(x) + B(x)$ is positive definite. For any $d \in S^n$ and $x \in X$, we have

$$d^T\Big(H(x) + B(x)\Big)d = d^T H(x)d + \sum_{i=1}^{n} p_i\frac{\partial h(x)}{\partial x_i}d_i^2 \geq \lambda_0 + \sum_{i=1}^{n}\beta_i d_i^2,$$

where

$$\beta_i = \begin{cases} p_i m_i, & p_i > 0 \\ p_i M_i, & p_i < 0 \end{cases}.$$

Thus, if for every $i = 1, \cdots, n$, there exists a p_i such that either $p_i m_i \geq -\lambda_0$ for a $p_i > 0$ or $p_i M_i \geq -\lambda_0$ for a $p_i < 0$, then we have

$$d^T[H(x) + B(x)]d \geq 0.$$

Thus, $h_p(y)$ is a convex function on Y_p for $p_i \in A_i^0$, $i = 1, 2, \ldots, n$, if all $A_i^0 \neq \emptyset$, $i = 1, \cdots, n$. Furthermore, $h_p(y)$ is strictly convex on Y_p for $p_i \in \bar{A}_i^0$, $i = 1, 2, \ldots, n$, if all $\bar{A}_i^0 \neq \emptyset$, $i = 1, \cdots, n$. □

Remark 2.1 *We can assume, without loss of generality, that $m_i < M_i$ for all $i = 1, \cdots, n$.*

1^0 *If $m_i > 0$, then* $A_i^0 = \Big([\max\{0, -\frac{\lambda_0}{m_i}\}, +\infty) \setminus \{0\}\Big) \cup [-\frac{\lambda_0}{M_i}, 0)$.

2^0 *If* $M_i < 0$, *then* $A_i^0 = (0, -\frac{\lambda_0}{m_i}] \cup \left((-\infty, \min\{0, -\frac{\lambda_0}{M_i}\}] \setminus \{0\}\right)$.

3^0 *If* $m_i < 0$ *and* $M_i > 0$, *then* $A_i^0 = (0, -\frac{\lambda_0}{m_i}] \cup [-\frac{\lambda_0}{M_i}, 0)$.

4^0 *If* $m_i < 0$, $M_i > 0$ *and* $\lambda_0 < 0$, *then* $A_i^0 = \emptyset$.

5^0 *If* $m_i = 0$ *and* $\lambda_0 \geq 0$, *then* $A_i^0 = (0, +\infty) \cup [-\frac{\lambda_0}{M_i}, 0)$.

6^0 *If* $M_i = 0$ *and* $\lambda_0 \geq 0$, *then* $A_i^0 = (0, -\frac{\lambda_0}{m_i}] \cup (-\infty, 0)$.

It becomes clear from Remark 2.1 that the monotonicity implies hidden convexity. If $h(x)$ is strictly monotone on X, more specifically, if there exists a set $I \subset \{1, \cdots, n\}$ such that $m_i > 0$ for any $i \in I$ and $M_i < 0$ for any $i \in \bar{I} = \{1, \cdots, n\} \setminus I$, then $h_p(y)$ is convex on Y_p when p_i is sufficiently large for any $i \in I$ and when p_i is sufficiently small for any $i \in \bar{I}$. If the primal function h is nonconvex $(\lambda_0 < 0)$ and there exists an $i \in \{1, \cdots, n\}$ such that $m_i < 0$ and $M_i > 0$, then $A_i^0 = \emptyset$.

3. Equivalent Convex Programming Problem

Theorem 2.1 provides us a sufficient condition to identify a class of hidden convex function. By adopting the variable transformation (17.3), we can convert the primal problem (17.1) into the following formulation:

$$\begin{aligned} \min \quad & g_{0,p}(y) = g_0\left(t_p(y)\right) \\ \text{subject to} \quad & g_{i,p}(y) = g_i\left(t_p(y)\right) \leq b_i, \quad i = 1, \cdots, m \\ & y \in Y_p, \end{aligned} \tag{17.17}$$

where Y_p and $t_p(y)$ are given by (17.5) and (17.3), respectively. The equivalence between (17.1) and (17.17) is obvious.

Theorem 3.1 *A solution y^* is a global or local minimum of (17.17) if and only if $x^* = t_p(y^*)$ is a global or local minimum of (17.1).*

Proof. Notice that the transformation

$$t_p(y) = (-\frac{1}{p_1} \ln y_1, \cdots, -\frac{1}{p_n} \ln y_n)^T$$

is a one-to-one mapping from Y_p to X. Obviously, both t_p and t_p^{-1} are continuous. Thus we can prove the theorem easily by following Sun et al. (2001). □

If there exists a parameter vector p such that problem (17.17), an equivalent transformation of problem (17.1), is a convex minimization problem, then the primal problem (17.1) is called a hidden convex minimization problem.

Let M_i^k and m_i^k be upper and lower bounds of $\frac{\partial g_k(x)}{\partial x_i}$ over X, respectively, $i = 1, \ldots, n$, $k = 0, 1, \ldots, m$, i.e.,

$$m_i^k \le \min_{x \in X} \frac{\partial g_k(x)}{\partial x_i}, \tag{17.18}$$

$$M_i^k \ge \max_{x \in X} \frac{\partial g_k(x)}{\partial x_i}. \tag{17.19}$$

Denote by λ_k a lower bound of the minimum eigenvalue of the Hessian of g_k over X, $k = 0, 1, \ldots, m$,

$$\lambda_k \le \min_{x \in X, ||d||=1} d^T H_k(x) d \tag{17.20}$$

where $H_k(z)$ is the Hessian of $g_k(x)$ at z and $d \in R^n$. Let

$$A_i^k = \{p_i > 0 \mid p_i m_i^k \ge -\lambda_k\} \cup \{p_i < 0 \mid p_i M_i^k \ge -\lambda_k\} \tag{17.21}$$

$$\bar{A}_i^k = \{p_i > 0 \mid p_i m_i^k > -\lambda_k\} \cup \{p_i < 0 \mid p_i M_i^k > -\lambda_k\} \tag{17.22}$$

$$A_i = \cap_{k=0}^m A_i^k \tag{17.23}$$

$$\bar{A}_i = \cap_{k=0}^m \bar{A}_i^k \tag{17.24}$$

$$A = \{p = (p_1, \cdots, p_n) \mid p_i \in A_i,\ i = 1, \cdots, n\} \tag{17.25}$$

$$\bar{A} = \{p = (p_1, \cdots, p_n) \mid p_i \in \bar{A}_i,\ i = 1, \cdots, n\}. \tag{17.26}$$

Theorem 3.2 *Assume* $g_i \in C^2(X)$, $i = 0, 1, \cdots, m$, *in (17.17). If* $A \ne \emptyset$, *i.e.,* $A_i \ne \emptyset$ *for all* $i = 1, 2, \ldots, n$, *then the problem (17.17) is a convex programming problem when* $p \in A$. *If* $\bar{A} \ne \emptyset$, *i.e.,* $\bar{A}_i \ne \emptyset$ *for all* $i = 1, 2, \ldots, n$, *then the problem (17.17) is a strictly convex programming problem when* $p \in \bar{A}$.

Proof. If $A \ne \emptyset$, then for all $i = 1, \cdots, n$, $A_i \ne \emptyset$. This further implies that for any $k = 0, 1, \cdots, m$, $i = 1, \cdots, n$, $A_i^k \ne \emptyset$. From Theorem 2.1, we know that $g_k\big(t_p(y)\big)$ is convex on Y_p when $p_i \in A_i$, $i = 1, \cdots, n$. Thus, all the functions $g_k\big(t_p(y)\big)$, $k = 0, 1, \cdots, m$, are convex on Y_p when $p \in A$. We can conclude that the programming problem (17.17) is convex on Y_p when $p \in A$. Similarly, the problem (17.17) is strictly convex when $p \in \bar{A}$ if $\bar{A} \ne \emptyset$. □

From Theorems 3.1 and 3.2, we know that the problem (17.1) can be converted into an equivalent convex programming problem (17.17) when $p \in A$ if $A \ne \emptyset$.

Without loss of generality, we can assume that $m_i^k \neq 0$ and $M_i^k \neq 0$ for all $k = 0, 1, \cdots, m$ and $i = 1, \cdots, n$. Let for $i = 1, 2, \ldots, n$,

$$\begin{aligned} I_i &= \{k \mid m_i^k > 0,\ k = 0, 1, \cdots, m\} && (17.27)\\ J_i &= \{k \mid M_i^k < 0,\ k = 0, 1, \cdots, m\} && (17.28)\\ \bar{I}_i &= \{0, 1, \cdots, m\} \setminus I_i && (17.29)\\ \bar{J}_i &= \{0, 1, \cdots, m\} \setminus J_i. && (17.30) \end{aligned}$$

If $k \in I_i$, then A_i^k in (17.21) reduces to

$$A_i^k = \Big([\max\{0, \ -\frac{\lambda_k}{m_i^k}\}, +\infty) \setminus \{0\}\Big) \cup [-\frac{\lambda_k}{M_i^k}, 0);$$

If $k \in J_i$, then A_i^k in (17.21) reduces to

$$A_i^k = (0, -\frac{\lambda_k}{m_i^k}] \cup \Big((-\infty, \min\{0, -\frac{\lambda_k}{M_i^k}\}] \setminus \{0\}\Big);$$

If $k \in \bar{I}_i \cap \bar{J}_i$, then A_i^k in (17.21) reduces to

$$A_i^k = (0, -\frac{\lambda_k}{m_i^k}] \cup [-\frac{\lambda_k}{M_i^k}, 0);$$

Then

$$\begin{aligned} A_i = \quad & \Big([\max\{0, \max_{k \in I_i}[-\frac{\lambda_k}{m_i^k}]\}, \ \min_{k \in \bar{I}_i}\{-\frac{\lambda_k}{m_i^k}\}] \\ & \cup [\max_{k \in \bar{J}_i}\{-\frac{\lambda_k}{M_i^k}\}, \ \min\{0, \min_{k \in J_i}[-\frac{\lambda_k}{M_i^k}]\}]\Big) \setminus \{0\}. \quad (17.31) \end{aligned}$$

Note from Remark 2.1 that if there exists $k \in \bar{I}_i \cap \bar{J}_i$ such that $\lambda_k < 0$, then $A_i = \emptyset$.

By (17.31), we can easily obtain the following corollary:

Corollary 3.1 *If for all $i = 1, \cdots, n$, one of the following inequality holds:*

$$\max\{0, \max_{k \in I_i}[-\frac{\lambda_k}{m_i^k}]\} \leq \min_{k \in \bar{I}_i}\{-\frac{\lambda_k}{m_i^k}\} \quad (17.32)$$

or

$$\max_{k \in \bar{J}_i}\{-\frac{\lambda_k}{M_i^k}\} \leq \min\{0, \min_{k \in J_i}[-\frac{\lambda_k}{M_i^k}]\}, \quad (17.33)$$

then the problem (17.1) is a hidden convex programming problem when the feasible value of p_i to (17.32) or (17.33) is not a singleton $p_i = 0$,

$i = 1, 2, \ldots, n$. Furthermore, if, for each $i = 1, \cdots, n$, the inequality (17.32) or the inequality (17.33) is strict, then problem (17.1) is a hidden strictly convex programming problem.

By Corollary 3.1, the hidden convexity of the primal problem (17.1) can be determined by simply checking if the condition (17.32) or (17.33) holds for all $i = 1, \cdots, n$. For a hidden convex programming problem (17.1), its global minimum can be found by using any existing efficient local search algorithm in the literature.

Example 3.1 *The following is a nonconvex minimization problem,*

$$\begin{aligned} \min \quad & g_0(x) = x_1^2 + \frac{3}{2}x_2^2 - x_1 - 2x_2 \\ \text{s.t.} \quad & g_1(x) = -x_1^2 - \frac{1}{2}x_2^2 - 14x_1 - 15x_2 + 40 \le 0 \\ & g_2(x) = 2x_1^2 + \frac{5}{2}x_2^2 - 6x_1 - 8x_2 + 10 \le 0 \\ & 1 \le x_i \le 2, \ i = 1, 2. \end{aligned} \tag{17.34}$$

The following are obvious,

$$\begin{aligned} m_1^0 = 1 \le \frac{\partial g_0(x)}{\partial x_1} &= 2x_1 - 1 \le 3 = M_1^0 \\ m_2^0 = 1 \le \frac{\partial g_0(x)}{\partial x_2} &= 3x_2 - 2 \le 4 = M_2^0 \\ m_1^1 = -18 \le \frac{\partial g_1(x)}{\partial x_1} &= -2x_1 - 14 \le -16 = M_1^1 \\ m_2^1 = -17 \le \frac{\partial g_1(x)}{\partial x_2} &= -x_2 - 15 \le -16 = M_2^1 \\ m_1^2 = -2 \le \frac{\partial g_2(x)}{\partial x_1} &= 4x_1 - 6 \le 2 = M_1^2 \\ m_2^2 = -3 \le \frac{\partial g_2(x)}{\partial x_2} &= 5x_2 - 8 \le 2 = M_2^2. \end{aligned}$$

It is evident that

$$I_1 = \{0\}, \ \bar{I}_1 = \{1,\ 2\}, \ I_2 = \{0\}, \ \bar{I}_2 = \{1,\ 2\},$$

$$J_1 = \{1\}, \ \bar{J}_1 = \{0, 2\}, \ J_2 = \{1\}, \ \bar{J}_2 = \{0, 2\},$$

and $\lambda_0 = 2$, $\lambda_1 = -2$, $\lambda_2 = 4$. Thus, we have that for $i = 1$,

$$\min_{k \in \bar{J}_1} \frac{\lambda_k}{M_1^k} = \min\{\frac{2}{3}, \frac{4}{2}\} = \frac{2}{3} > \frac{1}{8} = \max\{0, \frac{-2}{-16}\} = \max_{k \in J_1}\{0, \frac{\lambda_k}{M_1^k}\}.$$

and for $i = 2$,

$$\min_{k\in \bar{J}_2} \frac{\lambda_k}{M_2^k} = \min\{\frac{2}{4}, \frac{4}{2}\} = \frac{1}{2} > \frac{1}{8} = \max\{0, \frac{-2}{-16}\} = \max_{k\in J_2}\{0, \frac{\lambda_k}{M_2^k}\}.$$

By Corollary 3.1, it can be concluded that Example (3.1) is a hidden strictly convex minimization problem. So its local minimum must be its global minimum. The global minimum is $x^* = (1.5739, 1.0000)^T$, *with* $g_0(x^*) = 0.4033$.

4. Conclusions

A hidden convex minimization problem has its equivalent counterpart in a form of a convex minimization problem in a different representation space. In this sense, convexity, in certain situations, is not an inherent property. It is rather a characteristic associated with a given representation space. It should be emphasized here that no actual transformation is needed when solving a hidden convex minimization problem. Any local search method can be applied directly to the primal hidden convex minimization problem to obtain a global optimal solution.

We emphasize here that the proposed variable transformation $t_p(\cdot)$ is adopted in this paper to identify certain sub-class of hidden convex functions. When compared to the general results in Li et al. (2003), we can find the sub-class of hidden convex functions identified in this paper is large enough to compare with the general class of hidden convex functions identified in Li et al. (2003).

Acknowledgments

This research was partially supported by the Research Grants Council of Hong Kong under Grants 2050291 and CUHK4214/01E and the National Science Foundation of China under Grant 10171118.

References

M. Avriel (1976), Nonlinear Programming: Analysis and Methods, Prentice Hall, Englewood Cliffs, N.J.

R. Horst (1984), *On the convexification of nonlinear programming problems: An applications-oriented survey,* European Journal of Operations Research, 15, pp. 382-392.

D. Li, X. L. Sun, M. P. Biswal and F. Gao (2001), *Convexification, concavification and monotonization in global optimization,* Annals of Operations Research, 105, pp. 213-226.

D. Li, Z. Wu, H. W. J. Lee, X. Yang and L. Zhang (2003), *Hidden convex minimization,* to appear in Journal of Global Optimization, 2003.

Sun, X. L., McKinnon, K. and Li, D. (2001), *A convexification method for a class of global optimization problem with application to reliability optimization,* Journal of Global Optimization, 21, pp. 185-199.

Chapter 18

ON VECTOR QUASI-SADDLE POINTS OF SET-VALUED MAPS

Lai-Jiu Lin* and Yu-Lin Tsai
Department of Mathematics
National Changhua University of Education, Taiwan, R.O.C.

Abstract In this paper, we prove some existence theorems of vector quasi-saddle point for a multivalued map with acyclic values. As a consequence of this result, we obtain an existence theorem of quasi-minimax theorem.

Keywords: Upper (lower) semi-continuous functions, Closed (compact) multivalued maps, Acyclic maps, C-quasiconvex functions, Quasi-saddle points.

MSC2000: 90C47, 90C30

1. Introduction

Let X and Y be nonempty sets and f be a real-valued function on $X \times Y$. A point $(\bar{x}, \bar{y}) \in X \times Y$ is called a saddle point on $X \times Y$ if

$$f(\bar{x}, y) \leq f(\bar{x}, \bar{y}) \leq f(x, \bar{y}) \quad \text{for all } (x, y) \in X \times Y.$$

Recently, some existence theorems of saddle point for vector-valued functions and loose saddle point for multivalued maps are established, for example, see Chang et al (1997); Kim and Kim (1999); Lin (1999); Luc and Vargas (1992); Tan et al (1996); Tanaka (1994) and references there in.

Let X and Y be two convex subsets of locally convex topological vector spaces E_1 and E_2, respectively, Z be a real topological vector space,

*E-mail address: maljlin@math.ncue.edu.tw

$C : X \multimap Z$ be a multivalued map such that for all $x \in X$, $\text{int}C(x) \neq \emptyset$. Let $S : X \multimap X$, $T : X \multimap Y$ and $F : X \times Y \multimap Z$ be multivalued maps. In this paper, we consider the problem of finding $(\bar{x}, \bar{y}) \in X \times Y$ with $\bar{x} \in S(\bar{x})$, $\bar{y} \in T(\bar{x})$ and $\bar{z}_1, \bar{z}_2 \in F(\bar{x}, \bar{y})$ such that

$$z_1 - \bar{z}_1 \notin -\text{int}C(\bar{x}) \text{ for all } z_1 \in F(x, \bar{y}) \text{ and for all } x \in S(\bar{x})$$

and

$$z_2 - \bar{z}_2 \notin \text{int}C(\bar{x}) \text{ for all } z_2 \in F(\bar{x}, y) \text{ and for all } y \in T(\bar{x}).$$

A point $(\bar{x}, \bar{y})$ satisfied the above property is called a vector quasi-saddle point of F (in short VSPP).

In this paper, we first establish the existence result of (VSPP) by using a fixed point theorem of Park (see Park (1992)).

As a consequence of the existence results of (VSPP), we establish the following minimax theorem of finding
$(\bar{x}, \bar{y}) \in X \times Y$ with $\bar{x} \in S(\bar{x})$, $\bar{y} \in T(\bar{x})$ such that

$$\min_{u \in S(\bar{x})} \max_{v \in T(\bar{x})} f(x, y) = \max_{v \in T(\bar{x})} \min_{u \in S(\bar{x})} f(x, y),$$

where $f : X \times Y \to \mathcal{R}$ is a function.

Our results on existence theorems of vector quasi-saddle point are different from the existence results of vector saddle point.

2. Preliminaries

In order to establish our main results, we first give some concepts and notations.

Throughout this paper, all topological spaces are assumed to be Hausdorff. Let A be a nonempty subset of topological vector space (in short, t.v.s.) X. We denote by $intA$ the interior of A, by co$A$ the convex hull of A. Let X, Y and Z be nonempty sets. Given two multivalued maps $F : X \multimap Y$ and $G : Y \multimap Z$, the composite $GF : X \multimap Z$ is defined by $GF(x) = G(F(x))$ forall $x \in X$.

Let X and Y be two topological spaces, a multivalued map $T : X \multimap Y$ is said to be compact if there exists a compact subset $K \subset Y$ such that $T(X) \subset K$; to be closed if its graph $Gr(T) = \{(x, y) \mid x \in X, y \in T(x)\}$ is closed in $X \times Y$; to be upper semicontinuous (in short, u.s.c.) if for every $x \in X$ and every open set V in Y with $T(x) \subset V$, there exists a neighborhood $W(x)$ of x such that $T(W(x)) \subset V$; to be lower semicontinuous (in short, l.s.c.) if for every $x \in X$ and every open neighborhood $V(y)$ of every $y \in T(x)$, there exists a neighborhood $W(x)$ of x such that $T(u) \cap V(y) \neq \emptyset$ for all $u \in W(x)$; and to be continuous if it is both u.s.c. and l.s.c.

A topological space is said to be *acyclic* if all of its reduced Čech homology groups vanish. For instance, any nonempty convex or star-shaped set is acyclic. A multivalued map $T : X -\circ Y$ is said to be *acyclic* if it is u.s.c. with acyclic compact values. We denote

$$V(X,Y) = \{T \mid T : X -\circ Y \text{is an acyclic multivalued map}\}.$$

Let Y be topological vector space, a nonempty subset $C \subseteq Y$ is called a convex cone if C is a convex set and $\lambda C \subset C$ for any $\lambda \geq 0$, $\lambda \in \mathcal{R}$. A cone C is called pointed if $C \cap (-C) = \{0\}$. Let $x, y \in Y$, we denote $x \leq y$ if $y - x \in C$ and $x \not< y$ if $y - x \notin \text{int}C$.

Definition 2.1 *(Luc (1989)) Let X be a nonempty convex subset of a t.v.s E, Z a real t.v.s.,and C a convex cone in Z. Let* $G : X -\circ Z$ *be a multivalued map, G is said to be C-*quasiconvex *(respectively, C-*quasiconcave*) if for any* $z \in Z$, *the set*

$$\{x \in X : \textit{ there is a } y \in G(x) \textit{ such that } z - y \in C\}$$

(respectively, $\{x \in X : \textit{ there is a } y \in G(x) \textit{ such that } z - y \in -C\}$)

is convex.

Definition 2.2 *(Luc and Vargas (1992)) Let Y be Hausdorff t.v.s and C be a pointed closed convex cone, then the function* $\xi : Y \to \mathcal{R}$ *is said to be monotonically increasing (respectively, strictly monotonically increasing) with respect to C if* $\xi(a) \geq \xi(b)$ *for all* $a - b \in C$ *(respectively,* $\xi(a) > \xi(b)$ *for all* $a - b \in \text{int}C$*).*

Lemma 2.1 *(Luc (1989)) Let A be a nonempty compact subset of a real t.v.s. Z, C be a pointed closed convex cone of Z such that* $\text{int}C \neq \emptyset$. *Then*
(1) $MinA \neq \emptyset$ *and* $A \subset \text{Min}A + C$; *and*
(2) $MaxA \neq \emptyset$ *and* $A \subset \text{Max}A - C$.

Remark 2.1 *If C is a pointed closed cone with* $\text{int}C \neq \emptyset$, *it is easy to see that* $\text{Min}_C A \subset w\text{Min}_C A$ *and* $\text{Max}_C A \subset w\text{Max}_C A$ *hold.*

Lemma 2.2 *(Luc and Vargas (1992)) Let Z be a real t.v.s., C a closed convex cone in Z with* $\text{int}C \neq \emptyset$. *Then*

(i) For any fixed $e \in \text{int}C$ *and any fixed* $a \in Z$, *the functions* $\xi_{ea} : Z \to \mathcal{R}$, $\xi'_{ea} : Z \to \mathcal{R}$ *defined by*

$$\xi_{ea}(y) = \max\{t \in \mathcal{R} : y \in a + te - C\}$$

and

$$\xi'_{ea}(y) = \min\{t \in \mathcal{R} : y \in a + te - C\}$$

are continuous and strictly monotonically increasing functions from Z to $\mathcal{R}$.

(ii) Let X be a nonempty convex subset of a t.v.s. E and $f : X \multimap Z$, *if f is C – quasiconvex (respectively, C–quasiconcave) then the composite mapping* ξf *is* $\mathcal{R}_+$*-quasiconvex (respectively,* $\mathcal{R}_+$*-quasiconcave), where* ξ *stands for* ξ_{ea} or ξ'_{ea}.

Lemma 2.3 *(Aubin and Cellina (1994)) Let X and Y be topological spaces, and* $T : X \multimap Y$ *be a multivalued mapping.*

(a) *If T is u.s.c. with closed values, then T is closed.*

(b) *If Y is a compact space and T is closed, then T is u.s.c.*

(c) *If X is a compact space and T is an u.s.c. map with compact values, then T(X) is compact.*

Lemma 2.4 *(Lin (1999)) Let X be a convex subset of a t.v.s. E,* $T : X \multimap \bar{\mathcal{R}}$, *then T is quasiconvex if and only if for all* $x_1, x_2 \in X$, $\lambda \in [0,1]$, $y_1 \in T(x_1)$, $y_2 \in T(x_2)$, *there exists* $y \in T[\lambda x_1 + (1-\lambda)x_2]$ *such that* $y \leq \max\{y_1, y_2\}$.

Lemma 2.5 *(Park (1992)) Let X be a nonempty compact convex subset of a locally convex topological vector space E, and let* $F : X \multimap X$ *be a multivalued mapping. If F is upper semicontinuous on X and if* $F(x)$ *is nonempty closed and acyclic for every* $x \in X$, *then F has at least one fixed point, i. e. there exists an* $\bar{x} \in X$ *such that* $\bar{x} \in F(\bar{x})$.

Lemma 2.6 *(Lee et al (1997)) Let q be a continuous function from topological space Z to* $\mathcal{R}$ *and F be a multivalued map from topological space X to Z.*
(i) If F is u.s.c., then $qF : X \multimap \mathcal{R}$ *is u.s.c.*
(ii) If F is l.s.c., then $qF : X \multimap \mathcal{R}$ *is l.s.c.*

Lemma 2.7 *(Lin and Yu (2001)) Let X be a nonempty subsets of topological space* E_1, Z *a real t.v.s. and C a closed pointed convex cone such that* $\text{int}C \neq \emptyset$, $F : X \times X \multimap Z$ *and* $S : X \multimap X$ *be multivalued maps. Let* $m : X \multimap Z$ *be a multivalued map defined by*

$$m(x) = w\text{Min}_C F(x, S(x)) \quad \textit{for} \quad x \in X$$

and $M : X \multimap X$ *be a multivalued map defined by*

$$M(x) = \{u \in S(x) : F(x,u) \cap m(x) \neq \emptyset\}.$$

If both F and S are compact continuous multivalued maps with closed values, then both M and m *are closed compact u.s.c. multivalued maps.*

3. Vector Quasi-Saddle Points

As a simple consequence of Lemma 2.7, we have the following Proposition.

Proposition 3.1 *Let X and Y be nonempty subsets of topological spaces* E_1 *and* E_2 *respectively, Z a real t.v.s. and C a pointed closed convex cone in Z such that* $\text{int}C \neq \emptyset$ *and* $F : X \times Y \multimap Z$, $S : X \multimap X$ *be multivalued maps. Let* $m : X \times Y \multimap Z$ *be a multivalued map defined by*

$$m(x, y) = w\text{Min}_C F(S(x), y) \quad \text{for} \quad (x, y) \in X \times Y$$

and $M : X \times Y \multimap Z$ *be a multivalued map defined by*

$$M(x, y) = \{u \in S(x) : F(u, y) \cap m(x, y) \neq \emptyset\}.$$

If both F and S are compact continuous multivalued maps with closed values, then both m *and M are closed compact u.s.c. multivalued maps.*

Proof. Let the multivalued maps $H : X \times Y \multimap X$ and $A : X \times X \times Y \multimap Z$ be defined by

$$H(x, y) = S(x) \quad \text{for } (x, y) \in X \times Y,$$

and

$$A(u, x, y) = F(u, y) \quad \text{for } (u, x, y) \in X \times X \times Y.$$

Suppose F and S are compact continuous multivalued maps with closed values. It is easy to see that A and H are compact continuous multivalued maps with closed values. We also see that

$$w\text{Min}F(S(x), y) = w\text{Min}A(H(x, y), x, y).$$

It follows from Lemma 2.7 that m and M are closed compact u.s.c. multivalued maps. □

As a consequence of Lemma 2.7, Proposition 3.1 and Lemma 2.5, we have the following theorem.

Theorem 3.1 *Let X and Y be two nonempty compact convex subsets of locally convex t.v.s.* E_1 *and* E_2, *respectively, Z a real t.v.s.. Let* $C : X \multimap Z$ *be a multivalued map such that for all* $x \in X$, $\text{int } C(x) \neq \emptyset$ *and* $C_+ = \overline{\text{co}}(\bigcup_{x \in X} C(x))$ *be a pointed cone in Z and Z be ordered by* C_+. *Suppose that* $S : X \multimap X$, $T : X \multimap Y$ *are compact continuous* *multivalued maps with closed values and* $F : X \times Y \multimap Z$ *is a multivalued map satisfying the following conditions :*

(i) *F is a continuous multivalued map with compact values.*

(ii) *For each* $(x, y) \in X \times Y$, *the sets*
$H(x,y) = \{u \in S(x) : F(u,y) \cap m(x,y) \neq \emptyset\}$ *and*
$G(x) = \{v \in T(x) : F(x,v) \cap p(x) \neq \emptyset\}$
are acyclic,
where $m(x,y) = w\text{Min}F(S(x),y)$ *and* $p(x) = w\text{Max}F(x,T(x))$.

Then there exists $(\overline{x}, \overline{y}) \in X \times Y$ *such that* $(\bar{x}, \bar{y})$ *is a vector quasi-saddle point of F.*

Proof. Since $S(x)$ is a closed subset of compact set *X*, $S(x)$ is compact for each $x \in X$. Since *F* is a continuous multivalued maps with compact values, it follows from Lemma 2.3 that $F(S(x), y)$ is compact for each $x \in X, y \in Y$ and there exist $u \in S(x)$ and $w \in F(u,y)$ such that $w \in \text{Min}F(S(x),y) \subset w\text{Min}F(S(x),y)$. Hence $F(u,y) \cap m(x,y) \neq \emptyset$, where $m(x,y) = w\text{Min}F(S(x),y)$. That is to say $H(x,y) \neq \emptyset$. Since *X* and *Y* are compact and F is a continuous multivalued map with compact values, *F(X, Y)* is compact and *F* is compact. By Proposition 3.1 and Lemma 2.7, *H* and *G* are closed compact u.s.c. multivalued maps. Hence $H : X \times Y \multimap X$ and $G : X \multimap Y$ are u.s.c. multivalued maps with compact acyclic values. Then by Kunneth formula (Massay (1980)) and Lemma 3 in Fan (1952), $W = H \times G$ is also an u.s.c. multivalued map with compact acyclic values. Hence $W \in V(X \times Y, X \times Y)$. It follows from Lemma 2.5 that *W* has a fixed point $(\bar{x}, \bar{y}) \in X \times Y$. Then there exist $\bar{x} \in H(\bar{x}, \bar{y})$ and $\bar{y} \in G(\bar{x})$. This shows that $\bar{x} \in S(\bar{x})$, $\bar{y} \in T(\bar{x})$, $F(\bar{x},\bar{y}) \cap m(\bar{x},\bar{y}) \neq \emptyset$, and $F(\bar{x},\bar{y}) \cap p(\bar{x}) \neq \emptyset$. Therefore, there exist $\bar{z}_1, \bar{z}_2 \in F(\bar{x},\bar{y})$ such that

$$z_1 - \bar{z}_1 \notin -\text{int}C_+ \quad \textbf{for all } z_1 \in F(x,\bar{y}) \textbf{ and for all } x \in S(\bar{x})$$

and

$$z_2 - \bar{z}_2 \notin \text{int}C_+ \textbf{ for all } z_2 \in F(\bar{x},y) \textbf{ and for all } y \in T(\bar{x}).$$

Since $C(x) \subset C_+$ for all $x \in X$, the conclusion of Theorem 3.1 follows. □

Corollary 3.1 *In Theorem 3.1, if* $Z = \mathcal{R}$, $C(x) = (0, \infty)$ *for all* $x \in \mathcal{R}$ *and we assume that* $S(x)$ *and* $T(x)$ *are convex and condition (ii) is replaced by* **(ii)′**

(ii)′ *for each* $x \in X$, $y \multimap F(x,y)$ *is* $\mathcal{R}_+$*-quasiconcave and for each* $y \in Y$, $x \multimap F(x,y)$ *is* $\mathcal{R}_+$*-quasiconvex.*

Then there exist $\bar{z}, \bar{z}_2 \in F(\bar{x}, \bar{y})$ such that $z_1 \geq \bar{z}_1$ for all $z_1 \in F(x, \bar{y})$ and all $x \in S(\bar{x})$ and $z_2 \leq \bar{z}_2$ for all $z_2 \in F(\bar{x}, y)$ and all $y \in T(\bar{x})$.

Proof. It suffices to show that both $H(x,y)$ and $G(x)$ are convex for all $x \in X, y \in Y$. Let $u_1, u_2 \in H(x,y)$, $\lambda \in [0,1]$, then $u_1, u_2 \in S(x)$, $m(x,y) \in F(u_1, y)$ and $m(x,y) \in F(u_2, y)$, where $m(x,y) = \min F(S(x), y)$. There exist $a \in F(u_1, y)$ and $b \in F(u_2, y)$ such that $a = b = \min F(S(x), y)$. By assumption, $S(x)$ is convex for each $x \in X$, therefore $\lambda u_1 + (1-\lambda)u_2 \in S(x)$. Since $x \multimap F(x,y)$ is $\mathcal{R}_+$-**quasiconvex** for each $y \in Y$, it follows from Lemma 2.4 that there exists $c \in F(\lambda u_1 + (1-\lambda)u_2, y)$ such that

$$c \leq \max\{a, b\} = a = \min F(S(x), y) \leq c.$$

Hence, we have $m(x,y) = \min F(S(x), y) = c \in F(\lambda u_1 + (1-\lambda)u_2, y)$. Therefore $\lambda u_1 + (1-\lambda)u_2 \in H(x,y)$ and $H(x,y)$ is convex for each $(x,y) \in X \times Y$. This shows that, for each $(x,y) \in X \times Y$, $H(x,y)$ is an acyclic set. Similarly, we can show that, for each $x \in X$, $G(x)$ is an acyclic set. Then the conclusion of Corollary 3.1 follows from Theorem 3.1. □

The following Theorem is another special case of Theorem 3.1.

Theorem 3.2 *In Theorem 3.1, if we assume that $S(x)$ and $T(x)$ are convex for all $x \in X$ and condition (ii) is replaced by (ii)′*

(ii)′ *for each $x \in X$, $y \multimap F(x,y)$ is C_+-**quasiconcave** and for each $y \in Y$, $x \multimap F(x,y)$ is C_+-**quasiconvex**.*

Then there exists $(\bar{x}, \bar{y}) \in X \times Y$ such that $(\bar{x}, \bar{y})$ is a vector quasi-saddle point of F.

Proof. Let ξ_{ea} be a continuous and strictly monotonically increasing function from Z to R as defined in Lemma 2.2. Then the multivalued map $\xi_{ea}F$ is a continuous multivalued map with compact values.

Since for each $x \in X$, $S(x)$ is convex and $y \multimap F(x,y)$ is C_+-**quasiconcave** and for each $y \in Y$, $x \multimap F(x,y)$ is C_+-**quasiconvex**, it follows from Lemma 2.2 that, for each $x \in X$, $y \multimap \xi_{ea}F(x,y)$ is $\mathcal{R}_+$-**quasiconcave** and, for each $y \in Y$, $x \multimap \xi_{ea}F(x,y)$ is $\mathcal{R}_+$-**quasiconvex**. By Corollary 3.1, there exist $(\bar{x}, \bar{y}) \in X \times Y$, $\bar{x} \in S(\bar{x})$, $\bar{y} \in T(\bar{x})$, and $\bar{u}_1, \bar{u}_2 \in \xi_{ea}F(\bar{x}, \bar{y})$ such that $u_1 \geq \bar{u}_1$, for all $u_1 \in \xi_{ea}F(x, \bar{y})$, and for all $x \in S(\bar{x})$, and $u_2 \leq \bar{u}_2$, for all $u_2 \in \xi_{ea}F(\bar{x}, y)$, and for all $y \in T(\bar{x})$. Hence there exist $\bar{z}_1, \bar{z}_2 \in F(\bar{x}, \bar{y})$ such that $\xi_{ea}(z_1) \geq \xi_{ea}(\bar{z}_1)$ for all $z_1 \in F(x, \bar{y})$, for all $x \in S(\bar{x})$ and $\xi_{ea}(z_2) \leq \xi_{ea}(\bar{z}_2)$ for all $z_2 \in F(\bar{x}, y)$, for all $y \in T(\bar{x})$. Therefore,

$$z_1 - \bar{z}_1 \notin -\text{int}C_+ \quad \textbf{for all } z_1 \in F(x, \bar{y}) \textbf{ and for all } x \in S(\bar{x})$$

and

$$z_2 - \overline{z}_2 \notin \text{int}C_+ \quad \text{for all } z_2 \in F(\overline{x}, y) \text{ and for all } y \in T(\overline{x}).$$

The conclusion follows from that $C(x) \subseteq C_+$ for all $x \in X$. □

If $Z = \mathcal{R}$ and $F = f : X \times Y \to \mathcal{R}$ is a single-valued function, then Corollary 3.1 is reduced to the following minimax theorem.

Corollary 3.2 *In Corollary 3.1, let $f : X \times Y \to \mathcal{R}$ be a continuous function satisfying the following conditions :*

(a) *for each $x \in X$, $y \mapsto f(x, y)$ is quasiconcave; and*

(b) *for each $y \in Y$, $x \mapsto f(x, y)$ is quasiconvex .*

Then there exists $(\bar{x}, \bar{y}) \in X \times Y$ with $\bar{x} \in S(\bar{x})$, $\bar{y} \in T(\bar{x})$ such that

$$\min_{u \in S(\bar{x})} \max_{v \in T(\bar{x})} f(u, v) = f(\bar{x}, \bar{y}) = \max_{v \in T(\bar{x})} \min_{u \in S(\bar{x})} f(u, v).$$

Acknowledgments

The authors wish to express their gratitude to the referees for their valuable suggestions.

References

Aubin, J. P. and Cellina A. (1994), *Differential Inclusion,* Springer, Berlin, 1994.

Chang S. S., Yuan G. X. Z., Lee G. H. and Zhang Xiao Lan (1997), Saddle points and minimax theorems for vector-valued multifunctions on H-spaces, *Applied Mathematics Letters,* Vol. 11, No. 3, pp. 101-107.

Fan, K. (1952), Fixed point and minimax theorems in locally convex topological linear spaces, *Proceedings of the National Academy of Sciences, U.S.A.,* Vol. 38, pp. 121-126.

Kim, I. S. and Kim, Y. T. (1999), Loose saddle points of a set-valued maps in topological vector spaces, *Applied Mathematics letters,* Vol. 12, pp. 21-26.

Lee, B. S., Lee, G.M. and Chang, S. S. (1997), Generalized vector variational inequalities for multifunctions, in *Proceedings of workshop on fixed point theory,* edited by K. Goebel, S. Prus, T. Sekowski and A. Stachura, Vol. L.I. Annales universitatis Mariae Curie-Sklodowska, Lubin-Polonia, pp. 193-202.

Lin, L. J. (1999), On generalized loose saddle point theorems for set valued maps, in *Nonlinear Analysis and Convex Analysis,* edited by W. Takahashi and T. Tanaka, World Scientific, Niigata, Japan.

Lin, L. J. and Yu, Z. T. (2001), On generalized vector quasi-equilibrium problems for multimaps, *Journal of Computational and Applied Mathematics,* Vol. 129, pp. 171-183.

Luc, D. T. and Vargas, C. (1992), A saddle point theorem for set-valued maps, *Nonlinear Analysis; Theory, Methods and Applications,* Vol. 18, pp. 1-7.

Luc, D. T. (1989), *Theory of Vector Optimization,* Lecture Notes in Economics and Mathematical Systems, Vol. 319, Springer, Berlin, New York.

Massay, W. S. (1980), *Singular homology theory,* Springer-Verlag, Berlin, New York.

Park, S. (1992), Some coincidence theorems on acyclic multifunctions and applications to KKM theory, in *Fixed Point Theory and Applications,* edited by K. K. Tan, World Scientific, Singapore, pp. 248-277.

Tan, K. K., Yu, J. and Yuan, X. Z. (1996), Existence theorems for saddle points of vector valued maps, *Journal of Optimization Theory and Application,* Vol. 89, pp. 731-747.

Tanaka, T. (1994), Generalized quasiconvexities, cone saddle points and minimax theorem for vector-valued functions, *Journal of Mathematical Analysis and Applications,* Vol. 81, pp. 355-377.

Chapter 19

NEW GENERALIZED INVEXITY FOR DUALITY IN MULTIOBJECTIVE PROGRAMMING PROBLEMS INVOLVING *N*-SET FUNCTIONS*

S.K. Mishra
Department of Mathematics, Statistics and Computer Science,
G. B. Pant University of Agriculture and Technology, India

S.Y. Wang
Institute of Systems Science, Academy of Mathematics and Systems Sciences,
Chinese Academy of Sciences, China

K.K. Lai
Department of Management Sciences,
City University of Hong Kong, Hong Kong

J.Shi
Department of Computer Science and Systems Engineering,
Muroran Institute of Technology, Japan

Abstract In this paper, we introduce four types of generalized convexity for an **n-set** function and discuss optimality and duality for a multiobjective programming problem involving **n-set** functions. Under some mild assumption on the new generalized convexity, we present a few optimality

---

*The research was supported by the University Grants Commission of India, the National Natural Science Foundation of China, Research Grants Council of Hong Kong and the Grant-in-Aid (C-14550405) from the Ministry of Education, Science, Sports and Culture of Japan.
Corresponding author: S.K. Mishra, email: shashikmishra@rediffmail.com

conditions for an efficient solution and a weakly efficient solution to the problem. Also we prove a weak duality theorem and a strong duality theorem for the problem and its Mond-Weir and general Mond-Weir dual problems respectively.

Keywords: multiobjective programming, n-set function, optimality, duality, generalized convexity

MSC2000: 90C29, 90C30

1. Introduction

In this paper, we consider the following multiobjective programming problem involving n-set functions:

$$(VP) \quad \begin{array}{ll} \textbf{minimize} & F(S) := (F_1(S), \cdots, F_p(S)) \\ \textbf{subject to} & G_j(S) \leqq 0, j \in M, \\ & S := (S_1, \cdots, S_n) \in \Omega^n, \end{array}$$

where Ω^n is the n-fold product of a σ-algebra Ω of subsets of a given X, $F_i, i \in P := \{1, 2, \cdots, p\}$ and $G_j, j \in M := \{1, 2, \cdots, m\}$ are real-valued functions defined on Ω^n. Let $\Omega_0 := \{S : S \in \Omega^n, G(S) \leqq 0\}$ be the set of all the feasible solutions to (VP), where $G := (G_1, G_2, \cdots, G_m)$.

Much attention has been paid to analysis of optimization problems with set functions, for example see Chou et al. (1985), Chou et al. (1986), Corley (1987), Kim et al. (1998), Lin (1990), Lin (1992), Morris (1979), Preda (1991), Preda (1995), Preda and Stancu-Minasian (1997), Preda and Stancu-Minasian (1999) and Zalmai (1991). A formulation for optimization problems with set functions was first given by Morris (1979). The main results of Morris (1979) are confined only to set functions of a single set. Corley (1987) gave the concepts of a partial derivative and a derivative of real-valued n-set functions. Chou et al. (1985), Chou et al. (1986), Kim et al. (1998), Lin (1990)-Lin (1992), Preda (1991), Preda (1995), and Preda and Stancu-Minasian (1997), Preda and Stancu-Minasian (1999) studied optimality and duality for optimization problems involving vector-valued n-set functions. For details, one can refer to Bector and Singh (1996), Hsia and Lee (1987), Kim et al. (1998), Lin (1990)-Lin (1992), Mazzoleni (1979), Preda (1995), Rosenmuller and Weidner (1974), Tanaka and Maruyama (1984) and Zalmai (1990).

Starting from the methods used by Jeyakumar and Mond (1992), Ye (1991), and Preda and Stancu-Minasian (2001) defined some new classes of scalar and vector functions called d-type-I, d-pseudo type-I and d-

quasi type-I for a multiobjective programming problem involving n-set functions and obtained a few interesting results on optimality and the Wolfe duality .

Recently, Aghezzaf and Hachimi (2000) introduced new classes of generalized type-I vector valued functions which are different from those defined in Kaul et al. (1994). For details, see Aghezzaf and Hachimi (2000). In this paper, we extend generalized type-I vector valued functions in Aghezzaf and Hachimi (2000) to n-set functions and establish optimality and the Mond–Weir type and general Mond-Weir type duality results for the problem (VP).

2. Definitions and Preliminaries

In this section, we introduce some notions and definitions. For $x := (x_1, x_2, \cdots, x_m)$ and $y := (y_1, y_2, \cdots, y_m) \in R^m$, we denote $x \leqq y$ iff $x_i \leqq y_i$ for each $i \in M$; $x \leq y$ iff $x_i \leqq y_i$ for each $i \in M$ with $x \neq y$; $x < y$ iff $x_i < y_i$ for each $i \in M$ and $x \not< y$ is the negation of $x < y$. We note that $x \in R^m_+$ iff $x \geqq 0$. For two real numbers a and b, $a \leqq b$ is equivalent to $a \leq b$, that is, $a = b$ or $a < b$.

Let (x, Ω, μ) be a finite atomless measure space with $L_1(X, \Omega, \mu)$ separable, and let d be the pseudo metric on Ω^n defined by $d(S,T) := \left[\sum_{k=1}^{n} \mu^2(S_k \Delta T_k)\right]^{1/2}$, where $S = (S_1, \cdots, S_n)$ and $T := (T_1, \cdots, T_n) \in \Omega^n$ and Δ denotes the symmetric difference. Thus, (Ω^n, d) is a pseudometric space which will serve as the domain for most of the functions in this paper.

For $h \in L_1(X, \Omega, \mu)$ and $S \in \Omega$ with the indicator (characteristic) function $I_S \in L_\infty(X, \Omega, \mu)$, the integral $\int_s h d\mu$ is denoted by $\langle h, I_S \rangle$.

The notion of differentiability for a real-valued set function was originally introduced by Morris (1979); and its n-set counterpart was discussed in Corley (1987).

A function $\varphi : \Omega \to R$ is differentiable at $S^0 \in \Omega$ if there exist $D\varphi(S^0) \in L_1(X, \Omega, \mu)$ called the derivative of φ at S^0, and $\psi : \Omega \times \Omega \to R$ such that, for each $S \in \Omega$

$$\varphi(S) = \pi(S^0) + \langle D\varphi(S^0), I_S - I_{S^0} \rangle + \psi(S, S^0),$$

where $\psi(S, S^0)$ is $o(d(S, S^0))$, that is, $\lim\limits_{d(S,S^0)\to 0} \frac{\psi(S,S^0)}{d(S,S^0)} = 0.$

Function $F : \Omega^n \to R$ is said to have a partial derivative at $S^0 := (S^0_1, S^0_2, \cdots, S^0_n)$ with respect to its k-th argument if the function $\varphi(S^0_k) =$

$F(S_1^0,\cdots,S_{k-1}^0, S_k^0, S_{k+1}^0,\cdots,S_n^0)$ has derivative $D\varphi(S_k^0)$. We define

$$D_kF(S^0) := D\varphi(S_k^0)$$

and denote

$$DF(S^0) := (D_1F(S^0),\cdots,D_nF(S^0)).$$

Function $F:\Omega^n \to R$ is said to be differentiable at S^0 if there exist $DF(S^0)$ and $\psi:\Omega^n\times\Omega^n\to R$ such that

$$F(S) = F(S^0) + \sum_{k=1}^{n}\left\langle D_kF(S^0), I_{S_k} - I_{S_k^0}\right\rangle + \psi(S,S^0),$$

where $\psi(S,S^0)$ is $o(d(S,S^0))$.

A feasible solution S^0 of (VP) is said to be an efficient solution of (VP) if there exists no other feasible solution S of (VP) such that $F_i(S) \leqq F_i(S^0)$, for all $i\in P$, with strict inequality for at least one $i\in P$.

A feasible solution S^0 of (VP) is said to be a weakly efficient solution of (VP) if there exists no other feasible solution S of (VP) such that $F_i(S) < F_i(S^0)$, for all $i\in P$.

Along the lines of Jeyakumar and Mond (1992) and Aghezzaf and Hachimi (2000), we define the following types of **n-set** functions, called **d-strong** pseudoquasi-type-I, **d-weak** strictly-pseudo quasi-type-I, **d-weak** strictly pseudo-type-I, **d-weak** quasi strictly-pseudo-type-I functions.

Definition 2.1 *(F, G) is said to be* ***d*-weak** *strictly-pseudo quasi-type-I at $S^0\in\Omega^n$ with respect to $\eta:\Omega^n\times\Omega^n\to R^n$ and $\gamma_i:\Omega^n\times\Omega^n\to R^+\backslash\{0\}$, $i=1,2,\cdots,p$ and $\delta_j:\Omega^n\times\Omega^n\to R^+\setminus\{0\}$, $j=1,\cdots,m$, if for every $S\in\Omega_0$*

$$\sum_{i=1}^{p}\gamma_i(S,S^0)F_i(S)\leq\sum_{i=1}^{p}\gamma_i(S,S^0)F_i(S^0)$$

$$\Rightarrow \sum_{i=1}^{p}\sum_{k=1}^{n}\eta_k(S,S^0)\left\langle D_kF_i(S^0), I_{S_k}-I_{S_k^0}\right\rangle < 0$$

and

$$-\sum_{j=1}^{m}\delta_j(S,S^0)G_j(S^0)\leqq 0$$

$$\Rightarrow \sum_{j=1}^{m}\sum_{k=1}^{n}\eta_k(S,S^0)\left\langle D_kG_j(S^0), I_{S_k}-I_{S_k^0}\right\rangle \leqq 0.$$

It is an extension of weak strictly-pseudo quasi-type-I functions defined in Aghezzaf and Hachimi (2000). The concept also extends the **d-pseudo-quasi-type-I** functions defined in Preda and Stancu-Minasian (2001). There exist non **n-set** functions which are weak strict pseudoquasi-type-I, but not strict pseudoquasi-type-I and not type-I with respect to the same η, see Example 2.1 in Aghezzaf and Hachimi (2000).

Definition 2.2 *(F, G) is said to be* **d-strong-pseudo** *quasi-type-I at $S^0 \in \Omega^n$ with respect to $\eta : \Omega^n \times \Omega^n \to R^n$ and $\gamma_i : S^n \times S^n \to R^+ \setminus \{0\}$, $i = 1, 2, \cdots, p$ and $\delta_j : \Omega^n \times \Omega^n \to R^+ \setminus \{0\}$, $j = 1, \cdots, m$, if for every $S \in \Omega_0$*

$$\sum_{i=1}^{p} \gamma_i(S, S^0) F_i(S) \leq \sum_{i=1}^{p} \gamma_i(S, S^0) F_i(S^0)$$

$$\Rightarrow \sum_{i=1}^{p} \sum_{k=1}^{n} \eta_k(S, S^0) \left\langle D_k F_i(S^0), I_{S_k} - I_{S_k^0} \right\rangle \leq 0$$

and

$$-\sum_{j=1}^{m} \delta_j(S, S^0) G_j(S^0) \leqq 0$$

$$\Rightarrow \sum_{j=1}^{m} \sum_{k=n}^{p} \eta_k(S, S^0) \left\langle D_k G_j(S^0), I_{S_k} - I_{S_k^0} \right\rangle \leqq 0.$$

Definition 2.3 *(F, G) is said to be* **d-weak** *quasi strictly-pseudo-type-I at $S^0 \in \Omega^n$ with respect to $\eta : \Omega^n \times \Omega^n \to R^n$ and $\gamma_i : \Omega^n \times \Omega^n \to R^+ \setminus \{0\}$, $i = 1, 2, \cdots, p$ and $\delta_j : \Omega^n \times \Omega^n \to R^+ \setminus \{0\}$, $j = 1, \cdots, m$, if for every $S \in \Omega_0$*

$$\sum_{i=1}^{p} \gamma_i(S, S^0) F_i(S) \leq \sum_{i=1}^{p} \gamma_i(S, S^0) F_i(S^0)$$

$$\Rightarrow \sum_{i=1}^{p} \sum_{k=1}^{n} \eta_k(S, S^0) \left\langle D_k F_i(S^0), I_{S_k} - I_{S_k^0} \right\rangle \leqq 0$$

and

$$-\sum_{j=1}^{m} \delta_j(S, S^0) G_j(S^0) \leqq 0$$

$$\Rightarrow \sum_{j=1}^{m} \sum_{k=1}^{n} \eta_k(S, S^0) \left\langle D_k G_j(S^0), I_{S_k} - I_{S_k^0} \right\rangle \leq 0.$$

Definition 2.4 *(F, G) is said to be* **d-weak** *strictly pseudo-type-I at* $S^0 \in \Omega^n$ *with respect to* $\eta : \Omega^n \times \Omega^n \to R^n$ *and* $\gamma_i : \Omega^n \times \Omega^n \to R^+ \setminus \{0\}$, $i = 1, 2, \cdots, p$ *and* $\delta_j : \Omega^n \times \Omega^n \to R^+ \setminus \{0\}$, $j = 1, \cdots, m$, *if for every* $S \in \Omega_0$

$$\sum_{i=1}^{p} \gamma_i(S, S^0) F_i(S) \leq \sum_{i=1}^{p} \gamma_i(S, S^0) F_i(S^0)$$

$$\Rightarrow \sum_{i=1}^{p} \sum_{k=1}^{n} \eta_k(S, S^0) \left\langle D_k F_i(S^0), I_{S_k} - I_{S_k^0} \right\rangle < 0$$

and

$$-\sum_{j=1}^{m} \delta_j(S, S^0) G_j(S^0) \leqq 0$$

$$\Rightarrow \sum_{j=1}^{m} \sum_{k=1}^{n} \eta_k(S, S^0) \left\langle D_k G_j(S^0), I_{S_k} - I_{S_k^0} \right\rangle < 0.$$

Remark 2.1 *The above definitions are extensions of the corresponding definitions in Aghezzaf and Hachimi (2000). These definitions are different from the other definitions such as in Kaul et al. (1994) and Hanson and Mond (1987), for various examples refer to Aghezzaf and Hachimi (2000).*

The following results from Zalmai (1991) will be needed in Section 4.

Lemma 2.1 *Let* S^0 *be an efficient (or weakly efficient) solution for (VP) and let* $F_i, i \in P$ *and* $G_j, j \in M$ *be differentiable at* S^0. *Then there exist* $\lambda \in R^p_+, \mu \in R^m_+, (\lambda, \mu) \neq 0$ *such that*

$$\sum_{k=1}^{n} \left\langle \sum_{i=1}^{p} \lambda_i D_k F_i(S^0) + \sum_{j=1}^{m} \mu_j D_k G_j(S^0), I_{S_k} - I_{S_k^0} \right\rangle \geq 0 \text{ for all } S \in \Omega^n,$$

$\mu_j G_j(S^0) = 0, j \in M.$

Definition 2.5 *A feasible solution* S^0 *is said to be a regular feasible solution if there exists* $\hat{S} \in \Omega^n$ *such that*

$$G_j(S^0) + \langle DG_j(S^0), I_{\hat{S}} - I_{S^0} \rangle < 0, j \in M_0.$$

Thus incorporating the above in Lemma 2.1, normalizing λ such that $\sum_{i=1}^{p} \lambda_i = 1$ and redefining $(\mu_1, \cdots, \mu_m)$, we can have the following result.

Lemma 2.2 *Let S^0 be an efficient (or weakly efficient) solution for (VP) and let $F_i, i \in P$ and $G_j, j \in M$ be differentiable at S^0. Then there exist $\lambda \in R^p_+$ with $\sum_{i=1}^{p} \lambda_i = 1$ and $\mu \in R^m_+$ such that*

$$\sum_{k=1}^{n} \left\langle \sum_{i=1}^{p} \lambda_i D_k F_i(S^0) + \sum_{j=1}^{m} \mu_j D_k G_j(S^0), I_{S_k} - I_{S_k^0} \right\rangle \geqq 0, \text{ for all } S \in \Omega^n,$$
$$\mu_j G_j(S^0) = 0, j \in M.$$

3. Optimality Condition

In this section, we give a sufficient optimality condition for a weakly efficient solution to (VP) under the assumption of new types of generalized convexity introduced in Section 2.

Theorem 3.1 *Let S^0 be a feasible solution for (VP). Suppose that*

(i1) there exist $\lambda_i^0 \geqq 0, i \in P$, with $\sum_{i=1}^{p} \lambda_i^0 = 1$ and $\mu_j^0 \geqq 0$, $j \in M_0 := M(S^0) := \{j \in M : G_j(S^0) = 0\}$, such that for all $S \in \Omega^n$

$$\left\langle D_k(\lambda^0 F)(S^0) + D_k(\mu^0 G)(S^0), I_{S_k} - I_{S_k^0} \right\rangle \geqq 0$$

and one of the following conditions is satisfied:

*(i2) $(F, \mu G)$ is **d-strong** pseudo quasi-type-I at S^0 with respect to η, γ_i, δ_j and $\lambda > 0$;*

*(i3) $(F, \mu G)$ is **d-weak** strictly pseudo quasi-type-I at S^0 with respect to η, γ_i, and δ_j;*

*(i4) $(F, \mu G)$ is **d-weak** strictly pseudo-type-I at S^0 with respect to η, γ_i, and δ_j;*

with η satisfying $\eta^\top \alpha < 0 \Rightarrow \alpha_k < 0$ for at least one k in $\{1, 2, \cdots, n\}$. Then S^0 is a weakly efficient solution to (VP).

Proof. Assume that S^0 is not a weakly efficient solution to (VP). Then there is a feasible solution to (VP) such that

$$F_i(S) < F_i(S^0) \text{ for any } i \in P. \tag{19.1}$$

According to (i2) there exist $\gamma_i : \Omega^n \times \Omega^n \to R^+ \setminus \{0\}, i = 1, 2, \cdots, p$ and $\delta_j : \Omega^n \times \Omega^n \to R^+ \setminus \{0\}, j \in M_0$ such that, for all $S \in \Omega_0$

$$\begin{aligned}&\sum_{i=1}^{p} \gamma_i(S, S^0)\lambda_i^0 F_i(S) < \sum_{i=1}^{p} \gamma_i(S, S^0)\lambda_i^0 F_i(S^0)\\ &\Rightarrow \sum_{i=1}^{p}\sum_{k=1}^{n} \lambda_i^0 \eta_k(S, S^0) \left\langle D_k F_i(S^0), I_{S_k} - I_{S_k^0}\right\rangle < 0 \text{ (since } \lambda > 0)\end{aligned} \tag{19.2}$$

and

$$\begin{aligned}&-\sum_{j\in M_0} \delta_j(S, S^0)\mu_j^0 G_j(S^0) \leqq 0\\ &\Rightarrow \sum_{j\in M_0}\sum_{k=1}^{n} \mu_j^0 \eta_k(S, S^0) \left\langle D_k G_j(S^0), I_{S_k} - I_{S_k^0}\right\rangle \leqq 0.\end{aligned} \tag{19.3}$$

From (19.1) and $\lambda_i^0 \geqq 0, i \in P$, with $\sum_{i=1}^{p} \lambda_i^0 = 1$ and $\gamma_i > 0$ for any $i \in P$, we get

$$\sum_{i=1}^{p} \gamma_i(S, S^0)\lambda_i^0 F_i(S) < \sum_{i=1}^{p} \gamma_i(S, S^0)\lambda_i^0 F_i(S^0).$$

Using (19.2), we get

$$\begin{aligned}&\sum_{i=1}^{p}\sum_{k=1}^{n} \lambda_i^0 \eta_k(S, S^0) \left\langle D_k F_i(S^0), I_{S_i} - I_{S_i^0}\right\rangle < 0\\ &\Rightarrow \sum_{k=1}^{n} \eta_k(S, S^0) \left\langle D_k F_i(S^0), I_{S_k} - I_{S_k^0}\right\rangle < 0.\end{aligned} \tag{19.4}$$

Since S^0 is a feasible solution to (VP) and $G_j(S^0) = 0$ for $j \in M_0$, we obtain

$$-\sum_{j\in M_0} \delta_j(S, S^0)\mu_j^0 G_j(S^0) \leqq 0.$$

This relation together with (19.3) implies

$$\begin{aligned}&\sum_{j\in M_0}^{p}\sum_{k=1}^{n} \mu_j^0 \eta_k(S, S^0) \left\langle D_k G_j(S^0), I_{S_k} - I_{S_k^0}\right\rangle \leqq 0\\ &\Rightarrow \sum_{k=1}^{n} \eta_k(S, S^0) \left\langle D_k(\mu^0 G)(S^0), I_{S_k} - I_{S_k^0}\right\rangle \leqq 0.\end{aligned} \tag{19.5}$$

By (19.4) and (19.5), we get

$$\sum_{k=1}^{n} \eta_k(S,S^0)\left\langle D_k(\lambda^0 F)(S^0) + D_k(\mu^0 G)(S^0), I_{S_k} - I_{S_k^0}\right\rangle < 0$$
$$\Rightarrow \left\langle D_k(\lambda^0 F)(S^0) + D_k(\mu^0 G)(S^0), I_{S_k} - I_{S_k^0}\right\rangle < 0,$$

for at least one k in $\{1,2,\cdots,n\}$ (because $\eta^\top \alpha < 0 \Rightarrow \alpha_k < 0$ for at least one k in $\{1,2,\cdots,n\}$), which contradicts (i1).

By (i3), there exist $\gamma_i : \Omega^n \times \Omega^n \to R^+ \setminus \{0\}, i = 1,2,\cdots,p$ and $\delta_j : \Omega^n \times \Omega^n \to R^+ \setminus \{0\}, j \in M_0$ such that, for all $S \in \Omega_0$

$$\begin{aligned} &\sum_{i=1}^{p} \gamma_i(S,S^0)\lambda_i^0 F_i(S) \leq \sum_{i=1}^{p} \gamma_i(S,S^0)\lambda_i^0 F_i(S^0) \\ &\Rightarrow \sum_{i=1}^{p}\sum_{k=1}^{n} \lambda_i^0 \eta_k(S,S^0)\left\langle D_k F_i(S^0), I_{S_k} - I_{S_k^0}\right\rangle < 0. \end{aligned} \tag{19.6}$$

and

$$\begin{aligned} &-\sum_{j\in M_0} \delta_j(S,S^0)\mu_j^0 G_j(S^0) \leqq 0 \\ &\Rightarrow \sum_{j\in M_0}\sum_{k=1}^{n} \mu_j^0 \eta_k(S,S^0)\left\langle D_k G_j(S^0), I_{S_k} - I_{S_k^0}\right\rangle \leqq 0. \end{aligned} \tag{19.7}$$

By (19.1) and $\lambda_i^0 \geqq 0$, $i \in P$, with $\sum_{i=1}^{p} \lambda_i^0 = 1$ and $\gamma_i > 0$ for any $i \in P$, we get

$$\sum_{i=1}^{p} \gamma_i(S,S^0)\lambda_i^0 F_i(S) \leq \sum_{i=1}^{p} \gamma_i(S,S^0)\lambda_i^0 F_i(S^0).$$

Using (19.6) and the above inequality, we get

$$\begin{aligned} &\sum_{i=1}^{p}\sum_{k=1}^{n} \lambda_i^0 \eta_k(S,S^0)\left\langle D_k F_i(S^0), I_{S_k} - I_{S_k^0}\right\rangle < 0 \\ &\Rightarrow \sum_{k=1}^{n} \eta_k(S,S^0)\left\langle D_k(\lambda^0 F)(S^0), I_{S_k} - I_{S_k^0}\right\rangle < 0. \end{aligned} \tag{19.8}$$

From the feasibility of S and (19.7), we get (19.5). By (19.8) and (19.5), we get

$$\left\langle D_k(\lambda^0 F)(S^0) + D_k(\mu^0 G)(S^0), I_{S_k} - I_{S_k^0}\right\rangle < 0,$$

for at least one k in $\{1,2,\cdots n\}$ (because $\eta^\top \alpha < 0 \Rightarrow \alpha_k < 0$ for at least one k in $\{1,2,\cdots,n\}$), which again contradicts (i1).

By (i4), there exist $\gamma_i : \Omega^n \times \Omega^n \to R^+ \setminus \{0\}, i = 1,2,\cdots,p$ and $\delta_j : \Omega^n \times \Omega^n \to R^+ \setminus \{0\}, j \in M_0$ such that, for all $S \in \Omega_0$, we get (19.6) and

$$
\begin{aligned}
&-\sum_{j\in M_0} \delta_j(S,S^0)\mu_j^0 G_j(S^0) \leqq 0 \\
&\Rightarrow \sum_{j=1}^{m}\sum_{k=1}^{n} \eta_k(S,S^0)\left\langle D_k G_j(S^0), I_{S_k} - I_{S_k^0}\right\rangle < 0.
\end{aligned}
\tag{19.9}
$$

By (19.1) and $\lambda_i^0 \geqq 0, i \in P$, with $\sum_{i=1}^{p} \lambda_i^0 = 1$ and $\gamma_i > 0$ for any $i \in P$, we get

$$
\sum_{i=1}^{p} \gamma_i(S,S^0)\lambda_i^0 F_i(S) \leq \sum_{i=1}^{p} \gamma_i(S,S^0)\lambda_i^0 F_i(S^0).
$$

Using (19.6) and the above inequality, we get (19.8). From the feasibility of S and (19.9), we get (19.5). By (19.8) and (19.5), we have

$$
\left\langle D_k(\lambda^0 F)(S^0) + D_k(\mu^0 G)(S^0), I_{S_k} - I_{S_k^0}\right\rangle < 0,
$$

for at least one k in $\{1,2,\cdots,n\}$ (because $\eta^\top \alpha < 0 \Rightarrow \alpha_k < 0$ for at least one k in $\{1,2,\cdots,n\}$), which contradicts (il). This completes the proof. □

4. Mond-Weir Duality

In this section, we consider the following Mond-Weir dual problem (MD):

$$
\begin{aligned}
\textbf{maximize}\ & F(T) \\
\textbf{subject to}\ & \left\langle D_k(\lambda F)(T) + D_k(\mu G)(T), I_{S_k} - I_{T_k}\right\rangle \geqq 0, \forall S \in \Omega^n, \\
& \sum_{j=1}^{m} \mu_j G_j(T) \geqq 0, \\
& \lambda_i \geqq 0, i \in P, \text{ and } \sum_{i=1}^{p} \lambda_i = 1, \\
& \mu_j \geqq 0, j \in M \text{ and } T \in \Omega^n.
\end{aligned}
\tag{19.10}
$$

Let D be the set of all feasible solutions to (MD).

Theorem 4.1 *(Weak Duality). Suppose that $S \in \Omega_0$ and $(T,\lambda,\mu) \in D$. If any one of the following conditions is satisfied:*

(a) $(F, \mu G)$ *is* **d-strong-pseudo** *quasi-type-I at T with respect to* η, γ_i,, δ_j *and* $\lambda > 0$;

(b) $(F, \mu G)$ *is* **d-weak** *strictly pseudo quasi-type-I at T with respect to* η, γ_i *and* δ_j;

(c) $(F, \mu G)$ *is* **d-weak** *strictly pseudo-type-I at T with respect to* η, γ_i *and* δ_j;

with η *satisfying* $\eta^\top \alpha < 0 \Rightarrow \alpha_k < 0$ *for at least one* k *in* $\{1, 2, \cdots, n\}$, *then* $F(S) \geq F(T)$.

Proof. We proceed by contradiction. Suppose that there exist $S \in \Omega_0$ and $(T, \lambda, \mu) \in D$ such that $F(S) < F(T)$. Since γ_i is in $R^+ \setminus \{0\}$, we have

$$\sum_{i=1}^{p} \gamma_i(S,T)F_i(S) \leq \sum_{i=1}^{p} \gamma_i(S,T)F_i(T). \tag{19.11}$$

Since $(T, \lambda, \mu) \in D$, it follows that

$$\sum_{j=1}^{m} \mu_j G_j(T) \geqq 0, \text{ that is, } -\sum_{j=1}^{m} \mu_j G_j(T) \leqq 0.$$

Because δ_j is in $R^+ \setminus \{0\}$, we have

$$-\sum_{j=1}^{m} \delta_j(S,T)\mu_j G_j(T) \leqq 0. \tag{19.12}$$

By condition (a), (19.11) and (19.12) yield

$$\sum_{i=1}^{p}\sum_{k=1}^{n} \lambda_i \eta_k(S,T) \langle D_k F_i(T), I_{S_k} - I_{T_k}\rangle \leq 0$$

and

$$\sum_{j=1}^{m}\sum_{k=1}^{n} \mu_j \eta_k(S,T) \langle D_k G_j(T), I_{S_k} - I_{T_k}\rangle \leqq 0.$$

Since $\lambda > 0$, the above two inequalities imply

$$\sum_{k=1}^{n} \eta_k(S,T) \langle D_k(\lambda F)(T), I_{S_k} - I_{T_k}\rangle < 0$$

and

$$\sum_{k=1}^{n} \eta_k(S,T)\,\langle D_k(\mu G)(T), I_{S_k} - I_{T_k}\rangle \leq 0.$$

By the above two inequalities, we get

$$\sum_{k=1}^{n} \eta_k(S,T)\,\langle D_k(\lambda F)(T) + D_k(\mu G)(T), I_{S_k} - I_{T_k}\rangle < 0$$

$$\Rightarrow \langle D_k(\lambda F)(T) + D_k(\mu G)(T), I_{S_k} - I_{T_k}\rangle < 0,$$

for at least one k in $\{1, 2, \cdots, n\}$ (because $\eta^\top \alpha < 0 \Rightarrow \alpha_k < 0$ for at least one k in $\{1, 2, \cdots, n\}$), which contradicts (19.10).

By condition (b), we get

$$\sum_{i=1}^{p}\sum_{k=1}^{n} \lambda_i \eta_k(S,T)\,\langle D_k F_i(T), I_{S_k} - I_{T_k}\rangle < 0$$

and

$$\sum_{j=1}^{m}\sum_{k=1}^{n} \mu_j \eta_k(S,T)\,\langle D_k G_j(T), I_{S_k} - I_{T_k}\rangle \leqq 0.$$

These two inequalities imply

$$\sum_{k=1}^{n} \eta_k(S,T)\,\langle D_k(\lambda F)(T), I_{S_k} - I_{T_k}\rangle < 0$$

and

$$\sum_{k=1}^{n} \eta_k(S,T)\,\langle D_k(\mu G)(T), I_{S_k} - I_{T_k}\rangle \leqq 0.$$

By these two inequalities, we get

$$\sum_{k=1}^{n} \eta_k(S,T)\,\langle D_k(\lambda F)(T) + D_k(\mu G)(T), I_{S_k} - I_{T_k}\rangle < 0$$

$$\Rightarrow \langle D_k(\lambda F)(T) + D_k(\mu G)(T), I_{S_k} - I_{T_k}\rangle < 0,$$

for at least one k in $\{1, 2, \cdots, n\}$ (because $\eta^\top \alpha < 0 \Rightarrow \alpha_k < 0$ for at least one k in $\{1, 2, \cdots, n\}$). This contradicts (19.10).

By condition (c), (19.11) and (19.12) imply

$$\sum_{i=1}^{p}\sum_{k=1}^{n} \lambda_i \eta_k(S,T)\,\langle D_k F_i(T), I_{S_k} - I_{T_k}\rangle < 0$$

and

$$\sum_{j=1}^{m}\sum_{k=1}^{n}\mu_j\eta_k(S,T)\langle D_kG_j(T),I_{S_k}-I_{T_k}\rangle<0.$$

These two inequalities imply

$$\sum_{k=1}^{n}\eta_k(S,T)\langle D_k(\lambda F)(T),I_{S_k}-I_{T_k}\rangle<0$$

and

$$\sum_{k=1}^{n}\eta_k(S,T)\langle D_k(\mu G)(T),I_{S_k}-I_{T_k}\rangle<0.$$

By these two inequalities, we get

$$\sum_{k=1}^{n}\eta_k(S,T)\langle D_k(\lambda F)(T)+D_k(\mu G)(T),I_{S_k}-I_{T_k}\rangle<0,$$

$$\Rightarrow\langle D_k(\lambda F)(T)+D_k(\mu G)(T),I_{S_k}-I_{T_k}\rangle<0,$$

for at least one k in $\{1,2,\cdots,n\}$ (because $\eta^{\top}\alpha<0\Rightarrow\alpha_k<0$ for at least one k in $\{1,2,\cdots,n\}$), which contradicts (19.10). This completes the proof. □

Theorem 4.2 *(Strong Duality). Let $S\in\Omega_0$ satisfy*

(b1) S^0 is a weakly efficient solution to (VP);

(b2) S^0 is a regular solution to (VP).

Then there exist $\lambda^0\in R^p$ and $\mu^0\in R^m$ such that (S^0,λ^0,μ^0) is a feasible solution for (MD) and the values of the objective functions of (VP) and (MD) are equal at these points. Furthermore, if the conditions of the weak duality in Theorem 4.1 hold for each feasible solution (T,λ,μ) of (MD), then (S^0,λ^0,μ^0) is a weakly efficient solution to (MD).

Proof. By Lemma 2.1, there exist $\lambda_i^0\geqq 0, i\in P$ with $\sum_{i=1}^{p}\lambda_i^0=1$ and $\mu_j^0\geqq 0$, $j\in M_0$ such that (S^0,λ^0,μ^0) is feasible for (MD) and the values of the objective functions of (VP) and (MD) are equal. The last part follows directly from Theorem 4.1. □

5. Generalized Mond-Weir Duality

In this section, we study a general type of Mond-Weir duality and establish weak and strong duality theorems under a generalized invexity assumption.

Consider the following general Mond–Weir type of dual problem:

$$
\text{(GMD)} \quad
\begin{aligned}
\text{maximize} \quad & F(T) + \sum_{j \in J_0} \mu_j G_j(T) \\
\text{subject to} \quad & \left\langle D_k(\lambda F)(T) + D_k(\mu G)(T), I_{S_k} - I_{T_k} \right\rangle \geqq 0, \\
& \sum_{j \in J_\alpha} \mu_j G_j(T) \geqq 0 \text{ for } 1 \leqq \alpha \leqq r, \qquad (19.13) \\
& \lambda \geqq 0, \mu \geqq 0 \text{ and } \sum_{i=1}^{p} \lambda_i = 1,
\end{aligned}
$$

where J_α, $0 \leqq \alpha \leqq r$ are partitions of set M.

Theorem 5.1 *(Weak Duality). Assume that for all $S \in \Omega_0$ and all feasible (T, λ, μ) for (GMD), one of the following conditions holds:*

*(a) $\lambda > 0$ and $\left((F + \sum_{j \in J_0} \mu_j G_j)(\cdot), \sum_{j \in J_\alpha} \mu_j G_j(\cdot)\right)$ is **d-strong** pseudo quasi-type-I at T with respect to η, γ_i, and δ_j for any α, $1 \leqq \alpha \leqq r$;*

*(b) $\left((F + \sum_{j \in J_0} \mu_j G_j)(\cdot), \sum_{j \in J_\alpha} \mu_j G_j(\cdot)\right)$ is **d-weak** strictly pseudo quasi-type-I at with respect to η, γ_i, and δ_j for any $\alpha, 1 \leqq \alpha \leqq r$;*

*(c) $\left((F + \sum_{j \in J_0} \mu_j G_j)(\cdot), \sum_{j \in J_\alpha} \mu_j G_j(\cdot)\right)$ is **d-weak** strictly pseudo-type-I at T with respect to η, γ_i, and δ_j for any α, $1 \leqq \alpha \leqq r$;*

with η satisfying $\eta^{\mathsf{T}} \alpha < 0 \Rightarrow \alpha_k < 0$ for at least one k in $\{1, 2, \cdots, n\}$ Then the following can not hold:

$$F(S) \leq F(T) + \sum_{j \in J_0} \mu_j G_j(T).$$

Proof. Suppose to the contrary that the above inequality holds. Since $S \in \Omega_0$ and $\mu \geqq 0$, we have

$$F(S) + \sum_{j \in J_0} \mu_j G_j(S) \leq F(T) + \sum_{j \in J_0} \mu_j G_j(T).$$

From (19.13), we have

$$-\sum_{j \in J_\alpha} \mu_j G_j(T) \leqq 0, \text{ for all } 1 \leqq \alpha \leqq r.$$

Since γ_i and δ_j are in $R^+ \setminus \{0\}$ from the above two inequalities, we have

$$\begin{aligned}&\sum_{i=1}^{p}\gamma_i(S,T)\left(F_i(S)+\sum_{j\in J_0}\mu_jG_j(S)\right)\\ &\leq\sum_{i=1}^{p}\gamma_i(S,T)\left(F_i(T)+\sum_{j\in J_0}\mu_jG_j(T)\right)\end{aligned} \tag{19.14}$$

and

$$-\sum_{j\in J_\alpha}\delta_j(S,T)\mu_jG_j(T)\leqq 0. \tag{19.15}$$

By condition (a), (19.14) and (19.15), we have

$$\sum_{i=1}^{p}\sum_{k=1}^{n}\eta_k(S,T)\left\langle D_k\left(F_i(T)+\sum_{j\in J_0}\mu_jG_j(T)\right),I_{S_k}-I_{T_k}\right\rangle\leq 0 \tag{19.16}$$

and

$$\sum_{j\in J_\alpha}\sum_{k=1}^{n}\mu_j\eta_k(S,T)\left\langle D_kG_j(T),I_{S_k}-I_{T_k}\right\rangle\leqq 0, 1\leqq\alpha\leqq r. \tag{19.17}$$

Since $\lambda > 0$, from (19.16) and (19.17), we have

$$\sum_{i=1}^{n}\eta_k(S,T)\left\langle D_k\left((\lambda F)(T)+\sum_{\alpha=0}^{r}(\mu_{J_\alpha}G_{J_\alpha})(T)\right),I_{S_k}-I_{T_k}\right\rangle<0. \tag{19.18}$$

Since $J_0, J_1, \cdots, J_r$ are partitions of M, (19.18) is equivalent to

$$\begin{aligned}&\sum_{k=1}^{n}\eta_k(S,T)\left\langle D_k(\lambda F)(T)+D_k(\mu G)(T),I_{S_k}-I_{T_k}\right\rangle<0\\ &\Rightarrow\left\langle D_k(\lambda F)(T)+D_k(\mu G)(T),I_{S_k}-I_{T_k}\right\rangle<0,\end{aligned} \tag{19.19}$$

for at least one one k in $\{1,2,\cdots,n\}$ (because $\eta^{\top}\alpha<0\Rightarrow\alpha_k<0$ for at least one k in $\{1,2,\cdots,n\}$), which contradicts (19.13).

Using condition (b), from (19.14) and (19.15), we get

$$\sum_{i=1}^{p}\sum_{k=1}^{n}\eta_k(S,T)\left\langle D_k\left(F_i(T)+\sum_{j\in J_0}\mu_jG_j(T)\right),I_{S_k}-I_{T_k}\right\rangle<0$$

and

$$\sum_{j\in J_\alpha}\sum_{k=1}^{n}\mu_j\eta_k(S,T)\left\langle D_kG_j(T), I_{S_k}-I_{T_k}\right\rangle \leqq 0, 1\leqq \alpha\leqq r.$$

Since $\lambda \geqq 0$, the above inequalities give (19.19) and then again we get a contradiction to (19.13).

Suppose now that (c) is satisfied. From (19.14) and (19.15) it follows that

$$\sum_{i=1}^{p}\sum_{k=1}^{n}\eta_k(S,T)\left\langle D_k\left(F_i(T)+\sum_{j\in J_0}\mu_jG_j(T)\right), I_{S_k}-I_{T_k}\right\rangle < 0$$

and

$$\sum_{j\in J_\alpha}\sum_{k=1}^{n}\mu_j\eta_k(S,T)\left\langle D_kG_j(T), I_{S_k}-I_{T_k}\right\rangle < 0, 1\leqq \alpha\leqq r.$$

Since $\lambda \geqq 0$, the above inequalities give (19.19) and then again we get a contradiction to (19.13). This completes the proof. □

Theorem 5.2 *(Strong Duality). Let $S^0 \in \Omega_0$ satisfy*

(b1) S^0 is a weakly efficient solution to (VP);

(b2) S^0 is a regular solution to (VP).

Then there exist $\lambda^0 \in R^p$ and $\mu^0 \in R^m$ such that (S^0, λ^0, μ^0) is a feasible solution for (GMD) and $\mu_{J_0}G_{J_0}(S^0) = 0$, and the values of the objective functions of (VP) and (GMD) at these solutions are equal. Furthermore, if the weak duality holds between (VP) and (GMD), then (S^0, λ^0, μ^0) is a weakly efficient solution to (GMD).

The proof of this theorem follows the lines of the proof of Theorem 4.2 in the light of Theorem 5.1.

Acknowledgments

The authors wish to thank an anonymous referee and Prof. Andrew Eberhard for their constructive comments and suggestions on an earlier version of the paper.

References

Aghezzaf, B. and Hachimi, M. (2000), Generalized Invexity and Duality in Multiobjective Programming Problems, *Journal of Global Optimization,* vol. 18, pp. 91-101.

Bector, C.R. and Singh, M. (1996), Duality for Multiobjective *B*-Vex Programming Involving n-**Set** Functions, *Journal of Mathematical Analysis and Applications,* vol. 202, pp. 701-726.

Chou, J.H., Hsia, W.S. and Lee, T.Y. (1985), On Multiple Objective Programming Problems with Set Functions, *Journal of Mathematical Analysis and Applications,* vol. 105, pp. 383-394.

Chou, J.H., Hsia, W.S. and Lee, T.Y. (1986), Epigraphs of Convex Set Functions, *Journal of Mathematical Analysis and Applications,* vol. 118, pp. 247-254.

Corley, H.W. (1987), Optimization Theory for n-**Set** Functions, *Journal of Mathematical Analysis and Applications,* vol. 127, pp. 193-205.

Hanson, M.A. and Mond, B. (1987), Convex Transformable Programming Problems and Invexity, *Journal of Information and Optimization Sciences,* vol. 8, pp. 201-207.

Hsia, W.S. and Lee, T.Y. (1987), Proper D Solution of Multiobjective Programming Problems with Set Functions, *Journal of Optimization Theory and Applications,* vol. 53, pp. 247-258.

Jeyakumar, V. and Mond, B. (1992), On Generalized Convex Mathematical Programming, *Journal of Australian Mathematical Society* Ser. B, vol. 34, pp. 43-53.

Kaul, R.N., Suneja, S.K. and Srivastava, M.K. (1994), Optimality Criteria and Duality in Multiple Objective Optimization Involving Generalized Invexity, *Journal of Optimization Theory and Applications,* vol. 80, pp. 465-482.

Kim, D.S., Jo, C.L. and Lee, G.M. (1998), Optimality and Duality for Multiobjective Fractional Programming Involving n-**Set** Functions, *Journal of Mathematical Analysis and Applications,* vol. 224, pp. 1-13.

Lin, L.J. (1990), Optimality of Differentiable Vector-Valued n-**Set** Functions, *Journal of Mathematical Analysis and Applications,* vol. 149, pp. 255-270.

Lin, L.J. (1991a), On the Optimality Conditions of Vector-Valued n-**Set** Functions, *Journal of Mathematical Analysis and Applications,* vol. 161, pp. 367-387.

Lin, L.J. (1991b), Duality Theorems of Vector Valued n-**Set** Functions, *Computers and Mathematics with Applications* vol. 21, pp. 165-175.

Lin, L.J. (1992), On Optimality of Differentiable Nonconvex n-**Set** Functions, *Journal of Mathematical Analysis and Applications,* vol. 168, pp. 351-366.

Mangasarian, O.L. (1969), *Nonlinear Programming,* McGraw-Hill, New York.

Mazzoleni, P. (1979), On Constrained Optimization for Convex Set Functions, in *Survey of Mathematical Programming,* Edited by A. Prekop, North-Holland, Amsterdam, vol. 1, pp. 273-290.

Mishra, S.K. (1998), On Multiple-Objective Optimization with Generalized Univexity, *Journal of Mathematical Analysis and Applications,* vol. 224, pp. 131-148.

Mond, B. and Weir, T. (1981), Generalized Concavity and Duality, in *Generalized Concavity in Optimization and Economics,* Edited by S. Schaible and W. T. Ziemba, Academic Press, New York, pp. 263-280.

Morris, R.J.T. (1979), Optimal Constrained Selection of a Measurable Subset, *Journal of Mathematical Analysis and Applications,* vol. 70, pp. 546-562.

Mukherjee, R.N. (1991), Genaralized Convex Duality for Multiobjective Fractional Programs, *Journal of Mathematical Analysis and Applications,* vol. 162, pp. 309-316

Preda, V. (1991), On Minimax Programming Problems Containing n-Set Functions, *Optimization,* vol. 22, pp. 527-537.

Preda, V. (1995), On Duality of Multiobjective Fractional Measurable Subset Selection Problems, *Journal of Mathematical Analysis and Applications,* vol. 196, pp. 514-525.

Preda, V. and Stancu-Minasian, I.M. (1997), Mond-Weir Duality for Multiobjective Mathematical Programming with n-Set Functions, *Analele Universitati Bucuresti, Matematica-Informatica,* vol. 46, pp. 89-97.

Preda, V. and Stancu-Minasian, I.M. (1999), Mond-Weir Duality for Multiobjective Mathematical Programming with n-Set Functions, *Revue Roumaine de Mathématiques Pures et Appliquées,* vol. 44, pp. 629-644.

Preda, V. and Stancu-Minasian, I.M. (2001), Optimality and Wolfe Duality for Multiobjective Programming Problems Involving n-Set Functions, in *Generalized Convexity and Generalized Monotonicity,* Edited by Nicolas Hadjisavvas, J-E Martinez-Legaz and J-P Penot, Springer, Berlin, pp. 349-361.

Rosenmuller, J. and Weidner, H.G. (1974), Extreme Convex Set Functions with Finite Carries: General Theory, *Discrete Mathematics,* vol. 10, pp. 343-382.

Tanaka, K. and Maruyama, Y. (1984), The Multiobjective Optimization Problems of Set Functions, *Journal of Information and Optimization Sciences,* vol. 5, pp. 293-306.

Ye, Y.L. (1991), D-invexity and Optimality Conditions, *Journal of Mathematical Analysis and Applications,* vol. 162, pp. 242-249.

Zalmai, G.J. (1989), Optimality Conditions and Duality for Constrained Measurable Subset Selection Problems with Minmax Objective Functions, *Optimization,* vol. 20, pp. 377-395.

Zalmai, G.J. (1990), Sufficiency Criteria and Duality for Nonlinear Programs Involving n-Set Functions, *Journal of Mathematical Analysis and Applications,* vol. 149, pp. 322-338.

Zalmai, G.J. (1991), Optimality Conditions and Duality for Multiobjective Measurable Subset Selection Problems, *Optimization,* vol. 22, pp. 221-238.

Chapter 20

EQUILIBRIUM PRICES AND QUASICONVEX DUALITY

Phan Thien Thach

Institute of Mathematics, Vietnam

Abstract Given an economy in which there is a commodity trading between two Sectors A and B. For a given vector of prices Sector B is interested in getting a maximal commodity worth under an expenditure constraint. Sector A is interested in finding a feasible vector of prices such that the level of trade allowance per one unit of commodity worth is maximized. The problem under consideration is a quasiconvex minimization. Using quasiconvex duality we obtain a dual problem and a generalized Karush-Kuhn-Tucker condition for optimality. The optimal vector of prices can be interpreted as equilibrium and as a linearization of the commodity worth function at the optimal dual's solution.

Keywords: Quasiconvex, Duality, Price, Equilibrium.

MSC2000: 90C26

1. Problem setting

It is well-known that convexity plays an important role in linearization and linear approximation approaches to nonlinear problems and therefore it has a broad application in economic theory (e.g., Debreu (1959)-Luenberger (1995)). Dual interpretations and variations of price concept have brought both interesting theoretical aspects and efficient computational issues to mathematical programming problems. For a generalized convexity such as quasiconvexity there have been great research attempts to extend dual interpretations which are well performed in the case of convexity (e.g., Crouzeix (1974)-Thach (1995)). In the streamline of those researches this article presents an application of quasiconvex duality in a problem of finding an equilibrium vector of prices.

Consider two trading sectors A and B. There are n commodities exchanged between A and B. The commodity flow from A to B is denoted by a vector $x = (x_1, x_2, \ldots, x_n)$ with the following sign convention

- $x_i > 0$: x_i units of the i-th commodity flows from A to B;
- $x_i < 0$: $-x_i$ units of the i-th commodity flows from B to A.

Each vector x of commodity flow from A to B is associated with a gain $f(x)$ of commodity worth for B. Since the flow passes from A to B, a gain $f(x)$ for B means a loss $f(x)$ for A. The function $f(\cdot)$ is assumed continuous, quasiconcave and

$$\sup_{x \in R^n} f(x) = +\infty.$$

For a given commodity flow x in order to compensate the loss $f(x)$ of commodity worth for sector A, the manager of sector A issues a vector p of prices: $p = (p_1, p_2, \ldots, p_n)$, such that he receives from sector B a trade allowance :

$$p \cdot x = p_1 x_1 + p_2 x_2 + \cdots + p_n x_n.$$

In general we do not restrict the sign of p_i and we adopt the following sign convention

- $p_i > 0$: A receives p_i monetary units from B for one unit of the i-th commodity flowed from A to B;
- $p_i < 0$: B receives $-p_i$ monetary units from A for one unit of the i-th commodity flowed from A to B.

A price vector p is called feasible if it belongs to a given set P in R^n that is assumed bounded, closed, convex and containing 0 in its interior.

For a given vector p of prices the manager of sector B wants to find a commodity flow x that maximizes the gain function $f(x)$ subject to an expenditure constraint

$$p \cdot x \leq e,$$

where e is a limit of expenditure level ($e > 0$). By scaling we can assume without loss of generality that $e = 1$, i.e., the expenditure constraint is as follows

$$p \cdot x \leq 1.$$

The problem of sector B is thus formulated as follows

$$\begin{array}{lll} f(x) & \longrightarrow & \sup \\ \text{s.t.} \;\; p\cdot x & \leq 1. & \end{array} \tag{20.1}$$

Denote by $h(p)$ the supremum value in the above problem. Since x is a decision variable of the manager of sector B, for a given vector p of prices he can under a solvability condition assign a commodity flow x such that sector B gains the commodity worth of $h(p)$, or equivalently, sector A loses the commodity worth of $h(p)$.

The problem of sector A is now to find an equilibrium vector p of prices in the sense that it minimizes the loss function $h(p)$ over the set P of feasible vectors of prices :

$$\begin{array}{lll} h(p) & \longrightarrow & \inf \\ \text{s.t.} \;\; p & \in P. & \end{array} \tag{20.2}$$

It can be seen that $h(\cdot)$ is a quasiconvex function. The problem (20.2) is a quasiconvex minimization. In case $f(x) > 0$ for $x > 0$, the value $h(p)$ is positive and the amount $1/h(p)$ represents the level of trade allowance per one unit of commodity worth. Minimizing $h(p)$ is equivalent to maximizing $1/h(p)$. Therefore, problem (20.2) can be interpreted as a problem of maximizing the level of trade allowance per one unit of commodity worth over the set of feasible price vectors.

2. A Dual Problem and Generalized KKT Condition

Define X the set of all commodity vectors x satisfying the expenditure constraint for all price vector p in P :

$$X = \{x \in R^n : p\cdot x \leq 1 \;\; \forall p \in P\}.$$

Since P is bounded, closed, convex and containing 0 in its interior, so is X (cf. Stoer and Witzgall (1970), Rockafellar (1970)). A dual of problem (20.2) is defined as maximizing the gain function $f(x)$ over the set X :

$$\begin{array}{lll} f(x) & \longrightarrow & \sup \\ \text{s.t.} \;\; x & \in X. & \end{array} \tag{20.3}$$

The following theorem tells that the infimum value of the loss function $h(p)$ over P is greater than or equal to the supremum value of the gain function $f(x)$ over X.

Theorem 2.1 inf(20.2) $\geq$ sup(20.3).

Proof. For any $p \in P$ and $x \in X$ one has $p \cdot x \leq 1$, hence

$$\begin{aligned} h(p) &= \sup\{f(y) : p \cdot y \leq 1\} \\ &\geq f(x), \end{aligned}$$

proving the theorem. □

Problem (20.3) is a quasiconcave maximization, so we can apply a generalized KKT condition (cf. Thach (1995)). A vector p is called a quasisupdifferential of $f(\cdot)$ at x if

$$f(y) > f(x) \implies p \cdot y > p \cdot x, \tag{20.4}$$
$$p \cdot x = 1. \tag{20.5}$$

Condition (20.4) tells us that p gives a linear approximation to the upper level set $\{y : f(y) > f(x)\}$ at x, and it was used in the literature (cf. Greenberg and Pierskalla (1973)). However, it can be seen that if p satisfies (20.4) then so does λp for any $\lambda > 0$. Therefore vector 0 always belongs to the boundary of the set of such vectors p. To overcome this difficulty condition (20.5) provides a kind of normalization of p. The set of quasisupdifferentials of $f(\cdot)$ at x is denoted by $\partial f(x)$. From (20.4) it follows that

$$f(y) \leq f(x) \quad \forall y : \; p \cdot y \leq p \cdot x.$$

This together with (20.5) implies

$$\begin{aligned} f(x) &= \sup\{f(y) : \; p \cdot y \leq 1\} \\ &= h(p). \end{aligned}$$

Thus if $p \in \partial f(x)$ then $f(x) = h(p)$.

Denote by $N(x, X)$ the normal cone of X at x $(x \in X)$:

$$N(x, X) = \{p \in R^n : p \cdot (y - x) \leq 0 \;\; \forall y \in X\}.$$

Theorem 2.2 *A generalized KKT condition which appears in the form of the following inclusion*

$$0 \in \partial f(x) - N(x, X). \tag{20.6}$$

is sufficient for the optimality of a vector x in X. Furthermore, if this condition is satisfied then the intersection

$$\partial f(x) \cap N(x, X)$$

is nonempty, and any vector p in this intersection is an optimal solution to problem (20.2).

Proof. From (20.6) it follows that the intersection

$$\partial f(x) \cap N(x, X)$$

is nonempty. Let

$$p \in \partial f(x) \cap N(x, X).$$

Since $p \in \partial f(x)$, one has $p \cdot x = 1$ and

$$f(x) = h(p). \tag{20.7}$$

Since $p \in N(x, X)$, one has

$$\begin{array}{lll} & p \cdot (y - x) \leq 0 & \forall y \in X \\ \Leftrightarrow & p \cdot y - p \cdot x \leq 0 & \forall y \in X \\ \Leftrightarrow & p \cdot y \leq p \cdot x & \forall y \in X \\ \Leftrightarrow & p \cdot y \leq 1 & \forall y \in X \quad (\text{because } p \cdot x = 1). \end{array}$$

This in turn is equivalent to $p \in P$ (cf. Stoer and Witzgall (1970); Rockafellar (1970)). Thus p is feasible to problem (20.2). This together with (20.7) and Theorem 2.1 implies that x solves (20.3) and p solves (20.2). □

Let us discuss the solvability of the inclusion (20.6) in the following theorem.

Theorem 2.3 *The inclusion (20.6) is solvable, i.e., there is at least a vector x in X satisfying (20.6).*

Proof. Since X is a bounded, closed set and $f(\cdot)$ is continuous, $f(\cdot)$ achieves a maximum value on X :

$$+\infty > \alpha = \max\{f(y) : y \in X\}.$$

Since

$$\sup_{y \in R^n} f(y) = +\infty,$$

the set

$$S = \{y \in R^n : f(y) > \alpha\}$$

is nonempty and open. Define

$$M = \{y \in X : f(y) = \alpha\}.$$

Then M is a bounded, closed, convex set. Denote by $d(y, S)$ the Euclid distance from y to 5 :

$$d(y, S) = \inf\{\| y - z \| : z \in S\}.$$

Let x be a vector in M that is closest to S :

$$d(x, S) = \inf_{y \in M} d(y, S).$$

If $d(x, S) = 0$, then take $S' = S$. If $d(x, S) > 0$, then take

$$S' = \{x + \lambda(y - x) : 0 < \lambda \leq 1, \; y \in S\}.$$

Since $d(x, S) > 0$ one has

$$d(x + \lambda(y - x)), S) < d(x, S) \quad \forall \lambda \in (0, 1] \quad \forall y \in S,$$

therefore $S' \cap X = \emptyset$. So, in any case one has an open convex set S' satisfying

$$S' \cap X = \emptyset \text{ and } x \in cl(S') \cap X,$$

where $cl(\cdot)$ stands for the closure. Since $0 \in R^n$ belongs to the interior of X, by the separation theorem there exists a vector p such that

$$p \cdot x = 1, \tag{20.8}$$

$$p \cdot y \leq p \cdot x < p \cdot z \quad \forall y \in X, \; \forall z \in S'. \tag{20.9}$$

The first inequality in (20.9) means that $p \in N(x, X)$. On the other hand, $S \subseteq S'$, hence from (20.9) it follows that

$$\begin{aligned} & p \cdot x < p \cdot z \quad \forall z \in S \\ \Leftrightarrow \; & p \cdot x < p \cdot z \quad \forall z : \; f(z) > f(x) \;\; (\text{since } f(x) = \alpha). \end{aligned}$$

This together with $p \cdot x = 1$ implies $p \in \partial f(x)$. So the inclusion (20.6) is satisfied at $x \in X$, proving the theorem. □

As a consequence of the above theorem one has the strong duality between problem (20.2) and problem (20.3).

Corollary 2.1 min(20.2) = max(20.3).

Proof. Let x be a vector in X at which the inclusion (20.6) is satisfied, and let

$$p \in \partial f(x) \cap N(x, X).$$

Then, x solves (20.3), p solves (20.2) and $f(x) = h(p)$, proving the corollary. □

3. Illustration: a numerical example

In our trading problem we are given two commodities x_1 and x_2, and the commodity worth function of the commodity flow $x = (x_1, x_2)$ from A to B defined as follows

$$f(x_1, x_2) = \begin{cases} x_1 x_2 & \text{if } x_1 > 0 \text{ and } x_2 > 0 \\ 0 & \text{if } x_1 \leq 0 \text{ or } x_2 \leq 0. \end{cases}$$

The manager of sector A is holding the set P of feasible prices given by

$$P = \{(p_1, p_2) \in R^2 : 4p_1^2 + p_2^2 \leq 1\}.$$

For $p = (p_1, p_2)$ with either $p_1 \leq 0$ or $p_2 \leq 0$ it can be seen that

$$\begin{aligned} h(p_1, p_2) &= \sup\{f(x_1, x_2) : p_1 x_1 + p_2 x_2 \leq 1\} \\ &= +\infty, \end{aligned}$$

and with $p_1 > 0$ and $p_2 > 0$ that

$$\begin{aligned} h(p_1, p_2) &= \sup\{f(x_1, x_2) : p_1 x_1 + p_2 x_2 \leq 1\} \\ &= 1/(4p_1 p_2). \end{aligned}$$

So the problem of sector A is as follows

$$\begin{array}{l} 1/(4p_1 p_2) \quad \longrightarrow \quad \inf \\ \text{s.t. } 4p_1^2 + p_2^2 \leq 1, \; p_1 > 0, \; p_2 > 0. \end{array} \tag{20.10}$$

It can be seen that

$$\begin{aligned} X &= \{(x_1, x_2) : p_1 x_1 + p_2 x_2 \leq 1 \; \forall (p_1, p_2) \in P\} \\ &= \{(x_2, x_2) : x_1^2 + 4x_2^2 \leq 4\}. \end{aligned}$$

So the dual of (20.10) can be written as follows

$$\begin{array}{l} x_1 x_2 \quad \longrightarrow \quad \sup \\ \text{s.t. } x_1^2 + 4x_2^2 \leq 4, x_1 > 0, \; x_2 > 0. \end{array}$$

For (x_1, x_2) with $x_1 > 0$ and $x_2 > 0$ it can be seen that the set of quasisupdifferentials $\partial f(x_1, x_2)$ reduces to a single vector $(1/(2x_1), 1/(2x_2))$:

$$\partial f(x_1, x_2) = \{(1/(2x_1), 1/(2x_2))\}.$$

So the inclusion (20.6) at (x_1, x_2) such that

$$x_1 > 0, \; x_2 > 0 \text{ and } x_1^2 + 4x_2^2 = 4 \tag{20.11}$$

becomes the following equation

$$2x_1 = \lambda/(2x_1),\ 8x_2 = \lambda/(2x_2)$$

with $\lambda > 0$. Solving this equation under the condition (20.11) we obtain the following roots

$$\lambda = 8,\ x_1 = \sqrt{2},\ x_2 = 1/\sqrt{2}.$$

Thus, the inclusion (20.6) yields the dual's solution :

$$x_1 = \sqrt{2},\ x_2 = 1/\sqrt{2},$$

and the primal's solution can be calculated by taking the quasisupdifferential of $f(\cdot)$ at the dual's solution :

$$p_1 = 1/(2\sqrt{2}),\ p_2 = 1/\sqrt{2}.$$

The optimal value is 1.

4. Discussions

Suppose that the commodity worth function $f(\cdot)$ of the flow x from A to B is linear :

$$f(x) = c \cdot x, \quad (c \neq 0).$$

It can be seen that

$$h(p) = \sup\{c \cdot x : p \cdot x \leq 1\} = \begin{cases} +\infty & \text{if } p \notin \{\theta c, \theta > 0\} \\ 1/\theta & \text{if } p = \theta c \text{ for some } \theta > 0. \end{cases}$$

Then the problem of sector A is as follows

$$\begin{array}{lcl} 1/\theta & \longrightarrow & \inf \\ \text{s.t. } \theta c & \in & P. \end{array}$$

Thus if $f(\cdot)$ is a linear function with a vector c of linear coefficients then the optimal vector of prices must be propotional to vector c.

For a general case in which $f(\cdot)$ is nonlinear, in order to find an optimal vector of prices we can solve the inclusion (20.6) for the dual (20.3). A vector would be a primal's optimal solution if it is in the intersection of the set of quasisupdifferentials and the normal cone at the dual's optimal solution (Theorem 2.2). However a quasisupdifferential of $f(\cdot)$ is related to a linear approximation of $f(\cdot)$ (cf. Thach (1995)). So the primal's optimal vector of prices can be interpreted as a kind of linear approximation of the commodity worth function at the dual's optimal solution.

In the connection to duality by minimax we define a bifunction $L(y, q)$ for each commodity vector y and price vector q :

$$L(y,q) = \begin{cases} f(y) & \text{if } q \cdot y \leq 1 \\ -\infty & \text{if } q \cdot y > 1. \end{cases}$$

Then the primal problem is

$$\inf_{q \in P} \sup_{y \in R^n} L(y,q),$$

while the dual problem is

$$\sup_{y \in R^n} \inf_{q \in P} L(y,q).$$

If x is a vector at which the inclusion (20.6) is satisfied and p is a vector in the intersection between the set of quasisupdifferentials of $f(\cdot)$ at x and the normal cone at x, then (x,p) is a saddle point of the bifunction $L(y,q)$:

$$\inf_{q \in P} \sup_{y \in R^n} L(y,q) = L(x,p) = \sup_{y \in R^n} \inf_{q \in P} L(y,q).$$

Thus by our approach a saddle point problem is reduced to solving an inclusion.

Acknowledgments

The author would like to thank Professor C. Le Van for valuable discussions on equilibrium prices. He also expresses his thanks to an anonymous referee for helpful comments and suggestions.

References

Debreu, G. (1959), *Theory of Value,* John Wiley and Sons, New York.

Schaible, S., and Ziemba, W.T., Editors (1981), *Generalized Concavity in Optimization and Economics,* Academic Press, New York, New York.

Avriel, M., Diewert, W.E., Schaible, S., and Zang, I. (1988) *Generalized Concavity,* Plenum Press, New York, New York.

Luenberger, D.G. (1995), *Microeconomic Theory,* McGraw-Hill, Inc., New York.

Crouzeix, J.P. (1974) Polaires Quasi-Convexes et Dualité, *Compte Rendus de l'Académic des Sciences de Paris,* Vol. A279, pp. 955-958.

Crouzeix, J.P. (1981), A Duality Framwork in Quasiconvex Programming, *Generalized Concavity in Optimization and Economics,* Edited *by* S. Schaible and W.T. Ziemba, Academic Press, New York, New York, pp. 207-225.

Diewert, W.E. (1982), Duality Approaches to Microeconomic Theory, *Handbook of Mathematical Economics 2,* Edited by K. J. Arrow and M. D. Intriligator, North Holland, Amsterdam, Holland, pp. 535-599.

Diewert, W.E. (1981), Generalized Concavity and Economics, *Generalized Concavity in Optimization and Economics,* Edited by S. Schaible and W.T. Ziemba, Academic Press, New York, New York, pp. 511-541.

Greenberg, H.J., and Pierskalla, W.P. (1973), Quasi-conjugate functions and surrogate duality, *Cahiers Centre Études, Rech. Opér.,* Vol. 15, pp. 437-448.

Martinez-Legaz, J.E. (1988), Quasiconvex Duality Theory by Generalized Conjugation Methods, *Optimization,* Vol. 19, pp. 603-652.

Oettli, W. (1982), Optimality Conditions for Programming Problems Involving Multivalued Mappings, *Applied Mathematics,* Edited by B. Korte, North Holland, Amsterdam, Holland, pp. 196-226.

Singer, I. (1986), A General Theory of Dual Optimization Problems, *Journal of Mathematical Analysis and Applications,* Vol. 116, pp. 77-130, 1986.

Passy, U., and Prisman, E.Z. (1985), A Convex-Like Duality Scheme for Quasiconvex Programs, *Mathematical Programming,* Vol. 32, pp. 278-300, 1985.

Penot, J.P., and Volle, M. (1990), On Quasiconvex Duality, *Mathematics of Operations Research,* Vol. 4, pp. 597-625.

Thach, P.T. (1995), Diewert-Crouzeix Conjugation for General Quasiconvex Duality and Applications, *Journal of Optimization Theory and Applications,* Vol. 86, pp. 719-743.

Stoer, J., and Witzgall, C. (1970), *Convexity and Optimization in Finite Dimensions I,* Springer Verlag, Berlin, Germany.

Rockafellar, R.T. (1970), *Convex Analysis,* Princeton University Press, Princeton, New Jersey.